DIFFUSE POLLUTION OF WATER RESOURCES

PRINCIPLES AND CASE STUDIES IN THE SOUTHERN AFRICAN REGION

BALKEMA-Proceedings and Monographs in
Engineering, Water and Earth Sciences

Diffuse Pollution of Water Resources

Principles and Case Studies in the Southern African Region

Editor

R. Hranova
University of Botswana, Gaborone, Botswana

LONDON/LEIDEN/NEW YORK/PHILADELPHIA/SINGAPORE

Published by: Taylor & Francis/Balkema
P.O. Box 447, 2300 AK Leiden, The Netherlands
e-mail: Pub.NL@tandf.co.uk
www.balkema.nl, www.tandf.co.uk, www.crcpress.com

ISBN 0 415 38391 9

Printed in Great-Britain

CONTENTS

CHAPTER 3
CHARACTERISTICS OF AN URBAN ENVIRONMENT IN THE CONTEXT OF DIFFUSE POLLUTION CONTROL 47
R. Hranova

CHAPTER 4
ASSESING AND MANAGING URBAN STORM WATER QUALITY 69
R. Hranova & M. Magombeyi

CHAPTER 12

Preface

Water is often argued to be essential to life, but in our everyday activities, we usually take it for granted. Natural water resources have a limited capacity, and in many cases, especially in arid climates, cannot provide for an ever-increasing demand. In addition, they are subject to pollution stress due to the return flows of used water. The restoration of a polluted water body is a very costly and difficult to enforce exercise, even for the resources and capacity of a developed country. Most developing countries face serious and urgent problems, related to poverty alleviation and the resolution of different types of social conflicts, and often, the protection of water resources' quality is neglected, or the authorities lack the financial, technical and human resources to control and restore it. For all these reasons, the implementation of water pollution prevention measures, based on proper information regarding the specific site conditions, should have priority during the development and implementation of water resources management plans and strategies. It is intended that this book will make a contribution in this direction.

During the twentieth century, considerable achievements have been made with respect to water pollution prevention and abatement. Efforts have been directed to point pollution sources, associated with pipeline discharges of municipal and industrial origin. At present, it is a common practice to provide treatment of wastewater, discharged to natural water bodies. However, after the implementation of point source pollution abatement measures, it has been shown that many of the natural water resources do not recover fully their natural water quality, due to other forms of pollution generated by diffuse (non-point) sources. Non-point source pollution is associated mainly with polluted runoff from urban and agricultural areas, is dispersed in nature, and reaches the natural water bodies in poorly defined ways. Therefore, this type of water pollution is much more difficult to monitor, control and regulate, compared to point sources pollution. Recent practice in Southern Africa shows that the quality control of water resources focuses on point sources of pollution mainly and there is a need to address diffuse sources of pollution as well.

This book presents a variety of case studies dealing with diffuse pollution of natural water resources in the Southern African region. In addition, it presents basic principles, tools and approaches for its abatement and management. This allows a wider audience to understand and be familiarized with the subject. The case studies were performed by MSc students from the region, as part of capacity building programs in the field of water resources management. Therefore, the investigations were limited in terms of time and financial resources. Also, it should be noted that the cases studied reflect the actual practice of water quality management, which in many cases is basic. As a consequence, in several studies the results were not conclusive. Despite these limitations, the gathered information could be applied in the process of diffuse pollution abatement and management, and also, could serve as illustration to support the principles and approaches applied at international level.

The objective is that the principles, reflecting the international practice in the field of diffuse pollution, together with the specific case studies, will provide useful knowledge, guidance and information regarding the process of monitoring, control and abatement of diffuse pollution to both academics and professionals in practice. Emphasis has been given to diffuse pollution problems typical for the region and for developing countries elsewhere. The book's objectives and importance in the context of sustainable development, may be summarized as follows:

- To present specific methods, tools and approaches for the detection and evaluation of diffuse pollution, and to help in the introduction of new approaches in the process of monitoring, control and conservation of water resources.
- To outline the common problems to be addressed in future practice.

- To fill the gaps on a topic which is relatively unfamiliar to the public and to many professionals in the field of water resources management.
- To present a ground for common understanding among different specialists working in the field of water quality management and diffuse pollution control.
- To acknowledge the role of sponsors who have financially supported the research projects.

Finally, the editor would like to thank all authors and the support personnel for their efforts during the investigations and during the preparation of the manuscripts. The current contact details of all authors, at the time of publication, are presented in the list on page xvii.

LIST OF MOST COMMON ABBREVIATIONS

Al	Aluminum
BOD	Biochemical Oxygen Demand
Ca	Calcium
Cd	Cadmium
CO_2	Carbon dioxide
COD	Chemical Oxygen Demand
Cr	Total Chromium
Cu	Copper
DO	Dissolved Oxygen
EC	Electrical Conductivity
FC	Fecal Coliforms
Fe	Iron
Hg	Mercury
K	Potassium
Mg	Magnesium
Mn	Manganese
Na	Sodium
Ni	Nickel
ortho-P	phosphate (ortho)
Pb	Lead
T	Temperature
TC	Total Coliforms
TKN	Total Kjeldahl Nitrogen
TN	Total Nitrogen
TP	Total Phosphorous
TS	Total Solids
TSS	Total Suspended Solids
TDS	Total Dissolved Solids
Zn	Zinc

BOOK OVERVIEW

	GENERAL	SURFACE WATER		GROUND WATER	SOILS
		Storm water	Natural water		
Diffuse pollution topics					
Sources	Chapters 1 & 12				
• Urban runoff & sanitation		Chapter 4	Chapters 5 & 11	Chapter 6	
• Sludge land disposal			Chapter 8	Chapter 9	Chapter 9
• Irrigation with effluents				Chapter 10	Chapter 10
• Solid wastes / cemeteries				Chapter 7	
Regulation	Chapters 1 & 12	Chapter 5		Chapter 8 Chapter 10	
Monitoring	Chapters 2 & 12	Chapter 4	Chapter 5		
Assessments					
• Quality		Chapter 4	Chapters 3, 5, 8 & 11	Chapters 6, 7, 9 & 10	Chapter 9 & 10
• Runoff	Chapter 1	Chapter 4	Chapter 5		
• Pollution loads					Chapter 8
Abatement and Management	Chapters 2 & 12	Chapter 4	Chapter 5 Chapter 11	Chapters 6 & 7	
Other related topics					
Wastewater treatment and reuse	Chapters 8 & 10				
Pollutants' transport	Chapter 9				
Population trends / Urban development	Chapters 3, 5 & 11				

Case study
Basic principles

LIST OF CONTRIBUTING AUTHORS

Amos, Amos
Zimbabwe Open University
PO Box MP 1119, Mt.
Pleasant, Harare, Zimbabwe
amosii2000@yahoo.co.uk

Love, David
WaterNet
PO Box MP600, Mt. Pleasant,
Harare, Zimbabwe
Department of Geology,
University of Zimbabwe
PO Box MP167, Mt. Pleasant,
Harare, Zimbabwe
davidlove@science.uz.ac.zw

Gandidzanwa, Pamhi
Department of Geography and
Environmental Science,
Bindura University for Science
Education
P Bag 984, Bindura, Zimbabwe
pamigandi@yahoo.com

Gwenzi, Willis
Save Valley Research Station
P. Bag 2037
Chipinge, Zimbabwe
Wgwenzi@yahoo.com

Hranova, Roumiana
Civil Engineering Department
University of Botswana
Private bag 0061
Gaborone, Botswana
Hranova@mopipi.ub.bw

Magadza, Chris
Department of Biological
Sciences, University of Zimbabwe
PO Box MP167, Mt. Pleasant,
Harare, Zimbabwe
profmagadza@utande.co.zw

Magombeyi, Manuel
Civil and Water Engineering
Department, National
University of Science and Technology
PO Box AC 939, Ascot,
Bulawayo, Zimbabwe
Mmagombeyi@nust.ac.zw

Mangeya, Pride
Department of Geology,
University of Zimbabwe
PO Box MP167, Mt. Pleasant,
Harare, Zimbabwe
pride@science.uz.ac.zw

Manjonjo, Muwengwa
CPP Botswana Consulting Engineers
Private bag BR 121
Broadhurst, Gaborone,
Botswana

Moyce, William
Department of Geology,
University of Zimbabwe
PO Box MP167, Mt. Pleasant,
Harare, Zimbabwe
moyce@science.uz.ac.zw

Musiwa, Kudzai
Department of Mining
Engineering, University of Zimbabwe
PO Box MP167, Mt. Pleasant,
Harare, Zimbabwe
kudzie@eng.uz.ac.zw

Mwandira, Samson
Lilongwe Water Board
PO Box 96, Lilongwe, Malawi
Smwandira@lwb.malawi.net

Nkambule, Stenley
Faculty of Health Science
University of Swaziland
PO Box 369, Mbabane,
Swaziland

Nyama, Zanelle
Department of Soil Science and Agricultural Engineering,
University of Zimbabwe
PO Box MP167, Mt. Pleasant,
Harare, Zimbabwe
zanell07@yahoo.com

Ravengai, Seedwel
Department of Geology,
University of Zimbabwe
PO Box MP167, Mt. Pleasant,
Harare, Zimbabwe
ravengai@science.uz.ac.zw

Wuta, Menas
Department of Soil Science and
Agricultural Engineering,
University of Zimbabwe
PO Box MP167, Mt. Pleasant,
Harare, Zimbabwe
mwuta@agric.uz.ac.zw

Zingoni, Emmanuel
Department of Geography and
Environmental Science,
Bindura University for Science
Education
P Bag 984, Bindura, Zimbabwe
ezingoni2002@yahoo.co.uk

CHAPTER 1

Diffuse pollution-principles, definitions and regulatory aspects

R. Hranova

ABSTRACT: Definitions and basic classifications of water pollution in general and diffuse pollution in particular have been presented. The "language" of water quality, or the basic characteristics, which define it and allow the identification, assessment and control of pollution have been listed and explained briefly. The process of diffuse pollution generation has been described and the main sources mentioned, which would be the basis for the implementation of source control measures. Diffuse pollution is associated in most cases with polluted runoff and its evaluation and control would rely heavily on the correct estimation of runoff characteristics, therefore basic principles and methods in this direction have been emphasized. Definitions and principles with respect to regulatory instruments used to control and manage pollution, have been presented and different approaches and philosophies applied internationally in this field, have been discussed.

1 WATER POLLUTION

Water collects impurities from the moment of its formation in the clouds. Some are harmless, others, may be offensive aesthetically or even dangerous with respect to the intended use of the water. To establish the quality of water or to compare one type of water with another, it is necessary to define a basis for evaluation and comparison. Usually, this basis is defined in terms of the quality requirements for a specific beneficial use of water. The basic types of beneficial uses could be defined as follows:

- Public use – domestic, commercial and institutional needs, irrigation of parks, washing of streets and other public uses.
- Industrial use – water used for specific industrial processes, including that for the main production process, and auxiliary processes such as cooling and washing of materials, equipment, floors, etc.
- Recreational (contact) use – water in lakes, reservoirs, rivers, estuaries and the ocean, used for water contact sports.
- Agricultural use – water used for irrigation of crops, pastures and other agricultural activities, and live-stock watering;
- Environmental use – water used for the propagation of fish and as a habitat for aquatic and wild life.

There are different definitions of water pollution, depending on the approach and field of specialization of the author, but we apply this term in all cases when natural water quality has been altered in such a way that subsequent uses have become less acceptable. Ellis (1989) formulates surface water pollution as "...the alteration in composition or condition of surface water, either directly or indirectly, as the result of the activities of men, which initiates modification of ecological systems, hazards to human health and renders the stream less acceptable to downstream users." There are three points in this formulation, which clarify the term:

- Pollution is defined as a consequence of human activities.
- We consider that natural water is polluted not only in the cases of public health hazard, but when the natural environment is modified.

- Pollution is an act leading to the impairment of subsequent water use for a stated purpose.

Although this definition is formulated for surface water only it can be applied to ground water resources as well, and it could be stated that:

⇒ *Water pollution* is the alteration in the composition or condition of natural water, either directly or indirectly, as the result of human activities, which initiates modification of ecological systems, hazards to human health and impairs the subsequent water use.

The process of evaluation of the pollution status of a water resource system requires information regarding its natural (background) quality, in order to evaluate how much it has been affected by human activities.

⇒ *Background pollution* represents the physical, chemical and biological composition of surface or ground water, which could result from natural causes and factors. It varies with climatic, geographic and soil conditions and it is an indicator of the level of contamination of water in its pristine status, uninfluenced by human activities. Surface natural water quality is more prone to variation compared to ground water, the natural composition of which is dependent mostly on the geological formation of the aquifer.

Pollution contributed to water resources could be expressed in terms of pollution loads received by the corresponding natural water body from a specific source.

⇒ *Pollution loads* represent the mass quantity of pollution discharged per unit time into water bodies and are equal to the product of pollution concentrations and the flow rates during a stipulated period of time. We use the term "*pollution load*" in cases where an external source (a wastewater discharge, a tributary of a river, or a river itself) contributes pollution to a given water body (natural stream, river, lake or aquifer). Pollution loads contributed by different media within a water body could be regarded as pollution flux. Therefore, during the process of assessment and evaluation of pollution, both quantitative and qualitative measurements and data are needed in order to assess and evaluate impacts and effects.

1.1 *Classification of pollution according to pollution pathways*

1.1.1 Point sources of pollution

These are sources associated with man-made discharges of pipelines and canals, and are also known as "end of pipe" sources of pollution. They are identifiable discrete discharges from municipal and industrial wastewater collection and treatment systems. They could be broadly subdivided into two categories, based on their composition and origin:

- Municipal wastewater – The characteristic quality parameters have been well studied, defined and are relatively easy to measure and control. However, their values vary in certain limits depending on the geographical and climatic conditions, level of economic development, mixture of influent sources (industrial/domestic), and size of community served.
- Industrial wastewater – The quality characteristics are extremely diverse and require individual assessment for each specific case. We could differentiate between "manufacturing" wastewater, generated during the production process and "service water", generated by canteens, ablution blocks, and other activities, complementary to the industrial process. The service industrial wastewater usually has the characteristics of municipal wastewater. Dealing with industrial wastewater as a source of pollution requires understanding of the production process and, in most cases, segregation of different types of wastewater flows.

1.1.2 Non-point sources of pollution

They are associated mainly with land drainage and surface runoff, which enters a water body by dispersed and poorly defined ways. For this reason, the term "diffuse pollution" is commonly used as well. Atmospheric waste loads are chemicals and particulate matter, which settle from the atmosphere or are scavenged by precipitation. They could be classified as possible diffuse sources of pollution if they reach water bodies. Municipal wastewater applied to land may become a non-point source of pollution in runoff, if not handled properly. Usually, we differentiate between diffuse pollution from agricultural sources and diffuse pollution in urban areas. Agricultural sources are associated mainly with runoff and leachate from agricultural lands and animal operations, and are generated by the excess application of fertilizers, pesticides, herbicides, as well as, increased salinity of return flows. Diffuse pollution in urban areas is associated mainly with polluted urban runoff (drainage), contaminated with materials washed up from streets, roads, roofs, open spaces, etc.

1.2 *Classification according to pollution behavior*

- Non-conservative pollutants – these include most organics, some inorganics and many microorganisms, which are degraded by natural self-purification processes and their concentrations change with time due to decomposition and different types of transformation processes.
- Conservative pollutants – these are the ones, which are not affected by natural processes and their concentrations do not change with time. Typical examples are inorganic substances that can only have their concentrations reduced by dilution.

1.3 *Classification of pollution according to pollution characteristics*

- The addition of biodegradable organic material – it will result in the depletion or complete removal of dissolved oxygen in the water body, with corresponding adverse effects on its environmental status. It has been widely recognized that this type of pollution is of the greatest importance. Most of the efforts in water pollution control and abatement have been implemented to control or reduce the biodegradable organic matter discharged to surface or ground waters.
- Toxicity and mutagenic pollutants – it is due to the presence of synthetic organic compounds, salts of heavy metals, acids and alkalis, radioactive and other materials, which have toxic or mutagenic effect on living organisms. There are a variety of toxic substances resulting from industrial and agricultural operations, municipal practices and even normal domestic activities, which inhibit the normal aquatic activities of a water body and most importantly, pose serious threats to public health if the receiving body is used as a drinking water source or as a recreational site.
- Enhanced eutrophication – this is a natural phenomenon of lake aging, which could be classified as pollution when the natural process has been largely accelerated by the activities of men. It is dependent on the increases of potentially soluble inorganic nutrient materials, both in the sediment and in the lake water itself. Eutrophication could be considered as a truly natural process, only in cases when the inflow of nutrients to a lake is consistently higher than the outflow, within the unaltered conditions of the catchment area. Enhanced eutrophication becomes apparent principally by the development of large blooms of algae as the increased potential productivity of the lake responds to sunlight, with the production of heavy algal blooms and the appearance of algal mats on lake surfaces together with the dense growth of certain macrophytes.
- Salinization – this is the process of increasing the naturally dissolved solids concentrations of water resources due to certain activities. This type of pollution is often found in arid regions. Portions of the water for irrigation are lost through seepage and evaporation. Part of it is consumed by plant evapotranspiration, and the excess passes through the soil to be collected by tile drainage and returned to the river. If tile drainage system is not available, it will percolate to the nearest aquifer. As the salt concentration of river and drained water is the same, but the volume of drained water is considerably diminished, the salt concentrations of the receiving water body would rise with time. More severe salinity problems can be met in cases when transient leaching of soil salts occur, and when treated wastewater is used for irrigation purposes.
- The addition of inert, insoluble mineral material – this type of pollution is mainly due to the erosivity of runoff, and is generally considered a minor source of surface water pollution. It is associated with excess sediment deposition, increased turbidity and general change of river morphology. However, it should be pointed out that a considerable portion of the diffuse pollution load with respect to different

constituents are attached to insoluble mineral particles, which serve as a transportation vehicle of these pollutants to surface water bodies.

- Oil pollution – oil spreading as a thin film on water has the effect of interfering with gas exchange and hence preventing or severely reducing the rate at which the atmospheric oxygen can be adsorbed into deoxygenated water. Most oils will rapidly spread to achieve a thickness of about 1 mm in a matter of minutes. Although the direct organic load exerted by oil may be limited, the effect of a spill is potentially catastrophic because it retards the natural self-purification process, as well as, enhances the oxygen depletion of the water body, which has an adverse effect on the struggling aquatic life. In addition, many oils have direct toxic effects on humans and aquatic life.

2 POLLUTANT CONSTITUENTS

2.1 *Physical characteristics*

- Solids – this characteristic represents the sum of all constituents in a water sample, in particulate or dissolved form, expressed as solids content or TS, which is determined by evaporating a sample and weighting the dry residue. In the practice of pollution control and water/wastewater treatment it is often important to know the form in which the solids are present. Thus, we could differentiate between the suspended fraction (TSS) and the dissolve fraction (TDS) of TS. TSS is determined by filtering a sample of water through a standard glass-fiber filter, followed by drying and weighing of the residue. TDS could be determined either by evaporating and weighing the filtrate residue, or as the difference between TS and TDS. TS and TDS concentrations in conjunction with a detailed chemical analysis, is used to assess the pollution status of water samples and their suitability for various water uses – public, industrial or agricultural. Additional characterization of the solids content, in terms of organic or inorganic components, is shown in Figure 1.1. Actually, only TSS is considered to be a physical component, as it represents all impurities in particulate or colloidal (large colloids) form, while TDS should be considered a chemical characteristic as it presents constituents in ionic or molecular form. Based on the size of the particles and their ability to separate from suspension by means of gravity, TSS fractions could be characterized as settable and non-settable. This parameter is measured in Imhoff cones, where the water sample is left at rest, and the volume of the sediment is recorded after a specified period of time.
- Turbidity – it is a parameter closely related to TSS. It reflects the presence of suspended and colloidal particles, which are too small to settle and stay in suspension, thus causing specific opalescence and decreasing the clarity of water. It is measured by turbidimeters – instruments, which measure the deflection of a beam of light due to the opalescence of the water sample, and compare it to the deflection of a standard solution. Clay, silt, soil particles and other colloidal impurities usually cause turbidity in natural water and the level of turbidity would depend on the fineness of the particles and their concentration.
- Aesthetic characteristics such as color, taste and odor are of importance in cases when natural water would be used for drinking purposes or recreation. Water sometimes contains a considerable color, resulting from certain types of dissolved and colloidal organic matter leached from soil or decaying vegetation. True natural color is due to dissolved impurities. Tastes and odors in water are caused by the presence of decomposed organic material and volatile chemicals and are measured by diluting the sample until the taste and odor are no longer detectable by a human test. Potable water should be practically free of color, tastes and odor.
- The temperature of water is important in terms of its intended use and treatment, as well as an environmental indicator. It depends on the source of the water.

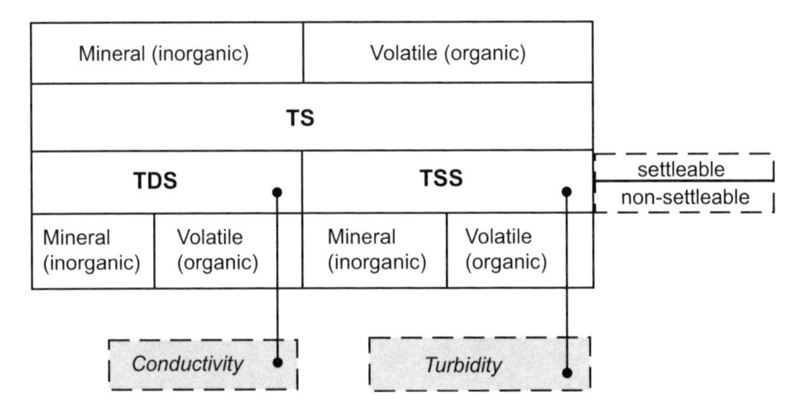

Figure 1.1. Solids classification in a water sample.

2.2 *Chemical characteristics*

2.2.1 *Inorganic constituents – specific ions and gases*

- The pH of water is taken as a measure of the acidic or basic nature and is defined as the logarithm of the reciprocal of the hydrogen ion concentration in moles per liter. Pure water at 24°C is balanced with respect to H^+ and OH^- ions and contains 10^{-7} mol/l of each type. Thus, the pH of neutral water is 7 and the range of neutrality is between 6 and 8. Acidic water has a pH lower than 6, basic water has a pH higher than 8.
- Dissolved cations and anions – the most common anions, which are naturally present in water are carbonates, bicarbonates, sulfates, and chlorides. The most common cations are Ca, Mg, Si, Na, Fe, Mn. The distribution of the specific species will depend on the source of the water. If a chemical analysis of a water sample is correct, the sum of the cations and anions expressed in terms of equivalents or milliequivalents per liter must be the same to satisfy the electroneutrality principle. This rule (ion balance) is applied to check the accuracy of laboratory analyses. The presence of selected ions in excess concentrations are connected with the following characteristics:
  - Electrical Conductivity (EC) – it is also known as "specific conductance", and is a measure of the ability of water to conduct an electric current. Measuring the electrical resistance between 2 electrodes and comparing it with the resistance of a standard solution of potassium chloride at 25°C determines the EC. It is sensitive to variations in TDS, mostly mineral salts. The degree to which salts dissociate into ions, the amount of electrical charge on each ion, ion mobility and the temperature of the water all have an influence on conductivity. There is a linear relationship between TDS (in mg/l) and EC (in µS/cm) with a coefficient varying between 0.55 and 0.75 (Chapman 1998). This coefficient is specific for each type of water body, but remains approximately constant, provided the ionic proportions of the water body remain stable. The conductivity of most freshwaters ranges from 10 to 1000 µS/cm but may exceed 1000 µS/cm in polluted waters or those receiving large quantities of land runoff (Chapman 1998). In addition to being a rough indicator of mineral content, conductivity can be measured to establish a pollution zone e.g. around an effluent discharge, or the extent of influence of runoff waters. It is usually measured in-situ by a conductivity probe.
  - Alkalinity of water is a measure of its capacity to neutralize acids. In natural waters the alkalinity is related to the concentrations of bicarbonate, carbonate and hydroxide ions. Total alkalinity usually is expressed in terms of equivalent calcium carbonate in mg/l.

- Hardness of water is a characteristic, reflecting the concentrations of selected ions, but the principle ones are Ca and Mg. Other bivalent and trivalent dissolved metal ions such as Al, Fe, Mn and Zn, also contribute to total hardness. "Hard" water contains excess concentrations of these metals, leading to specific properties, such as scale formation and low foaming capacity, resulting in increased soap consumption.

- Dissolved gases – the two most common gases present in water are carbon dioxide and oxygen.
  - Carbon dioxide is one of the minor gases present in the atmosphere and is the end product of both aerobic and anaerobic biological decomposition. Rainwater and most surface water supplies contain small amounts of CO_2 (less than 5 mg/l) but ground water may contain a significant amount resulting from the biological decay of organic matter. The presence of CO_2 is important because it affects the pH of water.
  - The oxygen content of natural waters varies with temperature, salinity, and turbulence, the photosynthetic activity of algae and plants and atmospheric pressure. The higher the temperature of water, the less oxygen it can hold. The large fluctuations in daily oxygen levels (high during midday from algal photosynthesis, low at night when both algae and microorganisms are respiring) are stressful and sometimes deadly to fish and other aquatic life. At DO levels less than 80% saturation in drinking water, odor and taste problems occur. In freshwaters, dissolved DO at sea level ranges from 15 mg/l at 0^0C to 8 mg/l at 25^0C. The determination of DO concentration is a fundamental part of a water quality assessment since oxygen is involved in, or influences, nearly all chemical and biological processes in water bodies. The DO can be used to indicate the degree of pollution by organic matter, the destruction of organic substances and level of self-purification of the water. It is usually determined based on electro-chemical principles by oxygen probes.

2.2.2 *Inorganic constituents – nutrients*

Nutrient enrichment (enhanced eutrophication) is a major theme in freshwater ecology. Some themes come and go, but the inevitable release of phosphorus and nitrogen that accompanies human presence seems to ensure that eutrophication will not soon become an outmoded subject of study (Lewis & William 1984). Eutrophication raises issues that range from the pressingly practical problems of phosphorus removal to the very fundamental ecological questions surrounding biological community regulation by resource supply and the conservation of the ecological habitat. Excessive levels of nitrogen and phosphorus in storm water could promote the growth of undesirable aquatic plants, including algae and floating macrophytes, which form a mat on the water surface. Eutrophied water reduces the possible recreational use of lakes and reservoirs, and increases the water treatment cost, when eutrophied water is used for potable purposes.

2.2.2.1 Nitrogen compounds

Nitrogen is essential for living organisms as an important constituent of proteins, including genetic material. Plant and microorganisms convert inorganic nitrogen to organic forms. In the environment, inorganic nitrogen occurs in a range of oxidation states as nitrate (NO_3) and nitrite (NO_2), ammonia in gaseous or ionic form, and molecular nitrogen (N_2). It undergoes biological and non-biological transformations in the environment as part of the nitrogen cycle. The major non-biological processes involve phase transformations such as volatilization, sorption and sedimentation. Biological transformations consist of:

i. Assimilation of inorganic forms (ammonia and nitrate) by plants and microorganisms to form organic nitrogen e.g., amino – acids;
ii. Reduction of nitrogen gas to ammonia and organic nitrogen by microorganisms;
iii. Complex heterotrophic conversions from one organism to another;
iv. Oxidation of ammonia to nitrate and nitrite (nitrification);
v. Ammonification of organic nitrogen to produce ammonia during the decomposition of organic matter;
vi. Bacteria reduction of nitrate to nitrous oxide (N_2O) and molecular nitrogen (N_2) under anoxic conditions (denitrification)

- Ammonia – it is formed by bacterial enzymes that hydrolyze urea as well as by oxidation of proteins, excretion by biota and from gas exchange with the atmosphere. It is also discharged into water bodies by some industrial processes (e.g. ammonia based pulp and paper production) and also as a component of municipal or community waste. At certain pH levels, high concentrations of ammonia are toxic to aquatic life and, therefore, detrimental to the ecological balance of water bodies. Moyo (1997) explains the massive fish death in Lake Chivero in 1996 as due to anoxia, presence of ammonia and other causes. In aqueous solution, unionized ammonia exists in equilibrium with the ammonium ion. Ammonia also forms complexes with several metal ions and may be adsorbed onto colloidal particles, suspended sediments and bed sediments. It may also be exchanged between sediments and the overlying water. Substantial losses of ammonia can occur via volatilization with increasing pH. Unpolluted waters contain small amounts of ammonia and its compounds, usually less than 0.20 mg/l expressed as nitrogen (Chapman 1998). A higher concentration could be an indication of organic pollution such as domestic sewage, industrial waste and fertilizer runoff. Ammonia at levels in excess of 1 mg/l contributes to the corrosion of copper pipes and brass plumbing fixtures adding copper salts to the water (Dean & Lund 1981). Ammonia also interferes with chlorine during the water treatment process, although low levels of monochloramine are considered desirable in water distribution systems.
- Nitrate – it is a readily available source of food for algae and higher plants, whereas nitrite is toxic to most animals and plants. Fortunately, the rate of oxidation of nitrite is faster than the rate of formation and concentrations of nitrite rarely exceed 1-2 mg/l in polluted water (Focht & Chang 1976). Nitrate, in contrast, is physiologically inert in adults at concentrations of 20 mg/l or more. It is well recognized as a curing agent for meats and leafy vegetables. However, some infants do not have sufficient acidity in their stomachs to inhibit the growth of bacteria, which can reduce nitrate to nitrites. Nitrites are objectionable, not only because they combine with hemoglobin, reducing its oxygen carrying capacity, but also, because they are believed to combine with organic amines in the body to produce carcinogenic nitro-amines (WHO 1996). The first effect can cause "blue babies" disease, if the nitrate concentration in water, which is used to prepare infant formulas, is in excess of 10 mg/l as nitrogen, equivalent to 45 mg/l as nitrate (Dean & Lund 1981). WHO recommended that the maximum limit for drinking water is 10 mg/l NO_3^--N, waters with higher concentrations represent a significant health risk (Chapman 1998). In lakes, nitrate levels of 0.2 mg/l nitrate nitrogen tend to stimulate algal growth and indicate eutrophic conditions (Chapman 1998).
- Organic nitrogen – it consists of protein substances (e.g., amino acids, nucleic and urine) and the products of their biochemical transformations (e.g. humic acids and fulvic acids). It is formed in natural water principally by phytoplankton and bacteria and cycled within the food chain. Organic nitrogen is usually determined using the Kjeldahl method, which gives a combined result of the total ammonia nitrogen plus total organic nitrogen (TKN).

2.2.2.2 Phosphorous compounds

There is a general agreement that the degree of phosphorus loading to water bodies, in terms of both dissolved and particulate species, determines the level of algal production or eutrophication. The relationship between phosphorus loading and algal biomass was assessed by Vollenweider (1968), and has resulted in remedial measures being taken to reduce external phosphorus inputs to many lakes suffering the severe effects of eutrophication.

In natural water and wastewaters, phosphorus occurs mostly as dissolved orthophosphates and polyphosphates, and organically bound phosphorous. Changes between these forms occur continuously due to decomposition and synthesis of organically bound forms and oxidized inorganic forms, at different pH values. Phosphorus in water bodies' sediments can be mobilized by bacteria and released to the water column. Sources of phosphorus include domestic wastewater, industrial effluents, detergents and fertilizers that could be washed out with the runoff from agricultural land. In most natural surface waters, phosphorus ranges from 0.005 to 0.020 mg PO_4-P/l (Chapman 1998).

2.2.3 Inorganic constituents – metals

"Heavy metals" is a generally collective term applying to the group of metals and metalloids with an atomic density greater than 6 g/cm^3. Although it is only a loosely defined term, it is widely applied to elements such as Cd, Cr, Cu, Pb, Zn, Ni and Hg, which are commonly associated with pollution and toxicity problems. In contrast, the ability of a water body to support aquatic life, as well as its suitability for other uses depends on many trace elements. Some metals e.g., Mn, Zn, and Cu present in trace concentrations, are important for the physiological functions of living tissue and regulate many biological processes. However, even small amounts of Hg and Cd can be toxic to both humans and aquatic life. The absence of Fe and Mn in some priority pollutants' lists results from their frequent classification as major ions (Chapman 1998). The effect of these metals on aquatic organisms depends upon the bioavailabilty of the metals to organisms. Bioavailability is influenced by water hardness, pH, life cycle (age) of organisms and water temperature. Water hardness diminishes the potential of dissolved metals to cause toxicity to aquatic life. Softer waters tend to make the metals even more toxic. In surface waters, at normal pH and redox conditions, most trace elements are readily adsorbed onto particulate matter. Consequently, the actual dissolved element concentrations are very low and the monitoring of trace metals on a routine basis is difficult due to the contamination of the sample from air, particularly lead. The toxicity of metals in water also depends on the degree of oxidation of a given metal ion together with the forms in which it occurs. For instance, the maximum allowable concentration of Cr^{+6} in USA is 0.001 mg/l whereas for Cr^{+3} it is 0.5 mg/l (Chapman 1998). As a rule, the ionic form of a metal is the most toxic form. However, toxicity is reduced if the ions are bound into complexes with, for example, natural organic matter such as fulvic and humic acids.

Considering the toxicity of different water quality constituents, normally the criteria used to limit their concentrations in order to prevent health or environmental hazards, are determined based on experimental or epidemiological studies. The last are long and costly investigations and in many cases the results are unreliable. Experimental studies are usually carried on animals whose reactions are known to be similar to the humans'. The results obtained are then extrapolated to conform to human conditions by means of models. The effect of toxic constituents on human health and aquatic life could be expressed in various ways. The following definitions (after Degremont 1991) could be useful in order to understand the effect of different pollutant constituents and corresponding criteria for their regulation:

- Acute toxicity – this is an indicator, which shows that the constituent leads to the death of the animal studied. The LD 50 value indicates the lethal dose, leading to death, in 50% of the individuals tested, for a specified time, e.g. 24h.
- Chronic toxicity – this is a dose, which, if ingested on a daily basis, leads to the premature death of the tested individuals. ADI (Acute Daily Intake) indicates a maximum dose ingested daily over a lifetime, which could be withstood by the metabolism of an individual without any risk.
- Cytotoxicity – this indicates doses of constituents, leading to the death of a certain percentage of cells. The tests are performed on cell cultures, but not on animals.
- Mutagenicity – this is a toxic effect, causing mutations in the tested individuals. The risk exists regardless of the dose of the constituent.
- Carcinogenicity – similar to mutagenicity, this is a toxic effect, which after ingestion or exposure causes the appearance of a malignant tumor.

2.2.4 Organic constituents

In general, the tests to determine the organic content of water and wastewater may be divided into two categories. The first category measures gross concentrations of organic matter, greater than 1 mg/l. It includes Biochemical Oxygen Demand (BOD), Chemical Oxygen Demand (COD) and Total Organic Carbon (TOC). The second category measures trace concentrations in the range of 10^{-12} to 1 mg/l. Normally, instrumental methods as gas chromatography and mass spectroscopy are used to determine trace organics.

The trace organic constituents found in water are derived from the breakdown of naturally occurring organic materials, domestic, commercial, industrial and agricultural activities. Naturally occurring organic materials include humic materials, microorganisms, and aliphatic and aromatic hydrocarbons. Organic materials from humans' activities often identified as "synthetic organic compounds" (SOCs) include constituents such as: pesticides, herbicides, degreasers, solvents, etc. A special class of SOCs known as "volatile organic compounds" (VOCs) is of concern in water pollution control because many of these compounds are possible or known human carcinogens.

- Biological Oxygen Demand (BOD) – by definition this is the quantity of oxygen consumed at 20°C and in darkness during a given period to produce the oxidation of the biodegradable organic matter present in water by biological means. The standard test is performed for a period of 5 days and is denominated as BOD_5. The BOD test is the most common test used in the field of wastewater control and treatment. If sufficient oxygen is available, the aerobic biological decomposition of an organic waste will continue until all of the waste is consumed.
- Chemical Oxygen Demand (COD) – this test is used to measure the oxygen equivalent of the organic material in wastewater that can be oxidized chemically using dichromate in an acid solution. In the vast majority of cases, COD of a given sample exceeds considerably its BOD value, because the chemical oxidation includes organic materials, which are difficult to biodegrade or are non-biodegradable.
- Total organic carbon (TOC) – this test is performed instrumentally, and it only takes 5 to 10 minutes to complete. It measures the organic carbon incorporated in organic materials present in a water sample, and as such represents the amount of the carbonaceous organic material, which is the most readily biodegradable. If a valid relationship can be established between results obtained with the TOC test and the results of the BOD test for a given wastewater, the use of the TOC test for process control is recommended. Typically, the TOC content of surface and ground water will vary from 1 to 20 mg/l and 0.1 to 2 mg/l, respectively.

TOC, BOD and COD reflect different fractions of the organic matter present in the sample and for a typical untreated municipal sewage the following indicative ratios could be mentioned:

$$TOC : BOD_5 : COD = 1 : 1.3 : 1.6.$$

Different types of water or wastewater samples would have different ratios. The lowest the ratio, the easier the available organic matter would be degraded by biological means.

2.3 *Biological characteristics*

Microorganisms are commonly present in surface water, but they are usually absent from most ground water because of the filtering action of the aquifer. A wide variety of microorganisms could be found in surface water, including viruses, bacteria, fungi, protozoa and nematodes. The most common microorganisms are bacteria. They vary in shape and size from about 1 to 4 μm. Disease-causing bacteria are called "pathogenic" bacteria. These bacteria, which require oxygen for their survival, are called "aerobic", while those, which thrive in an environment free of oxygen, are "anaerobic". "Facultative bacteria" are those that live either with or without free oxygen.

It would be extremely time consuming and would provide little useful information if we try to identify all types of microorganisms present in polluted water. All water – borne and many of the water – based diseases, are dependent on their transmission upon the access of feces from a contaminated individual, which contain pathogenic microorganisms, into a water supply source used for drinking or bathing purposes. Because the number of pathogenic microorganisms present in wastes and polluted water are relatively few and difficult to isolate and identify, the coliform bacteria, which are harmless, more numerous and more easily tested for, are commonly used as indicator organisms. *Escherichia coli (colon bacilli or coliforms)* are bacteria that inhabit the intestines of warm-blooded animals. The intestinal tract of humans contains countless rod-shaped bacteria known as coliforms. Each person discharges from 100 to 400 billion coliform organisms per day in addition to other kinds of bacteria. Thus, the presence of coliform organisms is taken as an indication that pathogenic organisms may also be present, and the absence of coliform organisms is taken as an indication that the water is free from disease-producing organisms. We could differentiate between "total coliforms" (TC) and "fecal coliforms" (FC). FC are capable to survive at 44 °C and originate exclusively from human feces, while other coliforms may occur naturally in unpolluted water and soils, as well as in feces. Their survival ambient temperature is 37 °C. TC are determined at this temperature and therefore include all types of coliforms, but only FC are the indicator organisms, which would show the presence of fecal contamination. However, Horan (1990) warns that in tropical conditions certain bacteria of non-fecal origin fulfill the definitions of FC, therefore results with respect to FC should be interpreted with caution.

Viruses are obligate parasitic particles consisting of a strand of genetic material with a protein coat. They do not have the ability to synthesize new compounds; instead they invade the living (host) cell where the viral genetic material redirects cell activities to the production of new viral particles at the expense of the host cell. When an infected cell dies, large numbers of viruses are released to infect other cells. Viruses that are excreted by human beings may become a major hazard to public health. The FC test indicates the possible presence of pathogen bacteria and viruses.

Algae are single-celled plants and can cause nuisance in surface waters because they reproduce rapidly under conditions of nutrient supply, warm temperature and sunlight, leading to the formation of large floating colonies in streams, lakes, and reservoirs, called algae blooms.

Other microorganisms would include plants and animals of importance, ranging in size from microscopic rotifers and worms to macroscopic crustaceans. Knowledge of these organisms is helpful in evaluating the condition of streams and lakes. Methods to control surface water quality by the use of biological indicator organisms (the procedure is known as "biomonitoring") are described briefly in Chapter 12.

3 DIFFUSE POLLUTION - GENERATION AND SOURCES

Diffuse pollution is generated during the process of rainfall and the corresponding runoff to natural water bodies and as such is difficult to be clearly evaluated and defined. In order to do it, we need a considerable database regarding the natural water quality status of the corresponding body, which needs to be analyzed, and trends of deterioration regarding specific parameters or benchmarks established. In this context the evaluation and detection of diffuse pollution is directly connected to the continuous and regular control, monitoring and assessment of the water quality status of natural water bodies. In addition, the level of diffuse pollution is directly related to the quantity and quality of runoff, which depends on the rainfall pattern, the land topography and land cover. Thus, in many cases of diffuse pollution, the major causes are associated with changed land use practices, caused by deforestation and large-scale construction.

Basically, we could divide the sources of diffuse pollution into two broad categories, depending on their location – in urban areas, and in rural (non-urban areas). However, in many cases we could find mixed patterns.

Most of the chemical composition of surface runoff is due to erosion from soil surfaces and the wash-off of particles from impervious surfaces. Organic matter, metals, different cations and anions may be transported and delivered to natural water bodies in particulate or dissolved form. Diffuse pollution of ground water occurs mainly in dissolved form. Based on the transportation pattern, diffuse pollution constituents in surface runoff could be found in:

- Solid phase – these are chemicals, which are strongly associated with particulate matter. These include organic nitrogen, particulate phosphorus and heavy metals. The assignment of metals to this category is arbitrary, since dissolved forms are often present under acidic conditions. Therefore, it is assumed here that the primary sources of metals in runoff are metal-based pesticides, which are tightly bound to solid particles. In urban runoff metals are also most often associated with particulate material, while metals in ground water might be in dissolved form due to soil acidification
- Dissolved phase – these are chemicals, which are dissolved in runoff. This group includes mainly in-organic nitrogen and soluble phosphorus. Inorganic nitrogen in the drainage is mostly

nitrate-nitrogen, and this ion does not adsorb to soil particles. Phosphorus is a special case. Most phosphorus in runoff is solid-phase, but dissolved phosphorus is directly available to plants and algae and hence cannot be neglected in eutrophication studies.

- Distributed phase – chemicals, which are transported in both solid phase and dissolved forms. Most organic pesticides could be included in this group.

3.1 *Atmospheric depositions*

Diffuse pollution from atmospheric depositions is associated with polluted precipitation (rainfall) and is also known as "acid rain". It is caused by polluted atmosphere, where rain droplets and atmospheric moisture absorb and dissolve oxides of atmospheric carbon dioxide, sulfur (emitted from burning fuels and specific industrial processes) and nitrogen (emitted from heavy traffic). These constituents make the precipitation acidic and damages man-made structures and the environment – vegetation cover, soils and water resources. The phenomenon of acid rain occurs most often in heavily industrialized and densely populated areas or in the neighborhood of heavy-industry enterprises.

3.2 *Sources of diffuse pollution in urban areas*

The most common sources of pollution within urban population centers are associated with polluted runoff from roofs, streets, highways, and open developed areas, but other sources could be mentioned as well. A brief description of such sources and the associated process of generation are discussed briefly.

3.2.1 Urban drainage

Storm water runoff in urban areas, known also as urban drainage, is usually much higher in volume and quantity, than rural runoff due to the distribution of impervious surfaces. In the case of separate sewers, main pollution loads are associated with conventional pollutants (SS, BOD_5, N, P), metals and organic compounds. They are attached to the sediments, carried with the runoff, and collected by urban drainage systems. In the case of combined sewer systems, any overflow during rains contributes a pollution load to the surface water, which has the characteristics of a weak sewage. Road runoff is an important source of contaminants, including oils, tar products, dioxins, metals, phenols and other materials mainly accumulated in sediments, which could have a toxic effect to several fresh water organisms. The BOD_5 values of urban runoff vary from 10 to 285 mg/l in USA, and from 100 to 350 mg/l in UK (Ellis 1989).

One of the most problematic areas, related to urban drainage pollution, is the presence of a wide variety of trace metals. Pb, Cd, Cu and Zn are the most frequently occurring metals, causing most concern in terms of their levels and toxic effects on aquatic life. The metals are derived mainly from vehicle exhaust emissions, roof and vehicle corrosion, tyre wear and atmospheric fallout. Deposition rates on street surfaces vary with the use of the developed area-industrial zones having greater deposition rates than commercial and residential areas.

The following points should be kept in mind when evaluating the risks of toxic metals in urban storm water. First, a large fraction will be attached to suspended particulates, which effectively reduces the level immediately available for biological uptake and bioaccumulation. Second, urban runoff events typically occur over shorter durations (1-6 hours) than the exposure intervals used in conventional bioassay toxicity tests (24-169 hours for chronic toxicity criteria). Also, urban storm water runoff will be subject to substantial dilution following mixing in the receiving water. Despite of these considerations, it is likely that trace metals in runoff would be toxic to the receiving water habitat, particularly for the more soluble species of Zn and Cu, if long retention times are involved (Ellis 1989).

The form and concentration of polluting materials in runoff will vary both with the density of population and with the type of activity associated with a particular area. In many cases, the surface runoff characteristics from industrial or commercial areas will differ from the runoff generated by purely residential areas. In general, the following classification of different types of land use practice in urban areas could be done, although it is not applicable to many cities, which have mixed patterns:

- Residential (low, medium and high density)
- Commercial
- Industrial (light and heavy industry)
- Other developed land (parks, sport complexes, large parking areas, cemeteries)
- Transportation (roads, streets, parking slots, highways, airports)

3.2.2 *Other sources in urban areas*

Urban drainage is considered by far, the most important source of diffuse pollution in urban areas and at present more efforts have been focused on abatement measures with respect to this source. However, other sources, as listed below, are also important, especially for developing countries.

- Solid waste disposal sites – unprotected solid waste disposal sites could be a major source of diffuse pollution to surface and ground water, characterized by high organic loads, toxic pollutants and pathogens. Protected disposal sites would require a special treatment of surface runoff and leachate, and could be classified as a point source of pollution.
- Illegal discharges and informal settlements – this category includes any point or non-point discharges, which are not authorized and accounted for. They are classified as diffuse pollution because of their possible large numbers and difficulty in location. Major sources of diffuse pollution are informal settlements, which are typical for developing countries. These are squatter camps at the outskirts of towns and cities, which most often have a very limited water supply and lack sanitation facilities. Other form of illegal discharges due to informal settlements are unauthorized buildings in developed areas, which discharge their effluents in the drainage system or directly on streets.
- On-site sanitation (septic tanks and pit latrines) – they could be a significant source of diffuse pollution only in the case of improper application practices, or when applied in areas with high ground water levels, which would be directly polluted by leachate from latrines or septic tank soak ways.

3.3 *Sources of diffuse pollution in rural areas*

The major cases of diffuse pollution in rural areas are associated with agricultural practices and animal operations. The most common cases are as follows:

- Application of fertilizers, pesticides and insecticides – excessive use of fertilizers, when application rates exceed the assimilative capacity of the crops, would lead to soil enrichments with nutrients. They could find their way to water resources in the form of polluted runoff, or in the form of dissolved fractions, could infiltrate to ground water. The same applies to pesticides and insecticides. This type of pollution is of higher concern, considering the fact that even low concentrations of SOCs have toxic effects on humans and the aquatic environment.
- Irrigation return flow – this is the excess irrigation flow, usually characterized by increased salts concentrations, compared to the original water used for irrigation. It percolates to ground water and might form the base flow in rivers, downstream of the irrigation fields. In cases where irrigation fields have drainage systems to collect and convey excess flows, they would be discharged into surface water. Irrigation return flows might be classified as diffuse pollution in specific cases of increased salt concentrations. Usually, this would be the case of irrigation in arid regions, where the evapotranspiration rates are very high. Another example is when the downstream users of irrigation return flows, reuse this water for irrigation again, and in this case we have a repeated cycle of salinization.
- Irrigation with wastewater/sludge – causes of pollution in such cases are associated with improper practice and excessive application rates. Pollutant constituents consist of organic matter, nutrients, microbiological contamination and heavy metals. Increased salinity of ground water is enhanced, because wastewater's salt concentrations, even after treatment, are much higher compared to natural water. It should be noted that this type of pollution source has been included under the rural areas

classification, but it is also applicable in many cases of urban or suburban developments, as it is illustrated by the case studies presented in Chapters 8, 9 and 10.

- Animal operations – diffuse pollution from barnyards, where animals are concentrated for milking or other operations might constitute a significant pollutant load through runoff, if not cleaned and managed properly. Pollution is severe in terms of organic matter, nutrients and pathogenic microorganisms.
- Mining operations – mine drainage and the wastewater from mineral processing could be defined as point sources of pollution from such activities. However, there is a significant amount of polluted water coming as a result of runoff from stockpiles, spoil heaps and surface drainage, which are typical non-point sources of pollution. Open mining activities are large contributors of diffuse pollution. Acid waters, together with dissolved iron compounds usually associated with them, are amongst the principal problems associated with mine discharges. Dissolved sulfates, together with chlorides, are frequently present in mine runoff to an appreciable concentration. In general, pollutant constituents depend on the geological formation of the site.

3.4 *Pollution concentrations and runoff events*

Given the nature of the generation of diffuse pollution, pollution concentrations vary considerably depending on: the rainfall characteristics, runoff erosivity, land cover, atmospheric depositions and other parameters. In addition, it is expected that a flush effect would lead to higher concentrations of pollutants at the beginning of the storm. The same applies to rainfall events at the beginning of a rain season. Many attempts have been made to obtain more uniform values, which could be used in models for the simulation of different scenarios and the prediction of undesirable effects. In general, it has been agreed that pollution concentrations of runoff should be determined as event orientated statistical values. This approach requires a considerable monitoring effort and a large data set in order to determine a statistically significant result with respect to mean pollutant concentrations of a given type of storm. Correlation between these values and specific land use practices has not been shown in all cases, but a statistically significant difference was found between urban areas and open spaces. Statistically significant results to confirm flushing effect in storm sewers were not found for the climatic conditions of USA (Novotny 2003). More information regarding the monitoring of runoff pollution and the determination of the concentrations of pollutant constituents is presented in Chapter 2.

4 METHODS OF SURFACE RUNOFF DETERMINATION

Diffuse pollution is closely related to the amount of rainfall and the portion of it, which would reach water bodies in the form of runoff. It has been mentioned that diffuse pollution of surface water bodies is associated mainly with the particulate material in runoff and the soil erosive capacity. Therefore, the correct estimation of surface runoff should be the basis of the estimation of pollution loads generated by surface runoff. It should be noted that not every rainfall would generate runoff or should contribute in the same way to diffuse pollution loads. Diffuse pollution abatement measures should focus on typical storms of medium magnitude with relatively high frequency and low return periods. These are the ones contributing the most significant pollution loads. However, in engineering practice, of most importance has been the determination of the "design storm", which is applied to determine design flows of drainage systems and other flood protection structures. Therefore, the analysis of the available rainfall data in the case of the estimation of pollution from surface runoff requires specific attention and rainfall data evaluation.

4.1 *Rainfall excess and surface runoff*

Surface runoff is that portion of the rainfall, which reaches natural or man-made drainage canals, after it falls on the ground and travels to the point of consideration, and, which remains after all water losses are satisfied. The most important rainfall losses along its way of the rainfall to the water body could be defined as follows:

- Interception storage – This is the part of the rainfall, which adheres to the surface of vegetation and other aboveground objects and is returned to the atmosphere by evaporation. Its value depends on the type of vegetation, intensity and volume of the rainfall and the growth stage of the vegetation. Grass and dense shrubbery could intercept 1.2-1.8 mm of the rainfall (Novotny 2003).
- Depression storage – This is the part of the rainfall, which is detained in surface depressions, which need to be filled before the runoff could be transported further. This water evaporates or percolates to the soil. The amount of depression storage depends on the moisture content of the ground cover and topography. In many cases, the interception storage and depression storage are grouped together and evaluated as one value, termed, "surface storage".
- Evapotranspiration – it represents water loss into the atmosphere by the combined effect of evaporation from soil and water surfaces and transpiration by plants. Transpiration denotes the water abstracted by plants from soil moisture and released to the atmosphere, as part of their life cycle.
- Infiltration – this is the process of percolation of the surface storage into the soil. It is a function of soil permeability, moisture content, vegetation cover and other factors. After reaching the aquifer, this portion of the rainfall is known as "groundwater runoff" and is the main source for ground water recharge. It is the source of the "base flow" in rivers and streams, maintained by springs and other forms of ground water discharges to the natural water bodies. The infiltration of ground water to sewer systems could be included in this category as well.

Considering the above-mentioned losses, we could differentiate between rainfall excess and surface runoff in the following way:

- Rainfall excess or "net rain" is used to denote that part of the rainfall, which is left after the subtraction of the above mentioned losses, and is expressed as the depth of water in "mm" over a given surface area for a specified period of time.
- Surface runoff – it represents that part of rainfall, which has been generated by the rainfall excess, and forms part of the surface flow in natural rivers and streams. It is expressed as flow rate (volume per unit time).

4.2 *Determining rainfall excess*

Rainfall excess could be determined by the curve number method, known as the Natural Resources Conservation Method (NRCM), developed first in the USA. It determines the rainfall excess as a function of the rainfall volume, surface storage and infiltration (Novotny 2003). Based on extensive rainfall/runoff data and variety of soil and cover conditions, the method results in the development of a set of curves, with a specific number, which links the 24-h rainfall with the corresponding rainfall excess. Each curve number is dependent on the type of land use, level of imperviousness, hydrologic conditions and type of soils. Different soil conditions are classified in four categories. This method was adapted for the Southern African conditions and is known as the SA-SCS method (Shulze et al. 1993). More detailed description of the procedure applied is given in Chapter 5, section 2.4. A simplified procedure, which could be used in cases where data is not available and the required accuracy is not high, is described in Chapter 4.

4.3 *Determining surface runoff*

4.3.1 The rational method

This method, also known as Lloyd-Davis method, is the oldest and most widely used method in engineering practice for determination of the design runoff quantity, during the process of drainage structures design for flood prevention. Its purpose is to determine a design surface runoff flow rate, which would be the base for the sizing of the conveying structure to transport the runoff from a given area to a point where it could be discharged safely into a natural water body or disposed on land. Thus, this method focuses on a selected rainfall event, which is the most probable one to cause flooding within a given period of time. In other words, the method determines the runoff from high intensity storms with a relatively low probability of occurrence. It should be applied with caution when used for the determination of pollution loads and the design of pollution abatement structures, as in most cases, diffuse pollution is attributed to medium storms with a high frequency of occurrence. Therefore, the selection of an appropriate "design" storm in terms of pollution control and abatement is very important. The method is based on the following assumptions:

- The peak rate of runoff at any point is a direct function of the average rainfall intensity during the time of concentration to that point.
- The frequency of the peak discharge is the same as the frequency of the average rainfall intensity.
- The time of concentration is the time required for the runoff to become established from the most remote part of the drainage area to the point under consideration. It includes the overland flow time (inlet time) and the time of flow along the channel, governed by channel hydraulics.

Reported practice generally limits the use of this method to urban areas of less than 13 km^2 (White 1978). For larger areas, the application of hydrograph methods is recommended. The rational method is represented in the following formula:

$$Q = C\,i\,A$$

where Q = the peak runoff rate; C = runoff coefficient, which depends on characteristics of the drainage area; i = the average rainfall intensity (i_{av}); and A = the drainage area.
The drainage area information should include the following:

- Land use – the present and predicted future practice-as it affects the degree of protection to be provided and the percentage of imperviousness.
- The character of soil and cover as they may affect the runoff coefficient.
- The general magnitude of ground slopes, which, with pervious items and shape of the drainage area, will affect the time of concentration.

The application of the rational method requires information with respect to rainfall data, as well as a clear understanding of the concepts and principles involved. The main parameters and procedures included in the determination of the i_{av} value are explained below:

- Time of Concentration – this is the time, required for the surface runoff flow to travel from the most remote part of the drainage area to the point of consideration along a conveying conduit. For urban drainage systems, the time of concentration consists of the inlet time plus the time of flow in the conduit from the most remote inlet point to the point under consideration. The time of flow may be estimated closely from the hydraulic properties of the conduit. The inlet time is the time for runoff to reach established surface drainage conduits. It would vary with surface slope, the nature of surface cover, and the length of the path of surface flow, as well as with variables such as the soil infiltration capacity and depression storage. Reported inlet times used for design purposes vary from 5 to 30 minutes, with 5 to 15 minutes most commonly used. In densely developed areas an inlet time of 5 minutes is often reported. In well-developed districts with flat slopes 10 to 15 minutes is common, and in flat residential districts 20 to 30 minutes are customary.
- Rainfall intensity – duration relationship – they are important characteristics of any rainfall event. Usually storm events have varying intensity along its duration. The curve representing the variation of the rainfall intensity during the storm duration is called "hyetograph". The i_{av} of the rainfall event would be equal to the cumulative depth of rainfall (in mm) divided by the storm duration. An "intensity-duration" curve for a given storm event would be constructed by rearranging and averaging the maximum consecutive rainfall intensities over an increasing storm duration, starting with the peak intensity. A graphical representation of the formation of intensity–duration curve for a specific storm event is shown on Figure 1.2. Different storms usually have different intensity duration curves. Rainfall events are statistical characteristics. The analysis of large data sets of storm records has found that the

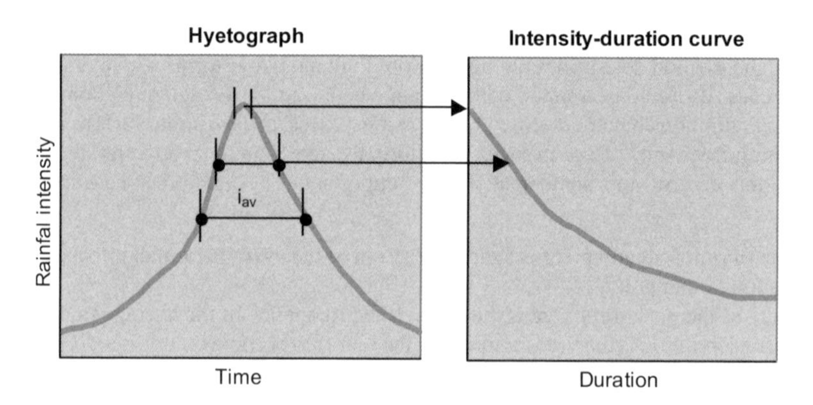

Figure 1.2. Hyetograph and corresponding intensity – duration curve for a storm event.

relationship between the i_{av} of a given storm and the storm duration are similar for similar geographic conditions. This has led to the development of unified intensity-duration curves for the specific geographic region and their linkage to the frequency of occurrence of the rainfall event.

- The frequency of a storm event with given i_{av}, is a statistical parameter, which shows how often this specific storm would be probable to occur, e.g. if a given storm event with i_{av} = 50 mm/h occurs once per two years, its frequency is 0.5. The frequency of a storm occurrence is reciprocally connected to the term "return period", also known as "recurrence interval", which shows the time period in which this specific storm would be most probable to occur. Considering the storm with a frequency of 0.5, its return period would be 2 years.

- Intensity-Duration-Frequency Relationships – these represent a set of intensity – duration curves, for different return periods, which are determined based on the statistical analysis of rainfall data for a given geographic region. The length of record, gage accuracy, placement, and density, significantly affect the reliability of rainfall data. Only the average intensity for a given duration may be determined from the rainfall intensity duration curve. This set of curves are used to determine the design value of i_{av}, which will be used in the rational formula, based on the determined time of concentration for the point under consideration. The method assumes that the peak flow would be formed when the storm duration is equal to the time of concentration. In practice, there are some cases when this assumption is not valid. Because of this, the hydrograph method, described in the following section is considered to be more accurate, but more data would be necessary for its application. One point, which would require specific attention during the process of determination of the value of i_{av}, is the choice of the correct "design storm", which would be used for the procedure. This choice would consider the possible effect and risks associated with the return period of this specific storm. In the case of flood protection structures, the "design storm" would have return periods of about 5, 10, 20 or 100 years. If the return period is 10 years, it means that there is a probability that a peak flow, higher than the design one, would occur once in 10 years. If the designer chooses a higher return period, then the inevitable risks associated with this choice are lower. However, the cost of the protection structure would rise considerably. Thus, the choice of the design storm with a corresponding return period is a compromise between risks and cost of the structure, which should be made after careful consideration of the possible undesirable consequences. The procedure is known as "risk assessment". In the case of diffuse pollution abatement measures (Chapter 2), and specifically, the provision of retention structures, two design storms must be considered-one with respect to the flood protection, and another with respect to pollution abatement. The last would have a much lower return period.

The Runoff Coefficient (C) is the variable least susceptible to precise determination. The range of coefficients, classified with respect to the general character of the tributary area reported in use, is very wide and could be found in numerous literature sources (White 1978, Novotny 2003). Actually, this coefficient gives a less accurate evaluation of the rainfall excess of a uniform rain with duration equal to the time of concentration, and reflects the level of impermeability of the drained area.

4.3.2 The unit hydrograph method

A hydrograph is a plot of flow rate versus time at a point of interest on the drainage system along a man-made conduit or along a natural stream or river. The development of the unit hydrograph theory serves to represent in mathematical form the transformation of rainfall excess into runoff. The unit hydrograph could be defined as the division of the excess rainfall into small pulses of the same duration as the unit rainfall input, and applied uniformly over the whole catchment basin. The unit hydrograph is the multiplier determining individual response hydrographs from the basin with the same duration. All individual hydrographs, based on the unit hydrograph applied for the specific conditions within the catchment area, are summed in order to obtain the final hydrograph, representing the total flow variation during a specified rainfall event for this specific basin.

The NRCS method for rainfall excess determination provides a methodology for the overland routing of excess rainfall. It proposes a triangular shaped dimensionless hydrograph, which allows for the peak flow determination. The time at which the peak flow will occur (peak time) would determine the shape of the hydrograph and would be different from the time of concentration. The value of the peak flow of the unit hydrograph will be determined by the watershed's area, storage and average slope. There is provision for adjustment of the result to incorporate the effect of imperviousness of the different types of areas. After the unit hydrograph for the area under consideration is constructed, it is used to convolute the excess rainfall into the runoff of the real storm event. The application of the method requires the determination of the storm hyetograph.

Urbanization leads to an increase in the impervious surfaces and the construction of drainage structures results in a fast transport of the surface runoff. Both of these reflect a considerable increase in the runoff peak flow, resulting in a changed urban streams morphology. Higher runoff peak flows lead to a higher erosion of the streambed and stream banks, lower base flow conditions and generally, causes damage to the natural habitat. In terms of diffuse pollution aspects, higher peak flows result in a higher possibility for pollution transport and change in the background water quality. In order to prevent this, in the USA, developers are obliged to provide measures and keep the peak runoff flows (or runoff volumes, or both), in relation to a predetermined design storm, at the pre-development levels (Novotny 2003). This could be achieved by different abatement measures and by careful urban planning practices, such as the reduction of impervious surfaces, provision of retention structures, infiltration facilities, etc. More details on the different types of abatement measures, with respect to diffuse pollution are presented in Chapter 2.

4.4 *Rainfall-runoff models*

The methods described above are event orientated. They allow for the determination of peak flow rates from selected storm events, defining specified conditions, which usually are connected with the design of flood protection structures. In addition, they assume constant values for the rainfall losses as evapotranspiration and infiltration. Continuous simulation models attempt to represent the entire hydrologic system, simulating the natural environment. They consider the whole runoff process, including surface runoff storage, infiltration, overland routing and channel flow on a continuous basis, considering the variation of parameters in time and space. The major advantage of models is the possibility to apply large data sets and to simulate and predict different scenarios at conditions closer to the naturally happening events.

Different models have been developed and applied in practice; some examples are the EPA storm water management model SWMM, which provides for qualitative aspects to be evaluated as well (Debo & Reese

2003, www.epa.gov/ceampubl, http://ccee.oregonstate.edu/swmm/); the ARM transfer function model (Novotny 2003); or the Rainfall flow erosion (RAFLER) model (Stephenson 2003). The practical application of these models in many cases is limited because of the large amount of data input needed, the complexity of application requiring advanced software and hardware facilities, and the need of highly trained personnel. In some cases, the application limitations might be related to the basic assumptions during the model development, connected with the mathematical tools applied for the problem solution.

Considering the sources and generation of diffuse pollution loads, simple techniques of quantification of runoff might lead to errors in the pollution loads estimation due to an inaccurate estimation of quantitative and qualitative parameters. A more detailed illustration of such possibility is presented in Chapter 4. An accurate pollution load determination should be based on the evaluation of event orientated actual pollution loads, rather than on average annual figures. Such an approach requires model application in order to incorporate the large amount of input data regarding the rainfall events, watershed and drainage system characteristics and water quality. A very useful tool in this direction is the application of geographical information systems (GIS), to form the basis for model application and data presentation.

GIS are data information systems, which deal with data referenced by spatial or geographic coordinates. It also contains components, allowing for data management and analysis procedures, and other functions, as the possibility for incorporation of more sophisticated models. Information can be stored in different layers, e.g. the topographic information is stored in one layer, the soil information on another, the ground water aquifer on a third layer, surface water on a fourth layer, etc. Each layer represents the spatial variation of the features, together with data for each specific item featured, e.g. the surface water layer could show the rivers and streams of a catchment basin together with monitoring data with respect to sampling points, as well as, qualitative and quantative data sets for each point. The system allows for the interaction of the information between different layers and also for processing the information by different mathematical expressions or by different models. Incorporating different models, databases, expert systems or other tools for data analysis into a GIS, requires a common framework between the different components in order to allow for data exchange in both directions, as input or output values. Thus GIS are very valuable tools because they allow for:

- Data storage and retrieval;
- Data processing, transformation and manipulation in an interactive way – spatially within the plane of the layer under consideration, and between different layers;
- Data visualization.

Recently, GISs find very wide applications in the field of water resources management in developed countries. Their applications in developing countries could be encouraged as well, but in many cases, it is restricted by the lack of data availability and the lack of culture to keep records of a different nature. The implementation of GIS in the water resources management practice in general, and the diffuse pollution control and abatement practice in particular, could become a valuable tool to support the decision-making process.

5 REGULATORY APPROACHES AND INSTRUMENTS

Considering its nature and generation, diffuse pollution is difficult to be evaluated, regulated and controlled. This process should be viewed as an integral part of the process of regulation and control of water quality in general. In order to understand it, basic definitions, principles and philosophies of water quality regulation is essential.

The practice of water pollution control in the past years was focusing on point sources of pollution only. Thus, in many countries, the regulatory basis was orientated in this direction, and did not have provision for the control of diffuse pollution sources. While in the USA and EU countries, new regulatory instruments have been developed in order to account, control and prevent water pollution from diffuse origin, in many countries such instruments do not exist, or are in an initial stage of development. In some cases the basic regulatory instruments are available but not implemented. For this reason, the proper understanding of existing regulatory instruments at local, national and international level, and their upgrading to include control and abatement of diffuse pollution sources, is a very important aspect of the contemporary development in the practice of water pollution control and water resources management.

5.1 *Basic definitions and concepts*

Water pollution regulatory instruments include a big variety of documents, which vary in different countries. One general classification could be made, as follows:

- Water Quality Standard – this is a document defining water quality objectives and restrictions, which is recognised in enforceable environmental control laws of a level of government. It is a definite rule, principle, or measure established by authority. Since established by authority, it is official or legal. However, because some documents are termed standards, it does not mean that they are rationally based on the best scientific knowledge and engineering practice.
- Criteria – this is scientific data, evaluated to derive the recommended limits for different water uses. A "criterion" should be capable of quantitative evaluation by existing analytical tests, and also should be capable of a definite resolution. In contrast to standards, criteria have no connotation of authority. When data is gathered in order to be used as a yardstick of water quality, "criterion" is the proper term. Water quality criterions are the benchmarks of the process of pollution control and they are internationally applicable in most cases. They are usually determined based on scientific investigation with respect to the effect of different constituents on human health, environment, crops, and others, e.g. toxicity tests.
- Guidelines – These are numerical values or could be narrative statements, recommended to support and maintain a designated water use. They define the required water quality, linked to the intended water use but are not legally enforceable. Usually, guidelines represent a set of criterions with respect to different constituents, formulated to provide a safe designated beneficial use of the water.
- Water Quality Objective (Goal) – This is a numerical value or narrative statement that has been established to support and protect the designated uses of water at a specified site. Usually, these types of documents address surface water bodies and establish the objectives to maintain their environmental integrity.

5.2 *Basic approaches*

Historically, the contemporary development of regulatory instruments for water quality control started with the need to protect public health from water-borne diseases and prevent their spreading. In this respect, the first priority and goal to be achieved, was the need to define and enforce in practice documents regarding drinking water quality and natural water bodies used for this purpose. Later on, it was to arouse the need to prevent the environment in general and natural water bodies in particular, in order to provide for a sustainable use of natural resources and protect them for the future generations. In this regard, we could differentiate between two basic philosophies in setting regulatory documents – the Uniform Effluent Standard (UES) approach and the Receiving Water Quality Objective (RWQO). The UES approach stresses on the quality of effluent discharges into receiving water bodies, while the RWQO approach emphasizes on the quality of the receiving water bodies and their designated beneficial use. This approach is not explicitly detailed regarding the effluent discharges directed to the receiving water bodies.

5.2.1 The Uniform Effluent Standards approach

The UES approach has been widely applied in the USA and the vast majority of the EU countries and has contributed significantly to the reduction of pollution from point sources of pollution. Its characteristic aspects could be described as follows:

- Aims to control the input of pollutants to the water environment by requiring that effluents comply with uniform standards.
- The underlying philosophy to the UES approach is that zero pollution (from point sources) is a desirable, ultimate goal (Van der Merwe & Grobler 1990).

- Standards are usually set so as to achieve pollutant concentrations in effluents, using the "Best Available Technology Not Entailing Excessive Costs" (BATNEEC) to treat them. The term is taken to mean the latest stage of development (state-of-the-art) of processes, of facilities or of methods for wastewater treatment and operation, which indicate the practical suitability of a specific measure for limiting the quality of effluent discharges. The choice of BATNEEC is based on the following:
  - Comparable alternative processes, facilities or methods of wastewater treatment and operation, which have recently been successfully tested;
  - The most recent technological advances and changes in scientific knowledge and the understanding of basic principles and concepts;
  - The economic feasibility of implementing such technology;
  - The local limits of application in terms of financial, technical and human resources;
  - Preference should be given to non-waste/low-waste technologies.

The UES approach has the advantage of simplicity in the process of enforcement, as it provides relatively easy to comprehend criteria to all parties involved in the process of water pollution control and regulation. However, it does not consider the different conditions at different locations and it disregards the assimilative capacity of receiving water bodies. As such, it is costly to implement, because all polluters are required to treat at high level their wastes before discharging them into water bodies. However, in many countries in the Southern African region, surface water streams and rivers are ephemeral, which means that for about half of the year, these natural channels convey effluents only. Correspondingly, this fact does not allow for any provisions in terms of effluents dilution or pollution assimilation and an uniform effluent standard could be justified in terms of environmental and public health protection.

One of the most serious disadvantages of the UES approach is the fact that it is applicable to point sources of pollution only, and fails to allow the evaluation and control of diffuse pollution sources. For this reason, the latest developments in the regulation and control of water quality worldwide are orientated towards the RWQO approach or the combination of both – the UES and RWQO approaches.

5.2.2 The Receiving Water Quality Objective approach

Historically, the RWQO approach has been developed and applied first in the UK. In many cases, the goals or objectives formulated for a given catchment basin or river consist of narrative statements, which are not supported by specific water quality criteria. Additional interpretation by the specific managing authorities, which are supposed to enforce them, would be necessary. This requires a higher level of education and professional expertise of the managing authority that would enforce them in practice. The RWQO approach could be characterised by the following points:

- The beneficial uses, for which a water body is suitable, determine the water quality objective (goal).
- Specification of water quality requirements (concentrations of water quality variables, or criteria) in receiving waters is based on the goals determined.
- The control of point and non-point sources of pollution should ensure that the specified water quality requirements have been met.
- Setting of "site-specific" effluent standards controls water pollution. They should take into consideration the contribution from point and non-point sources in conjunction with the defined goals, and also could vary from site to site, given the specific conditions of the catchment in terms of environmental status and economic development.

It could be stated that the drawbacks of the UES approach have been overcome by the RWQO approach, but at the expense of more complicated enforcement procedures. It requires a thorough investigation and understanding of the fate of pollutants in the water environment and their impact on water use. Also, it requires the application of site-specific effluent standards to be defined and administered; therefore it is not as straightforward as the UES approach to apply in practice. Also, it requires a higher standard and more complex skills from the regulatory organizations and consequently its application is a more costly procedure.

5.2.3 The Waste Assimilative Capacity Concept

One example of the combination of both approaches, which is applied in the USA (Novotny 2003), is the Waste Assimilative Capacity Concept (WACC). This concept combines both the UES and RWQO approaches and its main advantage is that it incorporates specific procedure for evaluation, assessment and control of diffuse pollution, together with the consideration of point sources pollution and the environmental health, not only of receiving water bodies, but also of all environmental elements. The main goal of this concept is the determination of the waste assimilative capacity (WAC) of the environment (water, air and soil). WAC represents the amount of pollution load, which could be absorbed and assimilated, without altering its integrity. The term "loading capacity" (LC) is introduced, which represents the sum of the load from background pollution (BL) and WAC. The procedures used during the application of the WACC could be summarized as follows:

- Establish background load (BL) – this procedure requires information regarding background quality and flow rates;
- Establish receiving body criteria, based on beneficial use – as per RWQO approach;
- Establish loading capacity (LC) – this procedure requires the estimation of the total assimilative capacity of the water body, which usually is based on the application of specific models, with respect to specified parameters, e.g hydrological models for water quantities estimation, qualitative models representing the DO assimilation, or toxicity evaluation of given toxic compounds;
- Estimate admissible load – WAC=LC-BL.
- Establish the actual pollution load for each specific case, from point and diffuse pollution sources;
- If WAC > pollution load – RWQO approach to be applied
- If WAC < pollution load – ES approach to be applied

It should be noticed that this concept is very sound in its logic and provides a possibility for more accurate evaluation of the available resources and the potential for their sustainable exploitation. Also, a very important consideration is the fact that decisions with respect to the use of available water resources could be done, based on scientific information. However, this approach requires very extensive data sets with respect to quantitative and qualitative characteristics not only for surface water, but for ground water, soils and air, which is a costly and time consuming exercise. The practical application of the concept requires the development of sound models, reflecting with enough accuracy, the assimilative processes in the environment. Correspondingly, the application of such models requires an advanced level of the technical support in terms of hardware and human resources. All this makes its practical implementation difficult even for developed countries and prohibitive for developing countries. However, it should be noted that developing countries could consider this concept as a future development and could orientate towards a data collection and monitoring strategy, which could help in the future implementation of the WACC.

5.3 *Setting regulatory instruments*

Setting regulatory instruments is a complicated and sensitive process, which needs to be undertaken carefully in order to obtain the final goal of protecting public health and the environment together with a sustainable economic and social development. Regulations might allow industries or municipalities to dump a certain amount of pollutants into water bodies, in order to support development needs. However, they should not be set at minimum levels so that the natural water quality should be destroyed for future generations. Also, they should prevent a polluter from profiting at the expense of other water users, and in addition, they should allow a certain level of flexibility, compensating, for example, for differences in summer and winter and dry versus wet years. During the process of setting or evaluating standards, some basic considerations should be followed to provide for a successful enforcement practice and achievement goals. These could be summarized as follows:

- Criteria and requirements should be based on sound logical and scientific grounds. The application of risk assessment techniques with respect to public and environmental health hazards is strongly recommended, in order to protect the public from chronic and acute water borne diseases. Both the environment and the public health should be protected from toxic, carcinogenic and mutagenic pollutants.

- Actual water quality values are statistical variables, which vary randomly. Therefore, water quality standards must specify not only the limiting value, but also, how often this value could be exceeded and for how long. These recommendations are usually based on specific research methods for the evaluation of chronic and acute toxicity of prescribed chemicals with respect to different biological species.
- Formulation of attainable and affordable goals – this consideration refers to many cases in the past, when high-level requirements were set in the regulatory instruments, and because of the fact that these goals were too difficult to attain, the regulations were never implemented in practice. Thus, the process of setting the objectives or requirements should consider actual conditions in terms of time and implementation cost. The cost should include the collection, transport and treatment of polluted water (construction and operation), and the cost of monitoring and the enforcement of the regulatory instruments as well.
- A phased approach in the implementation of regulatory instruments is highly recommendable, especially for developing countries.
- All stakeholders' participation during the process of setting standards is very important in order to provide for a common ground of the decisions taken, for the consideration of different opinions and points of view, and consequently, for the acceptance of the measures to be undertaken as a result of the regulations.
- Public awareness of existing regulatory instruments and new or upgraded ones would allow the gaining of public support and would make the public aware of the costs involved. Also, it could stimulate the willingness to accept the costs for public and environmental protection.
- Political will to enforce standards is also an important aspect of the process and should be sought.

5.4 *Specific aspects of diffuse pollution regulation*

Regulation of diffuse pollution sources is a relatively recent practice in the development of pollution control instruments, which accounts for a major shift from the application of the UES approach in most countries to the application of RWQO approach in Europe and the WACC in USA. The major characteristic of this development is the introduction of an integrated catchment based management of water resources in general and water quality in particular, which is linked with specific management programs to achieve predetermined goals. They are region specific and include specified control and abatement measures, implementation strategies and means for financial support for the program execution. This shift in the development of regulatory pollution control mechanisms is characterized by an evolutionary principle, where existing regulatory instruments are incorporated in the new legislation, thus combining the UES (emission approach) into a broader framework, which includes the consideration of both – effluent discharges and natural water bodies quality, in terms of surface/ground water, soil and air. Also, it recognizes that the environment is a legitimate user of water and should be treated as such in terms of quantity and quality.

Regulatory instruments to protect surface water consider as major pollution sources storm water discharges from industrial activities, major urban centers and major construction sites. In developing countries informal settlements and illegal connections to the storm water system are important sources of diffuse pollution as well, which needs to be addressed. Usually, major discharges of storm water drainage systems are regarded as point sources and are included in the list of polluters, requiring a discharge permit. Municipalities are required to develop storm water regulations, which include instructions and measures for the management of urban drainage systems, and address both water quantity (flood prevention) and water quality (pollution). These regulations could form part of municipal by-laws and are usually not legally enforceable.

Diffuse pollution from agricultural areas is controlled as part of the receiving water quality objective approach. In some specific cases of concentrated animal operations, the effluents or the runoff could be regarded and controlled as point source pollution.

While surface water regulations, through the discharge permit principle, allows for some extent of pollutants to be discharged into the environment, based on the assimilative capacity of the water body, ground water regulatory instruments adopt the principle that it should not be polluted at all. This principle is based on the fact that ground water in general has a very limited self-purification capacity and any rehabilitation measures are very difficult and costly to implement. In the EU, the so-called "precautionary approach" is applied, which prohibits any direct discharges to groundwater bodies and monitoring of groundwater quality in the case of indirect discharges (groundwater recharge methods) is prescribed. Regulatory instruments for groundwater protection should emphasize the point that, in some cases, measures to protect surface water might lead to the pollution of groundwater. Recommendations for best management practices should consider this fact and avoid such cases.

REFERENCES

Chapman, D. 1998. *Water Quality Assessment: A Guide to Use of Biota, Sediments and Water in Environmental Monitoring*. 2nd ed. London: SPON Press.

Dean, R, B. & Lund, E. 1981. *Water Reuse: Problems and Solutions*. Copenhagen: Academic Press.

Debo, T. & Reese, J. 2003. *Municipal storm water management* 2nd ed. USA: CRC Press LLC, Lewis Publishers.

Degremont 1991 *Water Treatment Handbook,* 6th ed. Vol.1, Paris: Lavoisier Publishing.

Ellis, K.V. 1989. *Surface water pollution and its control.* London: MacMilliam Press.

Focht, D. D. & Chang, A.C 1976. Nitrification process related to wastewater treatment. *Advances in Applied Microbiology* 19, 153-186.

Horan, N.J. 1990. *Biological wastewater treatment systems – Theory and operation.* Chichester: John Wiley & Sons

http://ccee.oregonstate.edu/swmm/

Lewis, I. & William E. (1984) Eutroplication and Land use, Lake Dillon, Colozelo *Ecology studies* vol 64, New York: Springer-Verlag

Moyo, N.A.G. 1997. Causes of massive fish deaths in Lake Chivero. In: Moyo, N.A.G. (Ed), *Lake Chivero: a polluted lake*. 98-104. Harare: University of Zimbabwe Publications.

Novotny, V. 2003. *Water Quality: diffuse pollution and watershed management.* Hobohen, New Jersey: John Willey & Sons.

Schulze, R.E., Schmidt, E.J., & Smithers, J.C. 1993. SCS-SA User Manual: PC – Based SCS design flood estimates for small catchments in Southern Africa. In *ACRU Report No 40*, University of Natal, South Africa.

Stephenson, D. 2003. *Water resources management* A.A Balkema Publishers, Lisse, The Netherlands

Van der Merwe, W. & Grobbler, D.C. 1990. Water quality management in RSA: Preparing for the Future *Water SA* 16, 1.

Vollenweider, R.A. 1968. Scientific Fundamentals of the Eutrophication of Lakes and Flowing waters, with particular reference to Nitrogen and phosphorus as factors in Eutrophication. In O.E.C.D. Tech Report.DAS/CSI/68.27, vol 15, 34.

White, J.B. 1978. *Wastewater engineering* 2nd Ed. London: Edward Arnold.

WHO (World Health Organisation) 1996. *Guidelines for Drinking Water Quality, Volume 2: Health criteria and other supporting information.* 2nd ed. Geneva: World Health Organisation Press.

www.epa.gov/ceampubl

CHAPTER 2

Monitoring, abatement and management of diffuse pollution

R. Hranova

ABSTRACT: Diffuse pollution could be identified and assessed, based only on the general water quality status of a catchment basin, and as such the monitoring process, which allows for this assessment is essential. Basic definitions, procedures and methods are presented in the light of the closed-loop approach to the monitoring process. Emphasis is given to the quality assurance and records handling procedures. Abatement measures, including source control activities and appropriate treatment methods and facilities to reduce storm water pollution before it reaches the natural water bodies, are presented as well. Specific aspects and approaches of the management practice of diffuse pollution control are discussed, considering the planning and preparation phase, public/stakeholders' participation and involvement techniques, as well as enforcement provisions.

1 IDENTIFYING DIFFUSE POLLUTION

Dealing with diffuse pollution, means dealing with the water quality status of surface and ground water. If the runoff, which maintains the rivers flowing, fills lakes and recharges aquifers, is polluted, it means that the natural water resources quality would be damaged and their future beneficial use jeopardized. In contrast to point sources of pollution, which are visible and easy to identify, diffuse pollution in many cases could be unnoticed and neglected for years. Polluted runoff is usually the consequence of social habits, established management practices and legal arrangements and it is difficult to point out one single entity responsible for diffuse pollution problems. Point sources of pollution are well defined, they could be easily monitored and assessed, and corresponding abatement measures, consisting in the vast majority of the cases of treatment systems, could be implemented. In the case of diffuse pollution, the specific source is difficult to be pointed out directly and the corresponding authority held responsible. Also, abatement measures are much more costly and might need the restoration of the whole environmental system. Therefore, the only way to identify the presence of diffuse pollution is to monitor and control the qualitative status of environment – water, soil and air.

It should be acknowledged that the monitoring process is usually a costly and demanding process, which requires technical support, manpower and institutional arrangements. Thus, the necessary expertise with respect to the design of optimum monitoring programs, which would provide maximum information value for a minimum operational cost, is important, and would help considerably a sustainable use of the available natural resources of any country.

What we do not know does not exist for us. In many developing countries diffuse pollution problems have not been recognized as such, because there is no regular monitoring of water quality. In the case of point discharges, the pollution is visible and therefore, it could be identified and appropriate measures taken for its abatement; while diffuse pollution stays unidentified and correspondingly, is not addressed. Only the effects of diffuse pollution, when it becomes apparent (e.g. an eutrophied lake) could be appreciated. Thus the monitoring of water quality is the first step to be implemented in a pollution prevention program. It is important not only with respect to pollution prevention but because of the need to know the status of the natural resources in terms of their best utilization in a sustainable way. However, different pollution abatement strategies could be implemented parallel to a water quality monitoring program, which would help in reducing diffuse pollution and maintaining a clean and healthy environment. These are most valid for cases of source control measures and introductions of practices to minimize waste generation and apply cleaner production processes.

2 MONITORING AND CONTROL OF WATER QUALITY

2.1 *The closed-loop approach to the monitoring process*

Over the last decades a pressing need has arisen for a comprehensive and accurate assessment and evaluation of water quality. Reliable data, collected by means of sound and efficient monitoring programs is an essential basis for such assessments.

⇒ Monitoring – the term could be defined as "the process of repetitive observations for a defined purpose, of one or more elements or indicators of the environment, according to pre-arranged schedules in space and time, and using comparable methodologies for environmental sensing and description."

The purpose of water quality monitoring is directly related to the practice of water quality management for public health and environmental protection. The management activities are based on two major factors: the availability of water resources in terms of quantity and quality, and water demand in terms of present use and future projections. To support these activities assessment of water qualities of natural water bodies and the effluents discharged to them, as well as the provision of well-structured information flows, is essential. It needs to be mentioned that monitoring programs are designed not only to control and assess the water quality of natural bodies or effluent discharges, but also that they form the basis of the operation and control of water and wastewater treatment plants. In the case of diffuse pollution assessments all types of monitoring are important because the analysis of the problems requires a catchment-wide analysis and the consideration of background quality and point pollution sources.

The success of any monitoring program, even the simplest one, depends on a proper preparation and design process. Therefore, serious consideration should be given to this preliminary step in order to achieve sound results at a minimum cost and to make an optimal use of the available technical, human and financial resources. As a first principle, it should be remembered that the monitoring process is a cycling or "closed-loop" process. This cycled process could have different steps, but a commonly used one is illustrated in Figure 2.1.

Often in developing countries, the authorities involved in the every day running of monitoring programs are not satisfied by the results, obtained from their programs. In many cases, such programs have not been implemented at all or are in their initial phase. Usually, the unsatisfactory results are explained with lack of financial resources, or lack of well-equipped and maintained laboratories, or lack of trained personnel to use the equipment. These factors have been officially recognized and in a number of cases, new water quality laboratories have been established and measures taken to provide the necessary capacity building of such institutions. It needs to be emphasized, however, that even the best technical basis could not substitute the need for a thorough analysis of the local conditions, and an adequate design of the monitoring program, which should be based on the available resources. Also, special care should be taken to maintain the technical facilities in good condition, and to upgrade the available resources in terms of equipment and skills of the supporting personnel.

In numerous cases, considerable efforts and costs have been spent in order to implement a new monitoring program with no satisfactory results. This is usually due to the fact that the cycle has not been closed or because mistakes have been made at its different stages. Some problems often met in developing countries' practices are:

- No clear objectives have been defined
- The parameters, which have been chosen, do not correspond to the formulated objectives. In many cases, too many parameters are listed, leading to waste of efforts and resources;
- Methodologies for measurements do not match the standard requirements, leading to unreliable information output;

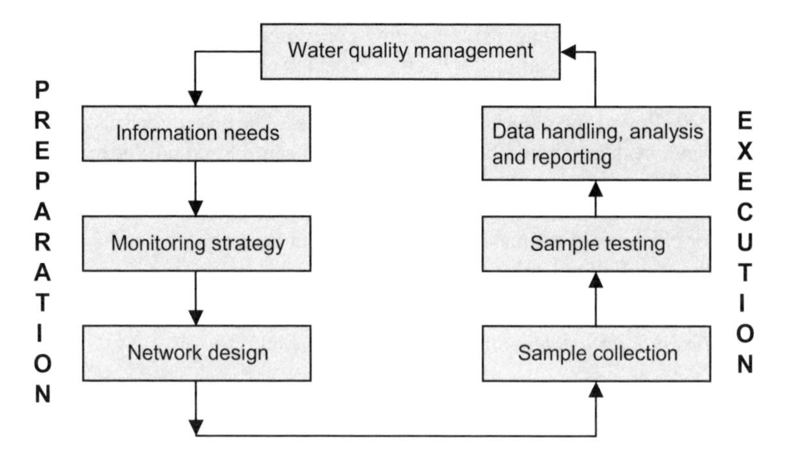

Figure 2.1. The closed-loop monitoring cycle.

- Wrong sample collection and preservation procedures, leading to unreliable information output.
- Lack of proper data storage, records handling, and unsuitable reporting format, leading to unsatisfactory information utilization.
- Lack of a quality assurance program, leading to unreliable information output.
- Lack of regular evaluation of the objectives and the different steps of the cycle, leading to unsuitability of the program to match changed site conditions.

2.2 *Formulation of objectives and goals*

Monitoring programs for water pollution assessment are mainly prepared to assess water quality and compare it with accepted and approved requirements or regulations, to detect trends and to provide information for the prediction and simulation of water quality changes. They may be applied to assess the status of natural water bodies, the quality of effluents, which are discharged into them, or both. We may be certain that a water body under consideration is polluted only in the case when its water quality is assessed and does not comply with prescribed regulations, guidelines or standards. Also, monitoring programs are applied to control and operate water and wastewater treatment plants, to choose a suitable treatment scheme during the preliminary stages of the design of these plants, or they may form part of a research project in the field of water quality management. Examples of different types of monitoring programs and discussion of problems encountered in the every day application are presented in Chapter 3.

The accepted and legislated standards and regulations play a key role in the pollution control process, and correspondingly in the formulation of the objectives of a pollution assessment program. If the main objective of a monitoring program is to control and prevent water pollution, the design of the program should be based on a thorough examination of the existing standards, regulations and guidelines.

It is important to define not only the purpose of the measurement but also the geographical scope of the measurement program and its duration. A map defining the geographical scope of the measurement program should also indicate all effluent source and tributary confluence points. Recently, geographical information systems (GIS) become a substitute of the maps, which allows to record and store information and to use this information as modules of more complex models for assessment, evaluation and control of water pollution.

2.3 *The network design*

Although some water bodies are monitored continuously, routinely and comprehensively, it is unreasonable to expect that every water body will be perpetually monitored at all locations. In general, the process of monitoring is expensive and time-consuming; therefore care must be taken to design an optimum network of sampling locations, reflecting expected changes of water quality. In all cases, the selected sample site locations will be determined by the objectives of the program and should be interconnected in a way that allows for comparison and data analyses. The heterogeneous distribution of quality parameters within water systems should be considered when sampling point locations are chosen. Generally, sampling locations near the boundaries of water systems are best avoided except when these regions are of direct interest. The sampling locations should generally be chosen so that the corresponding discharges are known or can be well estimated. This requirement becomes imperative for programs, which include the objectives of pollution loads determination.

2.4 *The choice of media, parameters and monitoring frequency*

Considering the integrated approach to water quality management and the need to protect all aspects of the environment (water, air and soil), the medium to be assessed may vary, or the objectives of the program may include monitoring of more than one media, e.g. in some cases both water quality and sediments quality should be sampled and tested. One other example is the case of effluents disposal on land, where effluents and soil quality, together with river and ground water quality might need to be tested, in order to examine environmental impacts of the disposal. For each specific medium, parameters to be tested, frequency and methods of monitoring should be identified and specified separately.

The choice of parameters to be monitored is very important and could influence considerably the cost of the program. The preparation of a list of selected parameters involves defining both the type of parameters (pH, dissolved oxygen, etc.) and the required level of measurement (mg/l, μg/l, and range). The selection of parameters should be based solely on the need to satisfy the objectives of the program and the appropriate media to be surveyed, and should consider the available resources. In the case of water quality monitoring, the selection process should include the following steps:

- Review the water quality criteria (standards) and tolerance limits applicable to the water body under investigation, or review the performance guidelines where the efficiency of a treatment operation is under consideration.
- Consult previously reported data, if available.
- Conduct a screening test on the water body under investigation, taking into account, thermal and material input from tributaries or domestic and industrial discharges in the case of rivers, and different flow streams in treatment plants.
- Develop a conceptual model of the system under investigation (a set of water quality characteristics).
- Identify parameters of particular interest to the purpose of measurement, based on thorough literature and documentary survey.
- Specify levels (range) of measurements for each parameter.

The values obtained by analysis are governed by the methods of analysis used and therefore all laboratories wishing to obtain comparable results should use the same method. When selecting a parameter it is necessary to specify the required range of concentrations. This information is significant in view of the fact that the procedures for the sampling and analysis of a given constituent in the mg/l range may be completely different from those used in the μg/l range. The specificity of a parameter is its ability to define selectively a given chemical or biological species. It is frequently necessary to provide an explicit statement of the degree of specificity required. For example, it may be important to state the parameter as "total phosphorus", "orthophosphate", "polyphosphate", or "organophosphorus compounds", which would necessitate structural identification and would require a specific testing procedure.

The choice of parameters to be monitored is a center point during the design process of a monitoring program and requires very good knowledge and understanding of the media to be studied, the available methodology and the study area. It also requires analysis of the conditions and resources available and ingenuity. This task should be delegated to experienced staff with proper knowledge of local conditions and resources.

The quality of a water body of just one instant is seldom of interest. Normally, information is required over a period of time during which the quality may vary. The times during this period at which samples are collected must be chosen so that they adequately represent the true quality and its variations. In addition, the number of samples must usually be kept to a minimum in order to save sampling and analytical effort. The problem is, therefore, to select the number of samples and the times at which they are taken, so that information of the required accuracy is obtained with minimum effort.

In most circumstances, the ideal technical solution would be to use automated instruments designed for the continuous measurement of the parameters. This approach can be extremely valuable but it is not generally applicable because suitable instrumentation for all parameters is not yet available. It could be executed regarding temperature, pH, dissolved oxygen, conductivity and other limited variables. At present the bulk of water analysis are made on discrete samples. Whatever the chosen sampling frequencies, data should be regularly revised to decide whether or not changes in the frequencies are necessary. In water systems regular cyclic variations of quality may occur, with periods of one day, one week or one year. For example, diurnal fluctuations can occur in rivers, lakes and effluents. Persistent cyclic variations with other periods may also occur (regular variations due to discharges in industrially exploited rivers). When cyclic variations occur, a biased estimation of quality will be obtained unless sampling times are carefully chosen. For example, when the dissolved oxygen content of the river varies diurnally, biased results for the mean daily value will be obtained if samples are always collected at the same time of the day.

2.5 *The sampling process*

The choice of a correct technique for collecting samples is most important; otherwise non-representative samples may be obtained. Some aspects of general importance in any sampling process are:

- The concentrations of parameters in the water entering the sample should clearly be the same as those in the water being sampled.
- The concentrations of parameters in the sample must not change during the process of collection and transport of the sample to the laboratory.

It is very important that the procedure, which will be used for collecting samples, should be carefully prescribed and followed. This is particularly true if the personnel involved are relatively unskilled scientifically. Collection of samples may involve hazards to the person sampling. A good description of such hazards and precautions should be available. When a laboratory receives samples, they should be preserved and stored, until analysis is started, under conditions prescribed by standard procedures, in order to prevent change in the original concentrations.

Representative samples might require different types of sample collection procedures, and we could differentiate among:

- Grab samples – a determined volume of sample collected at a given time.
- Depth-integrated samples – a mixed volume sample integrating identical portions of the total amount, taken as grab samples at different depth intervals, so that the integrated sample gives average concentrations for the whole depth column at this location.
- Time integrated or composite samples – similar to depth-integrated samples, but separate portions are taken at different time intervals, and the integrated sample represents average concentrations for the whole time interval, in cases where the flow variations are low.
- Discharge integrated samples – a mixed volume of sample, integrating grab samples at different time intervals, where each one of the grab sample is proportional in volume to the total discharge at this time interval. The integrated sample represents average concentrations of the total flow during the whole time interval, in cases where high flow variation is typical.

2.6 *Methods of testing*

In the majority of the water quality measurement programs the selection of methods of analysis is usually left to the experience of the analyst. A number of guidelines are presented to assist in making the most appropriate choice. The selection of measurement methods should be based on the following criteria:

- Total number of analysis.
- Frequency and geographical scope of measurement.
- Required rapidity of analysis.
- Sensitivity and detection limits.
- Selectivity and interference.
- Constraints on accuracy and precision.
- Available technical and human resources background

An assessment of the total number and frequency of analysis will serve to give a first estimate of the magnitude of the measurement program. This information will help to decide whether the analysis should be performed manually or by means of automated systems. Similarly, the required detection limit, selectivity, accuracy and precision will help the analyst to decide on the method to be used. The required rapidity of analysis will often by itself be enough to dictate the method of analysis. These would give an insight into whether the analysis should be done in the laboratory or in the field. Laboratory methods may be based on manual or instrumental techniques. If the requirements for method selection specify low detection limits or high selectivity, accuracy, and precision, it may be necessary to select an instrumental technique. For large numbers of measurements, it may be advisable to rely on automated systems, to achieve high precision and reduce the cost of measurement. Field analysis may be based on manual procedures, automatic water quality monitoring, or remote sensing systems.

The next step is the determination of the availability of equipment and manpower. This is particularly significant in the case of instrumental and automatic techniques, which require sophisticated equipment and specially trained personnel. If these prerequisites are not satisfied, it might be necessary to select manual methods of analysis, which may offer less desirable performance characteristics.

A recommendable (Degremont 1991) list of equipment for central and research laboratories would include infrared and ultraviolet spectrophotometers, polarographic electro analytical system, mass spectrometer, particle size distribution analyzer and carbon-hydrogen-nitrogen analyzer. In consequence of recent advances of technology, an increasing reliance on instrumentation and automated techniques for water pollution characterization is witnessed. However, it should be stressed that the reliance on instrumentation is a matter of convenience, rather than necessity.

Water analyses rely on instrumental methods of analysis to achieve high specificity and low detection limits. Usually instrumental techniques are economical for routine analysis whenever large numbers of samples are involved. Such automation not only permits simultaneous analysis of more than one parameter, but also frequently offers greater precision than manual methods.

2.7 *Quality assurance and evaluation*

Quality control is very important throughout the monitoring process, but these notes refer to quality assurance and control in the laboratory. In certified world class laboratories, about 20% of the allocated budget is designated for quality assurance and control (Chapman 1998). This process involves:

- Well-trained personnel, appropriate working conditions and equipment and well defined managerial structure;
- Preparation and a strict execution of Standard Operating Procedures (SOP) for the whole routine of sample handling (receipt, storage, analysis, disposal);
- Well-validated and robust standard procedures for analysis;
- Accurate and controlled reporting of the data;
- A frequent control of methods, instruments (e.g. equipment calibration) and personnel.

The quality assurance procedures would include activities, which could be subdivided into two categories – routine daily activities and specialised measures for quality assurance. The routine activities include using clean glassware and chemicals, strict adherence to the prescribed methods, and tidiness of the work place. The specialised measures include activities performed by the laboratory which is performing the analyses and may be termed "internal" control; as well as activities performed by an external laboratory as a reference material, which fall under "external control" (Chapman 1998). Internal control measures are usually carried out continuously, and include:

- Use of blank, duplicate and "spiked" samples, which should not be recognized by the staff executing the test;
- Checking the results and comparing them with well-known data from literature and theory, e.g. cation/anion balance;
- Use of "control charts", showing values exceeding certain control limits in the laboratory results of standard samples of known composition.

External control should be carried once or twice per year and includes:

- "Reference samples" of known composition could be used for critical check-up of the whole laboratory cycle;
- Inter-laboratory checks give insight into the general errors made in a group of laboratories.

The supervisor should create an open and transparent atmosphere, with a defined sense of professional commitment and responsibility. Errors should be discussed openly and measures for improving the quality of the analytical work suggested. It should be acknowledged that 100% perfect results are unobtainable and the quality goals will always be a compromise between the quality needed and the available resources.

2.8 *Data handling, analysis, storage and reporting*

This final stage of the measurement program is essential for the completion of its objectives. Poor record and storage of data obtained may jeopardize the efforts and achievements of a well-designed and well-executed program. Data record, storage and retrieval will be refereed as a "data storage system".

One can differentiate between manual and computer data storage systems, the first one being applied in small measurement programs. The latest developments of computer technologies make a "must" the application of computer based data storage systems for larger monitoring programs.

2.8.1 Data storage

Manual systems are traditional and most elementary. They involve reporting data in hand-written or type-written form. Examples for such reporting are the books in a treatment plant operation, the report of a water quality survey of a river, an inspection form or a waste discharge permit. The documents are stored in simple file cabinets or bookshelves, with a limited number of copies available. Even the simplest data storage system of a water quality monitoring program should include the following components:

- A map, showing clearly the geographical scope of the program, sample points' codes and locations. It would form the basis of the program's data system.
- A detailed description of the analytical methods used to determine parameter values;
- The data obtained during the monitoring process, with corresponding dates, sampling locations, parameters tested, units, and any remarks made during the process of sampling or testing.
- The description of the method of wastewater generation and the treatment technology, in cases of monitoring of effluent discharges.

It has to be mentioned that the stored data must be revised and checked periodically, in order to improve the program quality. Sample points locations and frequency of sampling may be changed or reduced, based on trends shown in the records.

2.8.2 Data analysis

The analysis of stored data would help to reach decisions regarding the status of pollution of the water body under consideration and to identify measures for its reduction, if necessary. Statistical tools are used to reach sound results and to demonstrate the trends. Data collected and stored may form the basis for the application of models describing different aspects of the pollution transport and transformation processes. Models help to simulate different scenarios and conditions or to predict events, and are very powerful tools for the decision making process in water quality management.

Water quality assessment is one important form of monitoring data analysis. In many cases the objectives of the program are associated with the evaluation and assessment of the current status or trends in the variation of different water quality characteristics. The term could be defined as follows (Novotny 2003):

⇒ To assess water quality means to evaluate the physical, chemical and biological characteristics of the body, which has been monitored, in relation to natural (background quality), health effects and intended beneficial use.

The data analysis stage of the monitoring process, together with the following stage – the preparation of the report, is an essential last step of the monitoring cycle, in which data obtained is structured, analyzed and conclusions drawn with respect to available regulatory instruments. It is also important in this stage to make trend analysis of water quality on a yearly basis. Regarding the assessment of the status of the examined body, by means of comparison with regulatory instruments, it should be remembered that stipulated values and criteria in the regulatory instruments should not be considered to be rigid values, but statistical variables. Well-designed regulatory instruments provide for flexibility by quoting statistical parameters as "percentiles". A 90th percentile concentration of 0.5 mg/l of any constituent, represented by a water quality data set, indicates that this concentration is exceeded in only 10% of the monitoring data. A more detailed discussion in this respect is presented in Chapter 12.

To enable a simpler analysis and interpretation of monitoring data, specifically regarding surface natural water quality, the use of water quality indices is often applied. Regarding physical and chemical parameters, the indices are based on lumping together different parameters. There are many different indices applied worldwide (Chapman 1998, Ellis et.al. 1993).

2.8.3 The reporting stage

The monitoring cycle aims at the generation of reliable data, which must be processed and presented in a way that is clear and understandable by non-specialists. Very often, water quality data is buried in annual reports, with data presented in tabular form only, without statistical assessment, interpretation or graphical presentation. Comparison with regulatory instruments should be made only after the statistical analysis of the data obtained has been done. In addition to the assessment of the status of the examined body, appropriate abatement measures and activities to be undertaken should be prescribed as well. Polluters should not be confined to a very strict scenario of pollution abatement measures but should be advised and given guidelines with a strict time frame for their implementation. The final choice of pollution abatement measures should be left to the responsible authority, which will implement it. A clear and understandable report is always cost effective, as it enables the implementation of decisions to protect, improve and restore water quality and to optimize the monitoring process.

Computer processing and presentation become an essential part of the data storage, handling and presentation process. Only for simple water quality projects, manual data processing, handling and reporting could be used. Care should be taken during the data processing operations to preserve and secure the integrity of the original data set.

Finally, the water quality monitoring report must show a clear structure and be understandable for all readers, including non-specialists. The minimum number of items in the report content should include:

- A short summary;
- An introduction to the objectives of the program;
- A description of the catchment and corresponding monitoring network;

- The methodologies applied, both in the field and in the laboratory, including quality assurance and control protocols;
- A clear presentation of results followed by analysis, including statistical processing and the interpretation of these results in the form of water quality assessment and evaluation;
- Conclusions and recommendations for the decision-making process.

2.9 *Specific aspects of diffuse pollution monitoring*

Considering the fact that in the vast majority of the cases, the monitoring of diffuse pollution means the monitoring of natural water bodies, it is unreasonable to expect or to plan, that every water body could be monitored continuously and at numerous locations. For this reason, a careful network design, based on a catchment principle, is imperative in order to obtain meaningful information at an acceptable cost. As a minimum requirement, locations at catchment boundaries, flow rate gauge stations, and major suspected pollution sources must be included. The frequency of sampling and parameters tested could vary for the different stations within a catchment, or among basins, based on specific conditions.

Another important consideration during the monitoring process of diffuse pollution is the need to identify the background pollution level, as diffuse pollution itself could be identified only if the background quality is known. Therefore, the network design must include points of pristine conditions, within a similar geographic region, which could serve as a control reference for background quality evaluation.

It has been emphasized that diffuse pollution is due to polluted runoff, and as such varies extremely in terms of quantity and quality, depending on the status of the washed up surface. Therefore, it is an event-orientated phenomenon, related to the type of land use practice in the area. In order to minimize monitoring efforts and costs, the monitoring practice of diffuse pollution is orientated towards the determination of "event mean concentrations" (EMC), which are statistical values of constituents, measured and statistically characterized for most common land uses within a given geographical boundary. EMCs are statistically related to specific rainfall events. Once EMCs are determined, it would allow the determination of characteristic "unit loads" for specified land use practices, which are expressed as the mass of pollution constituents per unit area for a specified time, e.g. kg SS/ha.year. Unit loads include background pollution loads. They are basic parameters for the application of models for pollution assessment and abatement, which forms part of the WAC regulatory concept. However, EMCs are site specific and depend on a variety of factors, and their determination should be carried on for each specific case as an individual exercise. Some examples of basic types of land use practice in agricultural areas are croplands, pastures and forests. In urban areas the different types of land use patterns are described in Chapter 1.

It should be noted that EMCs from agricultural areas might vary extremely, even within a geographic boundary, while EMCs in urban areas show a better correlation with the corresponding type of land use practice (Novotny 2003). For this reason, the monitoring network design should give more emphasis on urban areas, and the frequency of sampling should be event-orientated towards typical storm events with a low return period, which generates runoff. Also, the provision of discharge-integrated sampling is imperative. The selection of parameters depends on each specific case, but the basic recommended ones are: pH, DO, EC, turbidity, TS, TDS, TSS, general organic material (BOD, COD or TOC), different forms of nutrients, toxic metals and other listed compounds, which might be suspected to be present in the runoff.

In general, the EMCs of runoff are much lower compared to different types of wastewater. To illustrate this, Table 2.1 presents data from a nation wide regression analysis of urban runoff pollution in the USA, for different types of urban land use practice. However, the large volumes of untreated runoff result in a considerable pollution load to water bodies, which could have a substantial impact on the environment and public health and needs to be considered in the integrated assessment of water resources management.

Table 2.1. USA median values of runoff concentrations from different land use practices (in mg/l).

Pollutant	Residential	Mixed	Commercial	Natural open area
BOD	10	7.8	9.3	-
COD	73	65	57	40
TSS	101	67	69	70
Lead (total)	0.14	0.11	0.10	0.03
Copper (total)	0.03	0.03	0.03	-
Zinc (total)	0.13	0.15	0.23	0.19
TKN	1.90	1.30	1.18	0.97
Nitrate + Nitrite	0.74	0.56	0.57	0.54
Phosphorous (total)	0.38	0.26	0.20	0.12
Phosphorous (soluble)	0.14	0.06	0.08	0.03

After: Debo & Reese 2003

3 DIFFUSE POLLUTION REDUCTION AND ABATEMENT

3.1 *Source control*

Source control of diffuse pollution consists of measures to prevent its generation and in general, are aiming at the reduction of the quantity of the runoff (volumes and peak flows), its erosive capacity and pollutants concentrations washed up by the flow before it reaches the drainage system. Clearly, these are diffuse pollution prevention measures, which could achieve a significant reduction effect if applied adequately, and are mostly associated with diffuse pollution in urban areas. They should be applied where possible in new development projects, but are also important in the process of improvement of the existing practice of diffuse pollution management. In rural areas, diffuse pollution abatement measures are related to proper agricultural practices and avoidance of over-application of fertilizers, pesticides and herbicides. It should be emphasized and remembered that the implementation of preventative measures does not consist of specific engineering or scientific solutions only, but is a management philosophy based on the implementation of a good housekeeping practice in urban communities.

3.1.1 Solid waste management and street cleaning

Litter (street refuse) includes large items and particulate material, such as paper, plastics, food residues, glass bottles or particles, fallen leaves and other materials spread across streets, parking zones and open spaces in urban areas. An adequate solid waste management practice of regular waste collection, provision for waste collection containers at all required places and regular street and open spaces cleaning could reduce the level of diffuse pollution by 50%, especially in areas with a high percentage of impervious surfaces (Novotny 2003). A higher level of reduction could be achieved in developing countries, where improper solid waste collection and management practice is usual.

3.1.2 Control of pervious areas

In areas of unprotected bare soils, which are prone to erosion, the soil loss might be considerable, and would result in increased suspended solids concentrations in runoff and deposition of solids in the drainage system and the receiving natural channels. In urban areas, the suspended particles are usually the carriers of other pollutants as organic materials, nutrients and toxic metals. Large construction sites and open mining operations are major sources of such type of pollution. Pollution prevention activities consist in erosion control measures, such as stabilization of soils and temporary or permanent soil covers. Vegetation cover is an important erosion prevention measure and should be applied where possible.

3.1.3 *Pervious (porous) pavements*

In contrast to ordinary impervious pavements of roads, streets and parking slots, the pervious ones have incorporated porous modules, which allow for the infiltration of runoff into the road (street) sub base and after that to the ground. With these pavements, the volume of runoff is reduced drastically and directed to recharge ground aquifers. However, they have the disadvantages of reduced load bearing and more difficult maintenance, as the infiltration modules (sections) require a regular cleaning and sweeping in order to avoid clogging. The application of pervious payment is most recommendable for areas with permeable soils. If the street (road) base is impermeable, their application would require the addition of a sub drain system for the collection and discharge of the percolated runoff, which might increase the cost considerably. In general, the most appropriate applications of porous pavements are in the cases of parking slots and small streets (roads), with relatively low traffic load, where regular cleaning would be feasible, and the soil conditions allow for runoff infiltration. It should be noted that the infiltration rate of porous pavements is much higher than the soil infiltration rate, which allows for the accumulation of the infiltrated water in the sub-base and its gradual infiltration through the soil.

3.2 *Control during runoff transport*

3.2.1 *Filter strips and environmental corridors*

These are especially designated and vegetated areas, which have the purpose to retard and partially absorb the runoff from developed areas, to enhance infiltration and to retain suspended particles, which could be the carriers of pollutants. We could differentiate between:

- Filter strips – these are vegetated corridors of land parallel to road structures, which have the purpose of partial filtration and infiltration of runoff from upstream areas. Usually, they are covered by grass, which has the purpose of slowing down the runoff velocity, allowing for the increased rate of infiltration and sedimentation of suspended particles, thus preventing erosion and assimilating potential pollution constituents, associated with particulate material and nutrients. An illustration of a filter strip along a road junction is shown in Figure 2.2. Filter strips at a larger scale could serve as a buffer between different types of land uses, be landscaped to become aesthetically pleasant and provide groundwater recharge in areas with permeable soils.
- Environmental buffer zones are open vegetated areas or parks along natural water bodies, which have the purpose to retain runoff pollution before it is discharged into the water body. Often, they incorporate some of the treatment methods described in section 3.3, such as retention basins, ponds or wetlands.

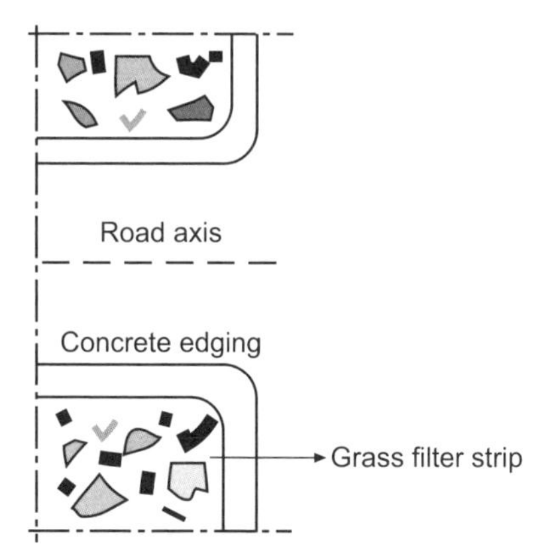

Figure 2.2. Schematic representation of a grass filter strip along a road.

3.2.2 *Catch basins*

A catch basin is a runoff inlet chamber, which receives the storm water from the street and conveys it into the drainage system. In contrast to ordinary inlet chambers, catch basins are equipped with a sump at their base, which allows for the partial sedimentation and retention of the heavy sand particles and debris, thus preventing the down stream transport system from clogging and eliminating part of the pollution carried by the runoff. The application of catch basins requires regular cleaning (desludging) of the sumps and removal of retained material.

3.2.3 *Grassed waterways and channel stabilization*

The drainage system (separated sewer system), which conveys the runoff, usually consists of open channels parallel to the roads/streets, which collect and transport the runoff. In its downstream ends, or in specific topographic locations, underground pipes might replace open channels. The design of the system requires the choice of an appropriate size and type of channel and channel lining, based on the hydraulic and hydrological data available, in order to transport the expected runoff quantity without flooding and to prevent channel erosion. The most economic solution is to apply grassed channels, where the grass provides for the necessary soil resistance to erosion. It should be noted, that grassed channels provide not only the protection of the channel cross-section, but also, as emphasized before, allows for partial infiltration and pollution retention and assimilation. However, the grass should be cut regularly to avoid clogging and the considerable reduction in the channel cross section. It is advisable to design the grass channels with triangular and trapezoidal cross-sections and low slide slopes in order to allow for mechanical grass removal, but this would increase the area requirements.

Measures to stabilize channels and to prevent erosion are necessary in specific locations as sharp bends, channel drops and flow energy dissipators at discharge points. The most common erosion prevention measures are ripraps and gabions. Ripraps consist of loose rock or concrete blocks placed over the erodible sols surface. Gabions are blocks of wire mesh filled with gravel or other appropriate material, which are usually used to stabilize the channel walls in the case of large channels. In the case of large channels, the dissipators of energy at the channel outlets require special hydraulic design.

3.3 *Runoff treatment and reuse*

The methods and structures discussed in this section are more suitable for the abatement of urban diffuse pollution sources. The most appropriate management practices of agricultural sources of pollution abatement are preventative measures to reduce pollution at the source, especially in cases of fertilizers and other chemical applications.

The choice of a specific treatment method is based on the location and purpose of the treatment facility. The location is closely related with the size and type of the drained area, and the purpose is related to the available regulatory instruments, which should indicate the downstream use and corresponding quality of the treated runoff. Thus, runoff treatment facilities might vary in size from small units for on-site treatment of roof runoff or parking areas to large ponds and even dams, which could control the quantity and quality of the runoff of the entire catchment area. The choice of an appropriate location should be made in the context of the integrated water resources management and catchment-wide analysis.

3.3.1 *Detention/retention basins*

Historically, runoff storage facilities were designed with the main purpose of runoff detention and flood alleviation, which could absorb the peak flow volume and release it into the stream at a later stage. Runoff treatment was not envisaged, except for the retention of coarse material and suspended solids. Detention facilities are those designed to retain the runoff for a short period of time and completely drain after the design storm event has passed. They are also referred to as "dry detention ponds". If the retained water is to be infiltrated into the ground, they are termed "recharge basins". Retention facilities are designed to retain a pool of water permanently, while discharging the runoff at a control rate, and are also known as "wet detention ponds". The design of detention facilities require a thorough knowledge of the drained area, the inflow/outflow hydrographs and requires a reservoir routing analysis in order to adequately fulfill

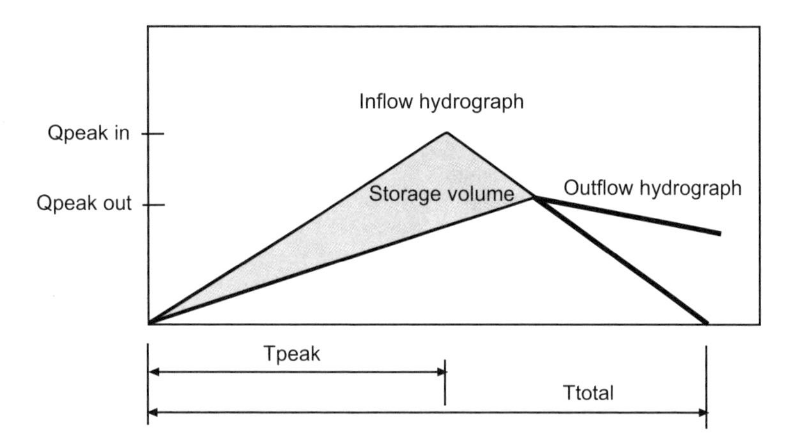

Figure 2.3. Schematic representation of storage volume determination, based on triangular hydrographs.

their purpose. A schematic representation of the principles of storage volume determination is shown in Figure 2.3.

Recently, storage facilities are designed as multipurpose basins, where in addition to storage, a substantial pollution reduction is achieved. These structures are retention type basins, named sometimes "dual-purpose" basins, where the wet pool is designed as a pond or wetland. During peak runoff events they retain the peak flow and release it to the stream at a delayed rate, while the lower part of the basin is always wet and perform the function of a treatment facility, as shown in Figure 2.4. Many different configurations could be developed, given the topographic and soil conditions.

Dual-purpose basins could be used as recharge facilities for ground water infiltration, as well. The application of storage facilities is most effective in cases where the topographic conditions allow for natural depressions, which could be easily transformed to storage basins.

3.3.2 Stabilization ponds

Stabilization ponds are widely used for wastewater treatment. The function and design principles are the same with respect to storm water treatment. However, it should be remembered that storm water usually has a lower organic content, compared to wastewater, and this consideration would influence the configuration and size of the pond system. They resemble closely the natural processes of stabilization of organic materials in natural water bodies, but are engineered in terms that adequate inlet and outlet facilities, retention time and depth are designed in advance, in order to allow for the provision of a proper process environment. Stabilization ponds are usually excavated basins, but if the topography allows, natural depressions could be used. There are three major types of ponds applied in practice–anaerobic, facultative and maturation ponds. With respect to the treatment of polluted runoff, which has relatively low pollutants concentrations, compared to wastewater, the last two types could be recommended for application.

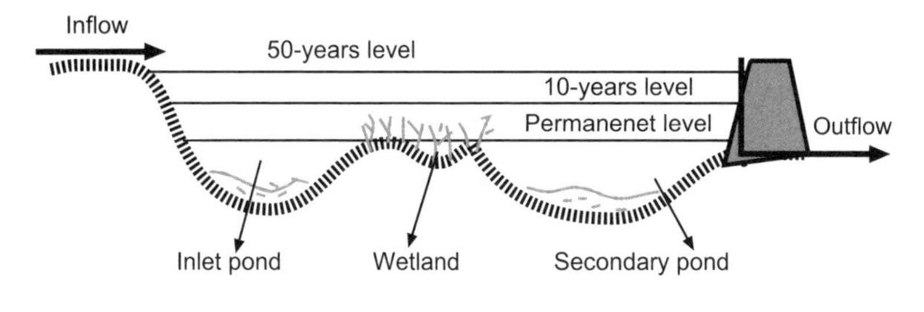

Figure 2.4. Schematic presentation of a dual-purpose basin.

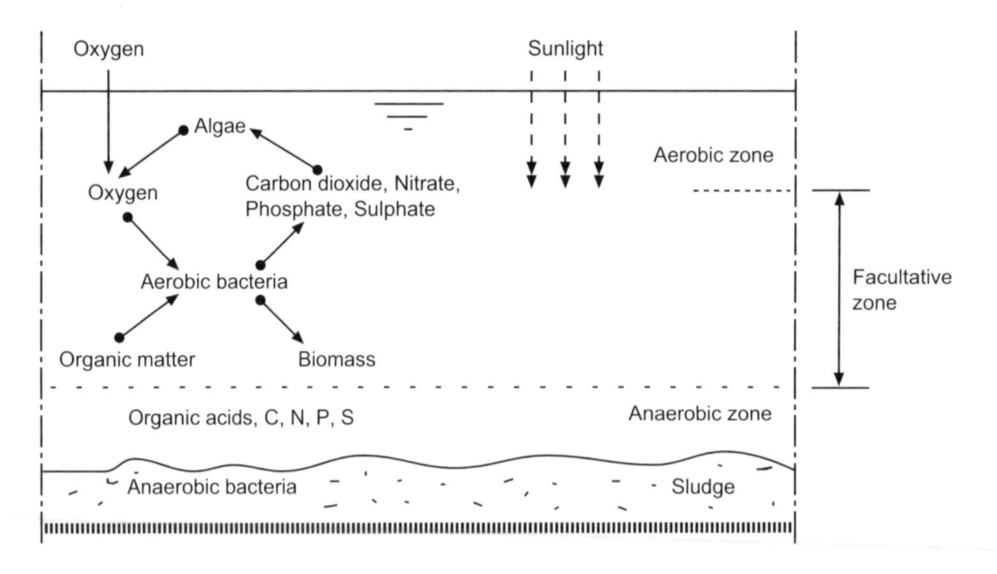

Figure 2.5. A scheme representing facultative pond reactions.

Facultative ponds have the purpose to remove solids by sedimentation, and organic material by biological degradation. The major processes, taking place in a facultative pond are shown in Figure 2.5.

Two distinct zones might be differentiated in a facultative pond – anaerobic zone, close to the bottom, and aerobic zone in the rest of the pond volume. The aerobic zone is of variable depth, depending on the volume of accumulated sediments, organic load, wind direction and algae concentrations. The zone at mid-depth, where both anaerobic and aerobic conditions might occur, is known as "facultative zone", and microorganisms in this zone are facultative – they can adjust and survive in both, aerobic and anaerobic conditions. It should be noted that the different zones are not well defined and change constantly in terms of depth variation. In addition, considerable interaction exists between the different zones in terms of products and materials released. The recommended depth for facultative ponds is between 1 m and 1.2 m, in order to allow for a relatively wide aerobic zone, where the major organic reduction occurs.

Algae play a significant role in the purification process, as shown in Figure 2.5, by acting in symbiosis with aerobic bacteria, consuming the end products of their activity and releasing oxygen in addition to the one supplied by reaeration from the atmosphere.

The design of facultative ponds is based on empirical values for recommended organic surface loading rates, expressed as kg BOD or COD/ha.day, which are strongly dependent on the average temperature conditions (Feachem et.al. 1977). The usual retention time of the water to be treated in the pond varies from 7 to 30 days, which explains the large area requirements.

Maturation ponds serve as a polishing step in the treatment process and provide mainly for pathogen removal. Empirical data from existing pond systems show a considerable pathogen removal efficiency, the second higher after disinfection. The process mechanism is not well studied but many authors mention the action of sunlight as the main disinfection agent. For this reason the recommended depth of maturation ponds is between 0.75 m and 1.2 m. Maturation ponds are usually located as a last stage of the pond system, after the facultative ponds, and it is strongly recommended that the ponds configuration should allow for several maturation ponds to be located in series (Horan 1990).

In general, stabilization ponds are robust treatment facilities, which require low operation cost and skills, and have a considerable buffer capacity under variable load conditions, which makes them suitable for runoff treatment. Their capacity to reduce nutrients concentrations is relatively low-up to 50%, and they retain heavy metals pollution, mainly through sedimentation.

3.3.3 Wetlands

Wetlands are shallow ponds, covered with intense vegetation. Natural wetlands are considered to be receiving water bodies and are subject to regulatory instruments to control their quality and maintain a healthy environment. If polluted runoff is to be discharged and treated into a natural wetland, it should comply with these regulations. Typical characteristics of natural wetlands include:

- The depth of water may vary throughout the year or be absent for some time.
- The saturation with water is the dominant factor determining the nature of soil development and the types of plant and animal communities living in it.
- Often they might be found in the boundary between terrestrial and truly aquatic systems (rivers or lakes).

Man-made wetlands are specifically constructed treatment structures, similar to the natural ones. However, selected hydraulic enhancements are provided in order to control the inflow and the hydraulic conditions, and in addition, to provide for certain retention time for biological reaction processes to take place. They have the same advantages as ponds, but have even higher land requirement, as their depth is lower. One specific advantage of wetlands, compared to ponds is their significant capacity of nutrients removal due to vegetative nutrients uptake. Therefore, their application for runoff treatment is strongly recommendable in cases of nutrients rich runoff.

Two major types of wetlands could be differentiated, based on the hydraulic conditions – free water system and submerged flow system wetlands. The free water system wetlands are basins or wide channels, which provide for a flow with a free surface among the grown vegetation and direct the flow movements. Their bottom is covered with impervious cover to prevent seepage to the ground water. Part of the basins and channel volumes are filled with soil to provide the basis for emerging vegetation. Submerged flow systems, known also as "constructed" wetlands, do not provide for free flow. A schematic representation is shown in Figure 2.6.

Wetlands are widely used options for treatment of polluted runoff at a large and at a small scale in on-site units. Large-scale wetlands are incorporated into the landscape design of the area and could form part of parks and recreational areas. It is recommendable to incorporate a pre-treatment facility for coarse and heavy particles removal before the wetland inlet, in order to avoid the accumulation of sludge and clogging. The design of a wetland system requires profound knowledge of the hydrology of the site, especially when the runoff water is the only source to feed and maintain the wetland. Care should be taken to provide for minimum water requirements throughout the year. If water is not supplied regularly and there are considerable periods of dry spells, the wetland might be lost. Wetlands require little operation and maintenance efforts, but should be regularly inspected to ensure that there is no disruption of the flow due to sediments deposition or vegetation overgrowth. Periodical cleaning and vegetation clearance would also be necessary.

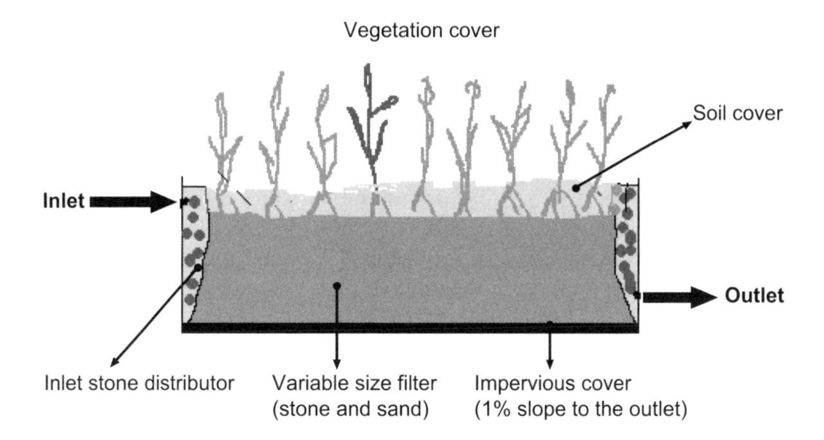

Figure 2.6. Schematic representation of a subsurface (constructed) wetland system.

3.3.4 Filters

Sand filters could be used as a polishing step for polluted runoff treatment. They have been applied mainly for small on site units or elements of the drainage system. Sand filtration removes pollutants based on adsorption and absorption processes. Filters remove this portion of the organic pollution, nutrients or heavy metals, which is associated with small-size suspended particles and colloidal material retained in the filter bed. In many cases, filters are incorporated as part of combined treatment units, which have compartments for sedimentation and/or other treatment methods.

Filter media might include different materials other than sand, such as anthracite, activated carbon or synthetic granules. The filter bed might be designed to contain layers of different materials with different size fractions. An important point to be remembered when filters are applied, is the fact that they have limited capacity to retain pollutants and need regular cleaning or replacement.

3.3.5 Infiltration structures

These structures could be classified more as facilitating structures, rather than treatment facilities. As in the case of detention reservoirs, infiltration structures allow for partial retention and infiltration of the runoff into the soil. They could be used at a small scale (on site structures), or at a large scale to allow for the runoff infiltration of a whole suburb or region.

Small-scale structures are usually designed as infiltration trenches, which have a similar construction as the well known "soak ways". These are trenches, filled with variable size inert porous material (stone, rubble, sand, etc), which are arranged to permit a smooth passage of the rainwater from the conveying drainage perforated pipes, through the drainage material, surrounding the pipe, into the soil. It is common practice to incorporate infiltration trenches within the drainage system, parallel to roads, streets or parking slots, in order to minimize the runoff volume, which needs to be transported downstream.

Large-scale structures include infiltration basins, which allow for the infiltration of the accumulated water through the bottom and walls of the reservoir. They might be designed in the shape of an embankment or an excavated structure as a shallow pond, which reaches the permeable soil layer. During most of the time these are dry structures, which could be incorporated in the landscape and form part of parks, recreational areas, etc. Care should be taken to clean periodically the debris, sludge and slime left after heavy storms, in order to prevent clogging of the surface. It is recommended that such types of basins be covered with grass, in order to help the biological regeneration of deposited organic material.

Infiltration structures usually have large land requirements, as the main design parameter is the infiltration capacity of the soil per unit area. The lower this parameter is, the larger infiltration areas will be needed. One important consideration for the design of infiltration basins is the retention time of the runoff volume, before its infiltration, which should not be excessively long in order to prevent the organic decomposition of debris, unpleasant odors and mosquito breeding.

3.3.6 Storm water reuse

Storm water is one of the basic sources of supply and replenishment of natural water resources. In many countries in arid and semi-arid regions, where natural water resources are limited, the practice of rainwater harvesting is applied. It consists in the collection and use of runoff from roofs and other open space areas at household level or small suburb levels. It is assumed that this water has a very low level of pollution and could be reused directly for different types of beneficial uses.

A relatively new development in the planning of urban water systems advocates decentralized wastewater systems, where the segregation of different types of water is recommended (Dean & Lund 1981). One possible form of combination of storm and gray water reuse is shown in Figure 2.7.

In decentralized systems, the transport of the different flows is minimized and all types of flows are treated and reused within a relatively small area. This allows for the utilization of the vast majority of the products, thus forming a close loop cycle with zero undesirable emissions. Storm water reuse forms part of the cycle. This practice requires the provision of storage reservoirs of adequate size. The "harvested" runoff could be reused again without treatment for irrigation or other beneficial uses, which do not have high water quality requirements. In other cases, the runoff from larger areas could be collected, and after treatment, could be reused for different purposes.

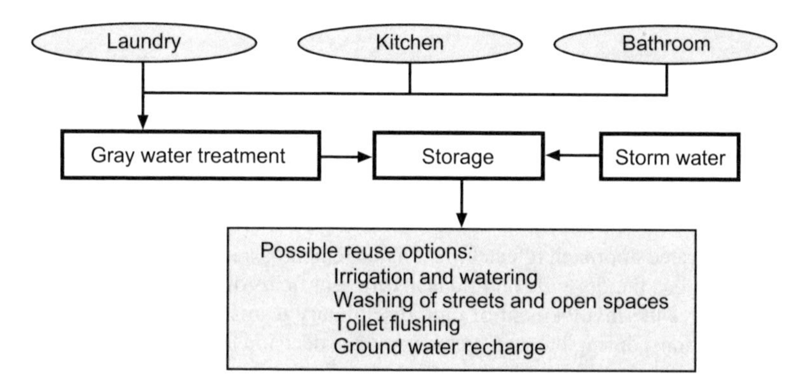

Figure 2.7. A schematic representation of storm water reuse in combination with gray water reuse.

There are numerous possible configurations of such type of systems at different scales; but the concept itself is still in the phase of development and research. The example, shown in Figure 2.7 envisages the combined storage and reuse for irrigation purposes of gray water and storm water, where gray water is domestic wastewater from household washing and bathing activities. This scheme does not show the black water released from toilets, which should be treated and reused in a separate line.

4 MANAGEMENT APPROACHES AND TOOLS

Historically, water resources management has dealt with the provision of water quantities necessary for the development of the society. Ancient civilizations have started to collect, store and transport water for different purposes. In ancient Egypt, the water levels of the Nile were controlled and the agricultural activities were linked to the seasonal flow variations. Romans are known for their advances in urban water supply, other people developed well-organized and constructed irrigation systems. Up to the 19th century, professionals involved in water engineering were thinking in terms of water quantities only. The link between water sources, their quality and public health issues was made in the second half of the 19th century after several disastrous epidemic events caused by water borne diseases. These events led to the implementation of sanitation structures and wastewater treatment facilities. During the last decades of the 20th century, characterized by extreme and unseen levels of urbanization and industrialization, a serious deterioration of whole geographic regions and corresponding natural water resources has been detected, especially in urban areas or regions with intensive agriculture. Thus, qualitative aspects and environmental protection become imperative for an adequate water resources management practice and the concept of integrated management of water resources was introduced.

4.1 *The integrated approach to catchment management*

Considering the character and generation of diffuse pollution, its assessment, regulation and control should be based on an adequate and proper water resources management practice, where the concept of sustainable economic and social development governs. This concept has different aspects in terms of economic, environmental, human resources and other conditions, but the general goal is to provide for the needs of present generations, without jeopardizing the needs of future generations. In order to achieve its objectives, the process of diffuse pollution management should be incorporated and should form part of the whole process of integrated water resources management. This approach requires consideration of factors and interactions in the following directions:

- All natural aspects of the system-surface and ground water and water's physical behavior, as well as environmental water requirements, impacts and interactions with other environmental media-soil and air;
- Consideration of quantitative and qualitative aspects;
- All sectors of the national economy that depend on water;
- The relevant national objectives and constraints (legal, institutional, financial, and environmental);
- The institutional hierarchy and arrangements at national, provincial, local levels and corresponding interactions;
- The spatial variation (upstream and downstream interaction, basin-wide analysis, interbasin transfer).

It is clear that an integrated approach to catchment management is a multi-objective task, which requires consideration of numerous factors and conditions in different fields of specialization. Therefore, its practical application requires the involvement of multidisciplinary teams and interactions between different disciplines and professions, during the management and the decision taking process. No single discipline, agency, institution or group could handle this task alone. Furthermore, if decisions are to be made on a sound and scientifically supported basis, the use of mathematical modeling and information processing becomes an imperative part of the process. It would allow for a more specific and precise determination of the present status, trend developments and impact/risk assessments.

4.2 *The planning stage*

The planning of any activities and measures for diffuse pollution control and abatement should be viewed as an indispensable and interrelated part of the integrated catchment plan for the development of water resources. The plan usually consists of separate programs, envisaging different aspects of the development of the catchment. It should be developed for a certain period of time (planning period), usually two to five years. The catchment principle of water resources management requires that the plans should be orientated towards a selected catchment area, but could be developed at different scales, ranging from a small stream to a large catchment basin, incorporating different streams, rivers or lakes. The most important constituents of any plan or program should include the following points:

- Formulation of goals and objectives;
- Translation of the objectives in measurable criteria or benchmarks;
- Formulation of activities to achieve the criteria;
- Preparation of the implementation plan;
- Monitoring of the plan implementation;
- Assessment of the results obtained at the end of the planning period, and the revision of the plan objectives (if necessary) for the next planning period.

Regarding the development of plans for diffuse pollution control and abatement, the formulation of objectives requires knowledge of the present status of the natural water resources. If preliminary investigations or historic records found that this status is satisfactory, then the objectives of the program would be oriented towards measures to maintain this status and prevent pollution from the future development of the area by implementing the best available and achievable practices of urban/rural development and environmental impact assessment. In cases where a deterioration of water quality has been found, the objectives should include the provision of specific treatment methods to reduce and alleviate pollution.

Planning objectives are usually narrative statements, which reflect the goals, based on stakeholders' desires and the mandate of the management agency. The translation of these statements into specific numerical values or well-defined criteria is the purpose of the next stage of the planning process, where available regulatory instruments must be applied, such as water quality standards and regulations, municipal by laws, ordinances, design manuals, etc. If these are not available, they must be developed for the specific objectives of the program. Based on the regulatory instruments adopted, the activities necessary for the implementation of the plan, should be defined. It is important to emphasize that in the past, activities related to water resources management, usually were limited to structural measures, such as engineering structures for water storage, supply and treatment. However, during the last decade of the twentieth century, activities to involve all stakeholders during the whole process of planning and implementation have become an imperative aspect. Thus, the development and implementation of programs for the public and all stakeholders' involvement in the water resources management process have a high priority in the planned list of activities. In addition, a successful diffuse pollution abatement program requires extensive water quality monitoring process and data collection; therefore, monitoring programs form a substantial constituent of the activities' list and provide the required information for the implementation of pollution abatement measures.

The planning stage of diffuse pollution abatement programs would require the preparation of an implementation plan, which would help in making it more realistic and enforceable. The plan should provide for well-formulated and concrete activities, clear benchmarks, or criteria for assessment of the work done, and time frames for the execution of the activities. Thus, a plan could include the following:

- Formulation of clear objectives and the corresponding activities to achieve them within a prescribed time frame;
- Listing the possible constraints for the implementation;
- Description of the mechanism of implementation with corresponding benchmarks;
- Providing for funding sources;
- Defining the responsibilities and level of involvement of all stakeholders participating in the program.
- Monitoring of the plan implementation – it would consist in regular control of the time schedule and the benchmarks for the plan implementation;
- Comparison of interim results and benchmarks to the initially formulated goals and criteria; if need arises, the plan might be reformulated or altered to reflect actual practical conditions;
- Final assessment at the end of the planned period – it should summarize the implemented activities and results, and lay the foundations for the next planning period.

Such an approach to the process of planning and the implementation of diffuse pollution control and abatement programs would be applicable to any program for water resources management in general. It is important to emphasize the fact that it should be viewed as a continuous succession of closed loop events for selected time periods. More detailed discussion of this topic is presented in Chapter 12.

4.3 *Public awareness programs and involvement*

Public awareness and involvement form an essential part of the efforts to control environmental pollution. The development of a public awareness program (PAP) as part of the activities of any organization, involved in water quality management, is a very important step, in order to streamline the efforts of all stakeholders, prepare a sound plan of the activities to be undertaken, and provide for the resources and financial backup. In a PAP we could differentiate two main streams:

- Continuous (on-going) public information/education programs;
- Specific public awareness campaigns related to a defined event or specific objective.

The first step in designing a PAP is to identify the term "public". There are different levels of involvement of different sectors of the public, based on technical expertise, job duties, levels of concern and willingness to invest time and effort. Therefore, different types of public education and involvement will be required to reach the different sectors or groups. A broad classification could include:

- Stakeholders-this group involves the parties directly concerned with the activities of the organization (legitimate stakeholders) and other interested parties as communities, environmental groups, and non-governmental organizations (NGOs) related to the PAP objectives.
- Press and media
- General public

4.3.1 The PAP planning stage

- During the process of planning and preparation of a PAP the following steps should be considered:
- Find out the historic background of the event or process;
- Formulate clearly your objectives and goals and derive from them the objectives of PAP. Some typical objectives of a PAP could be:
  - To inform and educate the stakeholders;
  - To seek input from the stakeholders and public and to involve them in the process;
  - To gain consensus on a specified problem;
  - To monitor the execution of your educational program;
- Identify the major potential stakeholders and their level and type of concern and the target audience, which could include: the respective community; local media; elected officials; government staff; industries related to the task; environmental groups; NGOs; women organizations; the general public;
- List and describe the different stages of the program;
- Outline specific activities and products within each stage and their interrelation;
- Identify ways or criteria to assess the program effectiveness;
- Include milestones for reassessment or correction of PEP;
- Estimate costs and levels of effort for each stage of the program.

4.3.2 Public involvement activities and techniques

The forms and activities of public involvement may vary considerably based on the specific conditions on the ground, the traditional forms used in the society and the available resources. Some possible activities for public involvement are listed below, but they should be considered as a guide to stimulate creativity and find the best option for each specific case.

- Advisory groups – different stakeholders are invited to participate and advice. Care should be taken for a balanced representation of different participants, who have the authority to make or enforce decisions. This will also prevent controversy, if the advice is not considered and applied.
- Hotline – a widely advertised telephone line to handle and answer questions. It could be used for two-way communication with the general public and provide information about the public reaction to a specific problem.
- Interviews – Face-to-face interviews with key stakeholders could be used to anticipate reactions, gain individual support and provide targeted education.
- Meetings – a widely used form to provide for the public to be heard on specific issues.
- Workshops – smaller meetings designed to complete a task or communicate detailed or technical information. They provide a maximum use of dialogue and consensus building, but is not appropriate for a large audience.
- Volunteer programs – involve citizens, who volunteer to provide services related to the task. It builds a sense of shared ownership and allows communities to experience some of the realities on the ground, but requires a significant input from professionals for supervision and guidance.
- Polls-carefully designed questions are asked of a statistically selected portion of the public. It provides a quantitative estimate of the general public opinion, but the information obtained is static, reflecting present conditions only and the procedure is costly.

4.3.3 Communicating technical information

PAP in the field of water quality management often requires communicating technical information to non-technical individuals and the public, which might create a lot of confusion, miss-understanding and have an adverse effect on the general objective of the program. For this reason, care should be taken to "translate" correctly the information, which needs to be communicated. The following points could be useful during this process:

- Anticipate in advance issues of public interest, based on previous results from interviews, workshops, and meetings.
- Invite public involvement during the preparation stage; a possible solution is to invite outside consultants as advocates for the citizens.
- Use an outside body (objective consultant) to review the technical aspects of the project.
- Present technical information in an understandable language, by hiring public relations experts to review the language in handouts, news releases, reports, etc.

4.4 *Enforcement of regulatory documents*

A successful enforcement of the regulatory instruments with respect to water quality management will determine the overall success of any program for water quality control, pollution prevention and water bodies' restoration. In the process of enforcement of the regulations, two phases could be differentiated:

- During the first stage, all supporting documentation must be completed and approved by the governing authority.
- During the second stage, inspection must be provided to ensure that the proposed activities and components of the execution body are installed and function as designed.

After ensuring that the approved regulatory instruments are backed up by sound institutional structure and financial support, the success of the enforcement process would depend on several factors, such as:

- Phased enforcement – it allows for warnings and advice, including technical recommendations, together with realistic time frames for the implementation of the activities by the polluters.
- Swift enforcement – it considers the time for the implementation of the prescribed activities and provides for the introduction of stringent penalties, if recommended time frames were not met. It should be emphasized that it is imperative to implement a swift enforcement of regulations regarding new developments. Failure to do so could result in costly reconstruction and irreversible damage to the environment.
- Effective enforcement – it is related to the fairness and consistency of the implementation of regulatory instruments. Each specific case should be handled in the same way without undue reliance on an individual inspector's subjective judgment. Also, authorities and political leaders, without undue interference, should back up the inspector's work.

REFERENCES

Chapman, D. 1998. *Water Quality Assessments A Guide to Use of Biota, Sediments and Water in Environmental Monitoring*. 2nd ed. London: SPON Press.

Dean, R, B. & Lund, E. 1981. *Water Reuse: Problems and Solutions*. Copenhagen: Academic Press

Debo, T. & Reese, J. 2003. *Municipal storm water management* 2nd ed. Boca Raton, Florida: CRC Press LLC, Lewis Publishers.

Degremont 1991. *Water Treatment Handbook, 6th edition*, Vol.1, Paris: Lavoisier Publishing.

Ellis, K.V., White, G.& Warn, A.E. 1993. *Surface Water Pollution And Its Control*. UK: Macmillan Press.

Feachem, R., McGarry, M., & Mara, D. 1977. *Water, Wastes And Health in Hot Climates* UK: John Wiley & Sons.

Horan, N.J. 1990. *Biological wastewater treatment systems–Theory and operation*. Chichester: John Wiley & Sons.

Novotny, V. 2003. *Water Quality: diffuse pollution and watershed management*. New Jersey: John Wiley & Sons.

CHAPTER 3

Characteristics of an urban environment in the context of diffuse pollution control

R. Hranova

ABSTRACT: Harare, the capital city of Zimbabwe, located in the Manyame River catchment basin, has been characterized, in terms of population growth characteristics, land use patterns and the water supply, sewerage and drainage systems. Specific attention has been given to the water quality of the Lake Chivero and tributaries, which receive the runoff from the city. A historic overview of the catchment water quality has been done, based on published data from previous research projects. Nutrients variations in urban drainage have been presented, based on data from the city's monitoring program for the period 1995-2000. Results show a sustained trend of increase in the concentrations of phosphates and ammonia. Comparison with earlier studies shows that this trend has been reported since the 1980's. The existing regulatory instruments and monitoring practice have been discussed with emphasis and implications with respect to diffuse pollution control and abatement, and corresponding recommendations for future development put forward.

1 INTRODUCTION

1.1 *Urbanization trends*

Throughout most history, the human population has lived in a rural life style, dependent on agriculture and hunting for survival. In 1800, only 3% of the world's population lived in urban areas, while in 1950, 30% of world's population resided in urban centers and the number of cities with over 1 million inhabitants had grown to 83. The world had experienced unprecedented urban growth in recent decades. In 2000, about 47% of the world's population, which is about 2.8 billion people, lived in urban areas, and with respect to developing countries, about 40% of the population resides in urban areas. It is expected that, by 2030, 60% of the world's population will be living in urban areas with the most growth occurring in the developing countries (UN Population Division 1997).

In Africa, at the start of the 20th century, 95% of Africans lived in rural areas. Even in the 1960's, Africa remained the least urbanized continent with an urban population of 18.85%. By 1996 this had more than doubled, and by 2010 at least 43% of the population is expected to live in urban areas. The average annual urban growth rate in Africa during 1970-2000 were the highest in the world at more than 4%, and it is projected to decrease slightly to about 3% during 2020-2025. In Northern Africa, more than half of the population now lives in cities, while in some countries of Southern, Western and Central Africa the percentage of population living in urban areas is between 33 and 37%. East Africa is the least urbanized sub-region with 23% of the population living in urban areas. (UN Population Division 1997).

At the time of colonial occupation in much of Sub-Saharan Africa, urban population settlements emerged as administrative and trading centers, not only on domestic trade routes but also in international trade. Urban centers further developed due to the establishment of infrastructure, especially railways, which have been developed to connect ports to their hinterlands. This legacy was followed after the independence of most of the African countries, which influenced the pattern of urban settlements, the

Table 3.1. Urban population growths in the South African region.

Country	Population in 1995 (Millions)	Urban population (%)			Urban population growth (%)			Total population growth (%)
		1980	1987	1995	1980	1987	1995	
Angola	10.8	21.0	26.0	32.2	6.0	5.7	5.7	3.1
Botswana	1.5	15.1	21.7	30.8	8.5	8.6	7.2	2.9
Lesotho	2.0	13.3	17.4	23.1	6.6	6.7	6.0	2.3
Malawi	9.8	9.1	11.0	13.5	6.7	5.9	5.6	3.0
Mozambique	16.2	13.1	22.4	34.2	11.6	9.7	7.8	1.9
Namibia	1.5	22.8	28.8	37.4	4.7	5.9	6.0	2.7
South Africa	41.5	48.1	48.7	50.8	2.6	2.6	2.8	2.2
Tanzania	29.6	14.8	18.9	24.4	11.0	6.9	6.3	3.1
Zambia	9.0	39.8	41.3	43.1	6.0	4.2	3.7	2.7
Zimbabwe	11.0	22.3	26.5	32.1	5.5	5.9	5.1	2.7

Source: World Bank (1997)

nature of urban built up, and the process by which urban development was administered. Recent patterns and trends in the urban population growth rate of some African countries are given in Table 3.1.

In Zimbabwe, urban growth has been very rapid since the 1970's following the relaxation of pre-independence population influx control. In recent years, urban areas accommodate about 30% of the total population and are growing at the rate of 5.4% per annum. Most of the larger urban areas in Zimbabwe show the characteristic manifestation of rapid urbanization in the third world, such as chronic unemployment levels, acute housing shortages and inadequate municipal service infrastructure. The result has been a tremendous pressure on municipal councils to provide adequate service, such as housing amenities and waste management (Drakakis-Smith & Kirell 1990)

1.2 *About Zimbabwe*

Zimbabwe is a landlocked country with an area of 389 000 km^2 extending from 15° 30 N to 22° 30 S and lies between 25° W and 33° E. The geographical location and the relief endow a sub-tropical climate. Three relief regions are generally recognized on the basis of their general elevation (Fig.3.1). These are the Lowveld with an altitude below 900m, the Middleveld with an altitude of 900 m to 1200 m, and the Highveld with an altitude between 1200 m and 2400 m, where the capital city, Harare is located. The Highveld includes a mountainous region, which has special characteristics not found in the rest of the country. It is known as the Eastern Highlands and consists of a narrow belt of mountains and high plateau ranging in altitude from 2000 m to 2400 m. Its considerable elevation gives a characteristic microclimate and associated vegetation.

The aim of this paper is to introduce and characterize this specific urban environment, in terms of development trends, location within the catchment basin, and corresponding influence on the water quality status, as well as to underline, and discuss the existing practice of monitoring and control of water quality in general and diffuse pollution in particular. This chapter serves as an introduction and the basis for a better understanding of the different case studies, presented in subsequent chapters.

2 THE URBAN ENVIRONMENT

The city of Harare is spread over a generally flat terrain, with a mean elevation of 1510 m above sea level. Its population is over 2 million people according to the 2002 census. The climatic conditions determine

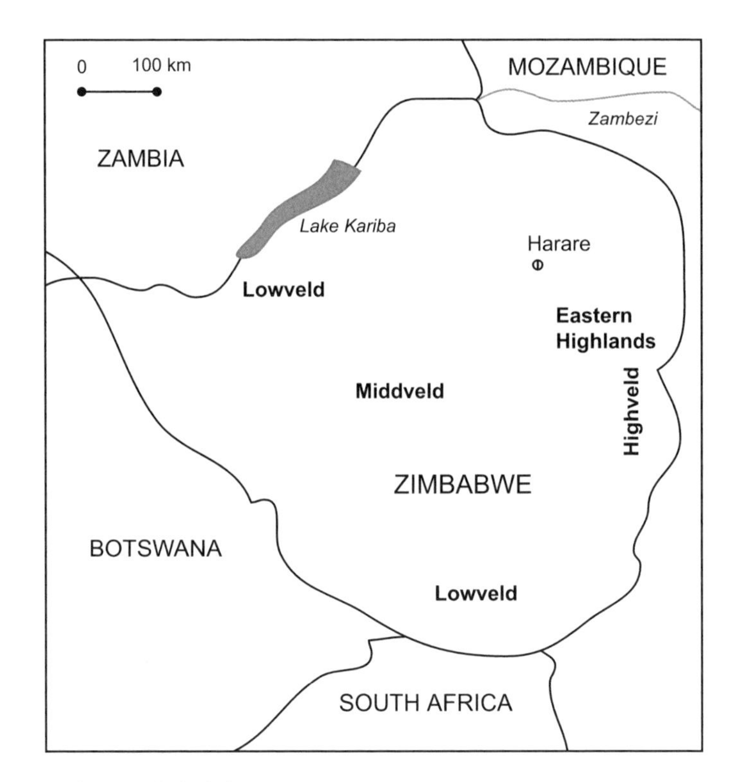

Figure 3.1. A schematic map of Zimbabwe.

two distinct seasons, with respect to rainfall – the hot summer period from October to March with an average temperature of about 27° C, and a dry weather for the remainder of the year. Annual rainfall figures vary within a range of 440–1220 mm. Most of the rainfall events are characterized by high intensities (AQUASTAT 2003). The mean annual rainfall is about 800 mm.

2.1 *Population growth trends*

Harare is a sprawling city covering 570 km^2 and is the largest urban center of Zimbabwe. The city has experienced a rapid growth since the attainment of independence in 1980. Harare's population had increased from 656011 in 1982 to 1.2 million in 1992, and it is now estimated to be above 2 million. According to the 1982 census, 33% of the urban population in the country lived in Harare alone; by 1992 the figure had increased to 36% and is likely to reach 40% before the year 2005, if the current trend prevails. The expansion of the capital city, which followed the repeal in the late 1970's of the influx control legislation that restricted the movement of the black population to the urban areas, has been accompanied by an unmet housing demand. The expansion of the City has been witnessed in the three major types of residential area developments – high, medium and low density. The most critical and problematic areas are the high-density urban developments. Sharp increases in the price of houses and rental accommodation have followed the miss-match between housing demand and supply, especially since 1991, after the relaxation of the rent control regulation. After the introduction of the Structural Adjustment and Liberalization Program (ESAP), rentals in low-income residential areas have increased by more than 300% between 1991 and 1996 and the urban influx has compromised the quality of life through a resultant densification and general over crowding (Potts & Mutambirwa 1991, Zinyama 1993). Drakakis-Smith & Kirell (1990) have confirmed the trend of excess population density, where owners rented rooms for coping with the housing shortage in Harare. This has had the effect of pushing the low-income groups into informal housing and the consequent emergence and proliferation of squatter (informal) settlements.

In general, two distinct patterns of overpopulation and undesirable urban development practices could be differentiated. The first one consists of overpopulation of existing high-density residential areas, which usually are provided with basic infrastructure. In order to increase their income, owners are renting available rooms, or are building additional shacks to accommodate as many tenants as possible. In Highfield for instance, which is the second high-density residential area to be built, a local plan survey revealed a total of 15 230 residential structures of which nearly half have been built without permission (Rakodi 1995). This results in over loading of existing infrastructure facilities, mainly the sewerage and drainage collection systems and the wastewater treatment facilities. The other pattern, resulting from overpopulation, is represented by the formation of informal settlements, which do not have any basic infrastructure, e.g. roads, streets, water supply and sewerage. These two patterns are typical for the vast majority of the countries in the Southern African region. In Harare, the predominant practice is the first one, but in the latest years, the development of informal settlements is also emerging.

Population growth and distribution have significant roles to play in the sustainability of the world's vast resources. The population growth, together with the social and consumption patterns, as well as the economic conditions, all have a direct effect on the environment. When we look at the impact of human activities, the situation is more complicated due to the diversity in the consumption patterns worldwide. A direct link between population growth and the status of the environment is difficult to be established. However, the view that population growth is solely responsible for all environmental ills, is highly objectionable. Population growth is only one of the several factors that place pressure on the environment in general and water resources in particular. Land distribution, inappropriate government policies, ineffective management practice, and inappropriate technologies, all of these and others could be mentioned as contributors to the environmental decline in over-populated areas. Thus, the population growth is one factor among many that exacerbates or multiplies the negative effects of other social, economic and political factors.

2.2 *The catchment area*

The city of Harare is located in the Manyame River catchment (Fig. 3.2), which comprises of different types of land use practice. It contains the highly urbanized areas of the capital city and the satellite town of Chitungwiza, as well as smaller population centers of urban type. Agricultural lands comprise of crops-growing fields and open-space animal farms. The total catchment area is 3930 km^2 and the sub-basin of lake Chivero and tributories, which recieves the runoff and the effluents of the city of Harare is 2250 km^2.

In terms of agricultural practice, a differentiation could be made between small-scale agricultural activities in the so-called "communal farms" and large-scale activities in "commercial" farms. Rural settlements are located in the upper reaches of the basin and consist of villages with a traditional African lifestyle. Natural pristine areas, as well as protected natural reserves, provide the habitat for diverse wild life and are relatively unharmed by human activities. Figure 3.2 presents a map of the Manyame River catchment basin and tributaries with water and wastewater treatment plants.

The city of Harare is located in the sub-catchments of the Mukuvisi and Marimba Rivers, the satellite town of Chitungwiza drains into the downstream reaches of Nyatsime River, while the suburban town of Ruwa drains into Ruwa River. Lake Chivero (formerly known as Lake McIlwaine) and Lake Manyame are man-made reservoirs, constructed with the main purpose of water supply for the capital city. Due to its downstream location, Lake Chivero is the natural sink of the surface run-off and effluents from the above-mentioned areas. Its major beneficial use is to supply a major part of the potable water demand for the City of Harare, but it is also a well-established recreational center and the natural environment for a wide variety of wildlife. This complex pattern requires an integrated approach to the water resources management practice, where water quality control and pollution abatement measures should be given priority. Decisions regarding the provision of safe potable water for the city of Harare, the proper management of effluents generated, and the protection of the environment should be made on the basis of sound

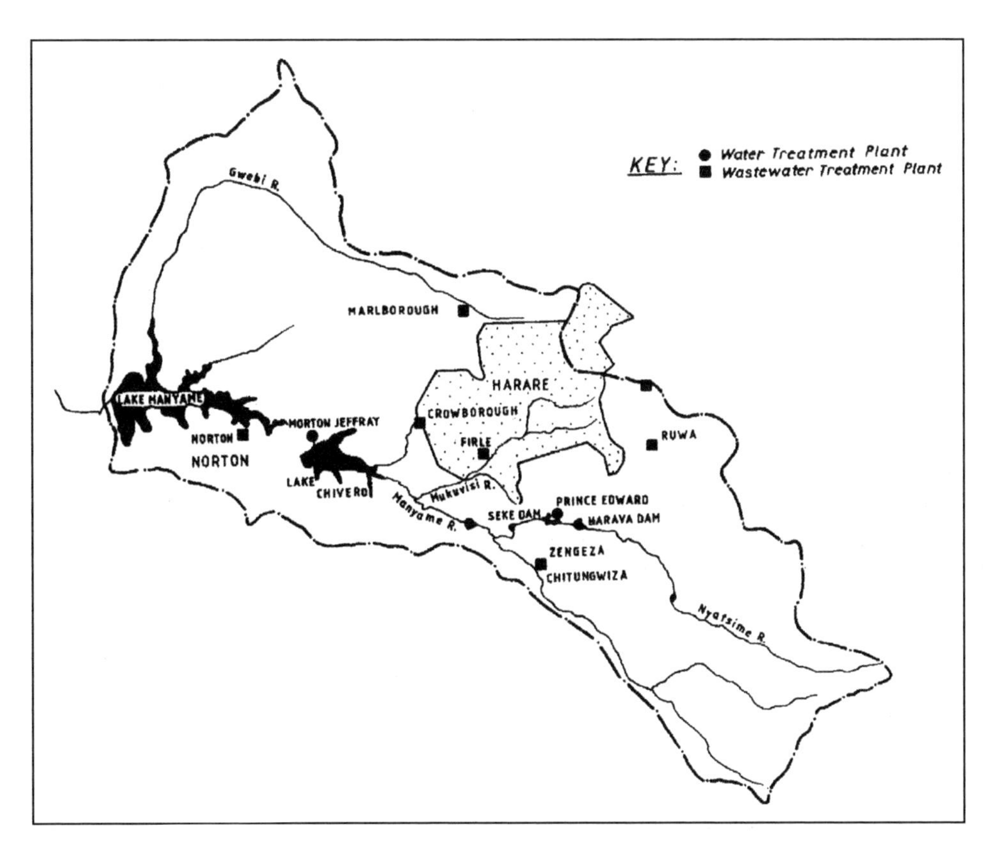

Figure 3.2. Manyame basin schematic map with water and wastewater treatment plants.

data regarding the corresponding water qualities. The availability of data regarding the quality of different types of water fluxes in the basin – natural water, storm water, effluents and potable water, should support the decision – making process, and should be the basis of optimal and cost-effective measures in order to provide for the development of the city and surrounding areas, in a safe and well-protected environment.

Earlier studies (Moyo 1997) show that the Lake Chivero is at an advanced stage of eutrophication. Typical signs of enhanced eutrophication, such as algae and other water aquatic vegetation blooms, are visible. Enhanced eutrophication results in periodical depletion of the dissolved oxygen of the lake's water and a general change of the environmental habitat. Several occurrences of massive fish deaths have been experienced during the last 30 years.

An additional effect of the lake's eutrophication is related to the increased operational and maintenance cost of the treatment of water for potable purposes, especially in dry years, when the water level in the lake subsides. The increase in the suspended solids concentration of the water, resulting from increased algae concentrations, necessitates a higher dosage rate of chemicals to be added, increases the WASHING frequency of the rapid sand filters, leading to increased electricity costs. During the treatment process the volumes of sludge and backwash water are increased considerably, thus leading to an increase in the cost of their conveyance and treatment. Also, a specific earthy taste is introduced to the treated water, as a result of algae blooms and the general decay of aquatic vegetation, which requires an additional treatment stage in order to obtain a good quality of the water supplied to the public.

Considering the importance of the Lake water quality, with respect to the water resources practice in general and the diffuse pollution management in particular, more extended information regarding the past development and trends of its quality is given in section 3.

2.3 *Urban land use patterns*

The city of Harare has a well-developed infrastructure and clearly defined types of urban development. Figure 3.3 represents graphically the city's urban planning scheme and the different types of land use practice. The urban planning provides a considerable percentage of developed, as well as natural open spaces and grounds, which could serve the purpose of environmental buffer zones for diffuse pollution reduction and alleviation. The city center comprises of high stories office buildings, used for institutional, governmental, commercial and business activities. The Central Business District (CBD) has a high level of impermeable cover and generates a considerable amount of runoff.

To the north of the CBD, residential areas are located, comprising of three-four stories blocks of flats, which gradually convert into single housing plots. The vast majority of the residential areas in the City comprise single story housing.

An approximate differentiation between the different types of residential areas, based on the average size of the plots could be done. Low-density areas comprise of housing plots with areas higher than 3000 m^2, medium-density plots vary in size from 800 m^2 to 3000 m^2 and high-density urban developments consist of plots between 300 m^2 and 800 m^2.

Low, medium and part of the high-density residential areas have well-developed infrastructure in terms of roads, electricity and water supply, sewerage and drainage structures, as well as solid waste management. Low-density areas use on-site sanitation systems, while medium and high-density areas are provided with gravity sewers. Housing developments in low and medium-density areas include well-developed and maintained gardens and green areas, which absorb runoff. Also, within residential areas, provision has been made for shopping centers, which contain commercial and business enterprises, as well as educational facilities, mainly primary and secondary schools. A considerable part of the high-density areas have basic infrastructure in terms of roads, water supply and sewerage, which in numerous cases, has not been maintained properly.

Industrial areas are located in the southern and south-east part of the city. It is a mixed land-use pattern, with predominantly light industry, food processing, cosmetics, automotive services and different types of commercial enterprises.

During the last decade, the city has expanded mainly with respect to its residential areas by new developments in the three different categories. However, the high rate of the population growth, as discussed earlier, has an adverse effect on the urban development, because the new-coming population in its vast majority consists of low income or unemployed population, who move from rural areas to the capital city in search of jobs and better life conditions. This has resulted in the emergence of high-density housing areas and small-scale industries, developed to the south-west of the city, to the north and south of the industrial areas, and to the east of the city at Tafara and Mabvuku, within the reach of Beverly. New high-density developments, such as Warren Park, Glen Norah, and Budiriro, were constructed during the eighties and nineties, in the vicinity and around existing high-density housing areas. The municipality has provided the land for such developments. The location of the largest high-density development at Kuwadzana, has been determined by the offer of two private farms to the city council. Typically, these are areas of small plots (324 m^2 or lower) with uniform low-cost, single story houses erected. However, the pressure on space and utilities in these areas continues, and many more people often occupy the rooms that were originally designed for single migrants. The overwhelming concentration of the new development to the southwest and west of the city has not been able to meet the demand. The growth of these high-density urban environments has remained steady over the years and continues to recent days. One major issue of concern has been the emergence of informal sector activities such as informal agriculture, hawking petty commodities and other servicing activities. (Drakakis-Smith & Kirell 1990). This high rate of population density leads to the overloading of the existing sewer lines, where blockages and overflows are typical. In addition, it provides for high percentage of the impermeable surfaces, thus increasing the runoff quantity. Road structures are not maintained properly, with numerous potholes, which enhance erosion by runoff. Collection and disposal of solid waste is irregular, allowing its spreading over open spaces, specifically at market places, thus becoming a source of diffuse pollution. In many cases, drainage channels are used to dispose solid waste.

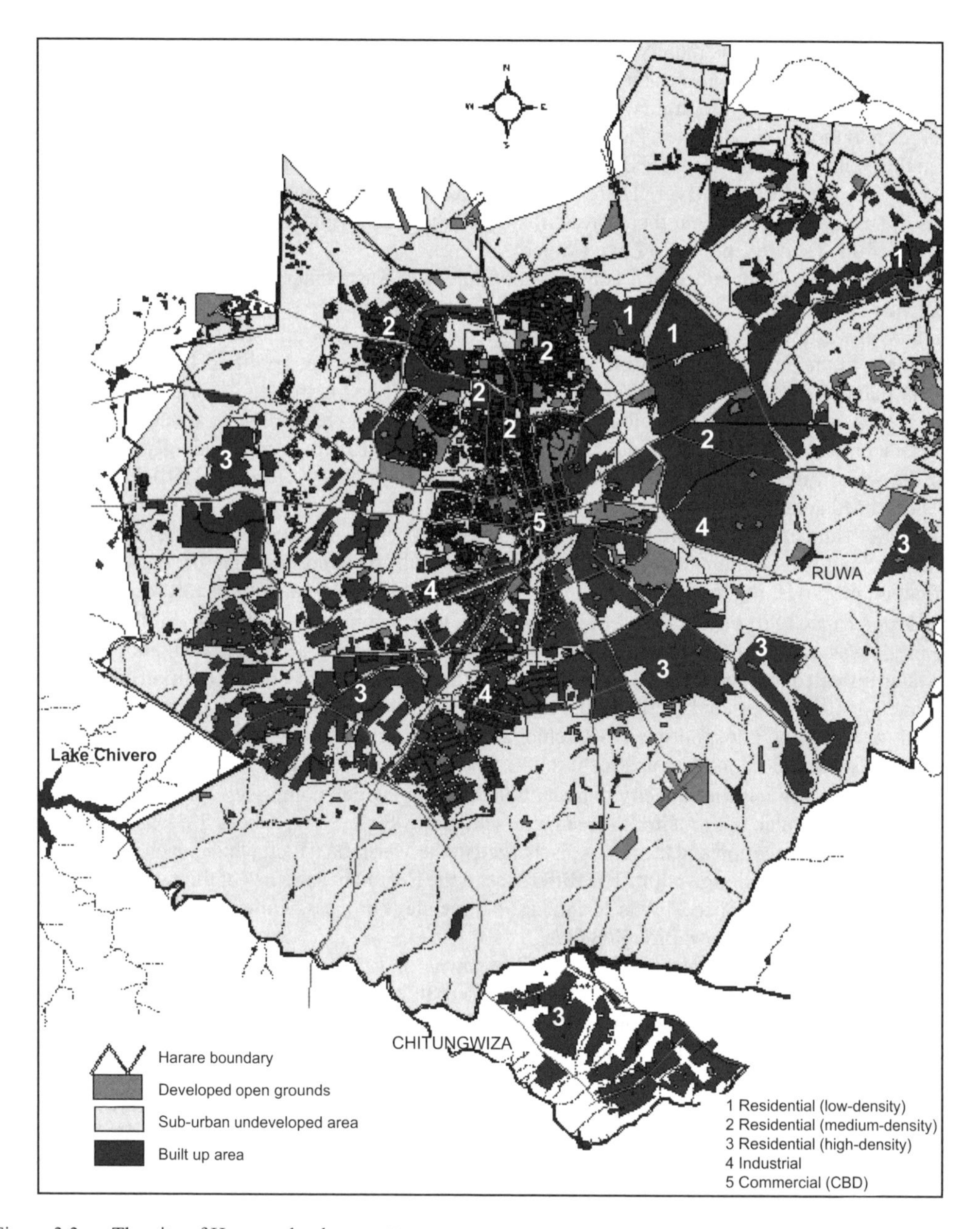

Figure 3.3. The city of Harare – land use patterns.

2.4 *Water supply, sewerage and drainage systems*

The city of Harare is supplied with water mainly through abstractions from the lakes Chivero and Manyame. After a conventional treatment at Morton Jeffrey water treatment plant, the water is directed to several reservoirs and distributed to the residents. The Prince Edward water treatment plant provides a relatively low percentage of the total demand. The city's formal residential areas are provided with treated reticulated water. The informal settlements, which have emerged due to the rapid increase of population in the last decades, are not provided with safe water, as described in Chapter 6.

The drainage system consists mainly of open ditches, parallel to the road structure throughout the urban development. At selected locations, the storm water is transported by pipelines and discharged into natural watercourses. In the CBD, the drainage system consists entirely of a subsurface pipeline system, which discharges into the Mukuvisi River. Part of the industrial areas and part of the medium-density residential areas in the southwest of the city, are drained by an open channel (Coventry Rd drainage channel), which discharges into the Marimba River. There is no provision for storm water storage or treatment before it reaches the natural water bodies; therefore, diffuse pollution from runoff is of concern, considering the proximity of the discharge points to Lake Chivero.

The sewer system is separated and conveys only domestic wastewater from residential areas and industrial wastewater, which is allowed to discharge into the municipal sewer. It should be mentioned that the municipal wastewater should be regarded as mixed industrial and domestic wastewater, because the vast majority of the industrial enterprises do not have local pretreatment facilities. It is treated in several treatment plants (Fig. 3.2). About 96% of the wastewater generated is treated in Firle and Crowborough sewage treatment works (STW). Biological nutrient removal (BNR) technologies are applied to this part of the effluent, which is discharged into Mukuvisi and Marimba Rivers correspondingly. The remaining 4% of the wastewater is treated in waste stabilization pond systems, except for Hatfield STW, which applies extended aeration technology.

The small treatment plants serve mainly high and medium density residential areas. The effluents from treatment technologies, which do not provide nutrient removal, are disposed on land or reused beneficially for irrigation purposes. Figure 3.4 shows the location of the different treatment plants.

The design capacity of Firle STW is 144,000 m^3/day, which comprises about 70% of the total volume of the city's wastewater. After the preliminary treatment and primary sedimentation, about 25% of the wastewater is treated in a separate line comprising of primary sedimentation, biofiltration and maturation ponds, and after that is reused directly for pasture irrigation. The rest is treated in several activated sludge units with nitrogen and phosphorous BNR removal and discharged into Mukuvisi River. The discharge point is located several kilometers upstream the River's confluence to Lake Chivero.

Crowborough STW has the capacity to treat 56,000 m^3/day, which comprises 26% of the total volume of the city's wastewater with a similar treatment scheme to Firle STW, where 73% of this volume is directed for pasture irrigation and the rest is discharged to the Marimba River, which reaches Lake Chivero 8 km downstream the discharge point. The difference in the treatment schemes of the two plants is related to the BNR units. Unit 5 of Firle STW is designed as a three-stage process, while units 3 and 4 are five-stage units as the BNR plant of Crowborough STW.

The wastewater treatment plants of the satellite towns of Ruwa and Chitungwiza comprise waste stabilization pond systems, which are heavily overloaded. In 2002, The Chitungwiza plant was upgraded with a BNR unit. The rapid population growth leads to a considerable overloading of the vast majority of the plants. Despite the fact that municipalities try to catch up with increased loads by upgrading plants' capacities in terms of volume and more advanced treatment technologies, they cannot cope with the volume of wastewater generated. The most overloaded plant at present is Crowborough STW. The hydraulic overload is highly pronounced during wet-weather conditions, when infiltration flows increase significantly the volume of the influent. During high intensity storm events, hydraulic loads are very high, forcing the plant operators to by-pass the plant and discharge a considerable amount of the untreated wastewater into the rivers in order to protect the plant from process failure. The overflow volumes are retained for a short period of time in designated ponds, and in many cases, discharged directly into the River. This practice has also been reported by Nhapi et.al. (2001). Thus, the wastewater treatment plants become a source of diffuse pollution through illegal discharges. The diffuse character of the pollution in this case is not related to the location of the discharge, which is well defined, but to the irregularity of the discharges and the variation of the effluents quality. In the cases of overloaded stabilization ponds systems, the result is a reduced quality of the effluent, leading to the overloading of the disposal sites and indirect pollution of the soil, ground water and surface water. A more detailed study of the impacts of diffuse pollution from storm water in different types of land use practices is presented in the consequent chapters.

3 ABOUT THE LAKE CHIVERO AND TRIBUTARIES

3.1 *About the lake water quality*

Since the year of its construction in 1953, Lake Chivero was a mesotrophic impoundment similar in many respects to the existing reservoirs along the Manyame River. Nitrogen and phosphorus concentrations were relatively low. Species diversity was great and the fish fauna of the lake had the predominantly riverine characteristics of the newly created man-made lake (Marshall 1982). Throughout this formative period (1960 – 1964), rainfall was good and the lake levels were reasonably constant, although drought years did occur in 1960 and 1964, when water levels dropped by several meters. During the same period, the volume of municipal wastewater being discharged by the local sewage works, rose considerably and the first manifestations of eutrophication were detected. (Balllinger & Thornton 1982).

Marshall and Falconer (1973) conveyed a study to investigate the Lake water quality, which reflected the trend of increased nutrients concentrations. The most important parameters, which were influencing the trophic status, were nitrates, ammonia nitrogen and soluble reactive phosphorus (SRP). Their concentration rose considerably during the late sixties due to point discharges from Crowborough and Firle STWs, which at this time were treating wastewater by biofiltration. As a result of this investigation, the municipality decided to divert the effluent from discharging into the rivers and to reuse it for pasture irrigation. Also, the water quality regulatory instruments restricted the discharge of nutrients rich effluents into natural water bodies. Later studies (Thornton 1980, Thornton & Nduku 1982c) reported a considerable reduction of SRP into the Lake water after these measures were implemented. It was found that nitrite nitrogen and particulate phosphorus were of lesser importance and were varying slightly during the year. The concentrations of these parameters, measured at mid-lake station, during the period 1969 to 1979 were usually less than 0.005 mg/l, whilst concentrations of nitrate, ammonia and SRP were generally in excess of 0.010 mg/l. (Thornton & Nduku 1982c). Calculations regarding the pollution load before and after the implementation of the new management strategy, and corresponding reduction in loads are presented in Table 3.2. It could be noted that positive changes are reported mainly with respect to SPR and ammonia. Also, this reduction is noted only at Marimba and Mukuvisi Rivers. Thornton & Nduku (1982b) point out that a tenfold increase of phosphorous loads, expressed as Total Reactive Phosphorous (TRP) during this period, was observed at Manyame River (see Table 3.2), which was attributed to diffuse pollution from urban run-off, mainly associated with informal settlements or excess population in the drainage area of the river upstream the Lake.

Thirteen common trophic indicators have been monitored both prior to and following the nutrient diversion scheme (Thornton & Nduku 1982c). Since the diversion of wastewater nutrients to pasture irrigation schemes, reductions in the mean concentrations of SRP and ammonia were observed in the order of 0.16 mg/l and 0.4 mg/l respectively in 1968-69 to 0.04 mg/l since 1976. Nitrite concentrations have remained largely unaffected. The apparent lack of effect of nutrient diversion on the nitrate and nitrite fractions is not entirely unexpected as nitrogen is virtually unaffected by pasture irrigation and may even be enhanced by some pasture crops (McKendrick 1982). Reduction in phosphorus concentration in the lake during 1977 is due to the more complete flushing of the lake during that year than during the pre-diversion study in 1967. Flow has been shown to play an important role in the phosphorus cycle, with phosphorus concentration and flow in Lake Chivero being inversely related at low flows and directly related at high flows (Thornton 1980). The studies confirm the link between water quality management strategies and implementation measures with corresponding effects on the status of water resources, as well as the need for an integrated approach. Lessons from past experience show that diffuse sources of pollution should be identified and controlled in conjunction with measures to reduce point sources of pollution.

Other investigations (Moyo 1997) show that during the nineties, the lake has been found to be in a state of advanced eutrophication due mainly to an accelerated urban population growth. Several occasions of massive fish kills have been reported, due to deoxygenation in periods of draughts and low flows. Another possible reason could be increased ammonia concentrations released from the Lake sediments. It has been reported that the main causes for the Lake's eutrophication are associated with overloaded or poorly

Table 3.2. Phosphorus, ammonia and nitrate loadings to Lake Chivero during 1967 and 1977.

Source	1967			1977		
	SPR	NO_3-N	NH_3-N	TRP	NO_3-N	NH_3-N
Inputs in t/year						
Manyame	3.2	3.6	0.9	31.8	93.1	6.5
Mukuvisi	183.5	121.4	139.1	20.5	58.1	2.7
Marimba	101.4	33.3	17.8	29.3	35.0	1.1
Sediments	-	-	-	16.1	-	-
Totals	288.1	158.3	157.8	97.7	186.2	10.3
Output in t/years						
Manyame	-	-	-	27.2	252.2	13.0
Water-works	-	-	-	1.2	2.2	0.4
Algal Uptake	-	-	-	3.4	-	-
Sedimentation	-	-	-	54.0	-	-
Totals				85.8	254.4	13.4
Change in solution in lake	-	-	-	–0.8	+10.9	–3.1
Mean amount in solution in lake	-	-	-	8.4	76.0	16.0
Loading rates						
Surface load (gm^{-2})	11.0	6.3		3.9	7.5	0.4
Volumetric load (gm^{-3})	1.2	0.6		0.4	0.8	0.1
Total inflow (10^6m^3)		128.3			433.8	

Note: "1967" data from Marshall and Falconer (1973); "1977" data from Thornton (1980)

maintained treatment facilities and runoff from sludge and solid waste disposal sites (Muthuthu et al. 1993, Nhapi et.al. 2001).

3.2 *Lake sediments*

Studies of nutrient cycling in temperate lakes have shown the importance of sediments as a source and/or sink of nitrogen and phosphorus. Investigations on Lake Chivero have suggested the existence of a sediment source/sink of nutrients in that lake (Marshall & Falconer 1973). Calculations presented by Robarts & Ward (1978), have suggested that the internal nutrient loading from this source might be considerable. It has been shown that sediment-water exchange processes do in fact have a significant effect on the lake nutrient budget, as shown in Table 3.2, but that the sediments act predominantly as a nutrient sink. This fact was supported by the high nutrient concentrations in sediments, where the observed concentrations of phosphorous were between 0.08 to 3.87 mg/l, nitrogen concentrations were between 0.01 mg/l and 0.11 mg/l with selected samples containing up to 10 mg/l, and organic carbon was between 0.5% and 16% (Nduku 1976, Robarts & Ward 1978). The sediments of Lake Chivero during this period were relatively high in nutrients, indicating to an eutrophic status of the lake. Non-eutrophic impoundments in Zimbabwe have sediment phosphorus concentrations of about 0.3 ppm of phosphorus whilst eutrophic lakes have concentrations in excess of 1.0 ppm. (Thornton & Nduku 1982a). Other studies, related to interstitial or pore water nutrient concentrations of sediments in the lake (Nduku 1976) show high concentrations of ammonia – 20 to 50 mg/l. SRP concentrations of 1.05-1.15 mg/l have been measured during the same period, while nitrogen concentrations are relatively low. These results confirm the eutrophic status of the lake.

3.3 *Water quality of the Lake Chivero tributaries*

Mukuvisi River is one of the three major rivers feeding Lake Chivero. Muthuthu et al. (1993) studied its water quality during the period 1991 – 1995. The most common pollution contributors identified in the Upper Mukuvisi River are: a fertilizer manufacturing plant (Zimphos); surface drainage from Graniteside, a light industrial area in the south-west of the City; seepage from solid waste disposal sites, stretching from Braeside to Graniteside along the south banks of the river; and surface drainage from the CBD and residential areas. The river water in this stretch is used for the irrigation of informal agriculture plots. Also, at some locations, squatters living in informal settlements along the riverbanks use its water for washing and drinking purposes. The results of the study during the wet season of 1991 regarding 3 sampling points are shown in Table 3.3. The first sampling point was located at the old Chiremba Road Bridge (MUK1) and served as a reference point; the second was located at Seke Road Bridge downstream from the Braeside (Arcadia) solid waste disposal site (MUK2); and the third point was located at Cripps Road Bridge downstream from the light industrial area and the Graniteside area (MUK3). The results show that conductivity increased progressively from MUK1 to MUK3 suggesting that the seepage from the landfill area are responsible for the rise in the ionic content of the river water. The pH was practically the same at all the sampling points and within acceptable WHO standards. Nitrates seemed to decrease steadily from MUK1 to MUK3, suggesting that the landfill seepage was diluting the nitrate in the river water. Phosphate levels at MUK2 and MUK3 were on the average double that at MUK1 and about three times the recommended limit of 0.5 mg/l. The DO levels were below the recommended lower limit of 6.0 mg/l. The lowest concentration at MUK3 could be due to possible organic pollution from the urban drainage or due to lower flow velocities in the River. Only the solid waste site showed an appreciable impact on the river water quality along this stretch, with respect to phosphorous contribution. The dry season results were similar in trend with lower temperature values around 16°C, and increased conductivity values due to lower flow rates.

Muthuthu et al. (1995) carried out a special investigation to evaluate the influence of the fertilizer plant (Zimphos), located in the southwest industrial area, near the banks of Mukuvisi River. The levels of various water quality parameters were measured at four sampling sites. The Mutare Road Bridge about 1 km upstream from the fertilizer plant served as a reference point (Point 1). Point 2 was located after the fertilizer plant, and point 3 was located at the Mutare Rail Bridge, about 3 km downstream from the plant. The other sampling point was a canal, which carries discharges from an underground drainage collecting system between the fertilizer plant and the river (Point 4). Water sampling was carried out periodically between January 1989 and September 1990. During the months of January to March 1990, samples were collected on a weekly basis. During the months of August and September 1989 and 1990, samples were collected biweekly. Intensive sampling on a daily basis was done from the 11th to the 16th of March 1990. The results of this study, with respect to nutrients are presented in Table 3.4.

The results show a high level of pollution released from the area near Zimphos in the form of surface runoff, as well as in the form of ground water and they confirm that the fertilizer plant is a major source of emission into the upper Mukuvisi River. The levels of nitrate and phosphates increased considerably

Table 3.3. Mean values and standard errors of the physical and chemical parameters for six sets of samples collected during the wet season of 1991.

Parameter	MUK1	MUK2	MUK3	WHO Limits
Temperature (°C)	23.5 +/ –1.5	23.2 +/ –1.4	30.0 +/ –0.6	-
Conductivity (μS/m)	568 +/ –95	660 +/ –161	697 +/ –85	500
pH	7.1 +/ –0.2	7.6 +/ –0.5	7.5 +/ –0.3	6.5 – 8.5
Nitrate (mg/l)	3.4 +/ –1.3	2.4 +/ –1.9	1.5 +/ –0.8	10
Phosphate (mg/l)	0.5 +/ –0.5	1.0 +/ –0.9	0.9 +/ –0.8	0.3
D O (mg/l)	4.7 +/ –1.2	6.0 +/ –1.4	3.0 +/ –1.7	> 6

Table 3.4. Mean values of the various parameters at Points 1 to 4.

PARAMETERS	POINT1	POINT2	POINT3	POINT 4
Temperature (°C)	19.0 +/ –3.6	18.8 +/ –0.5	15.5 +/ –0.6	21.5 +/ –4.7
Conductivity (μS/m)	190 +/ –72	1550 +/ –135	325 +/ –110	2140 +/ –970
pH	6.9 +/ –0.3	3.9 +/ –0.4	7.2 +/ –0.2	3.7 +/ –0.2
Nitrate (mg/l)	4.0 +/ –1.3	42.0 +/ –23.1	13.5 +/ –1.7	17.5 +/ –12.7
Phosphate (mg/l)	4.3 +/ –2.1	11 +/ –3.9	9.3 +/ –1.4	11.7 +/ –12.1

from Point 1 to Point 2. A decrease in the level of pollutants at Point 3 indicates some degree of natural recovery and self-purification, or dilution, in a distance of only 3 km down-stream of Point 2. A seasonal examination of the results obtained, show that the impact of Zimphos is more pronounced during the dry season, and most probably is due to the pollution from ground water recharge. Increased volumes of the river flow during the wet season lead to dilution of the contribution of the plant.

Jarawaza (1997) reported data about the water quality of Upper Manyame River and tributaries, collecting the effluents from the satellite town of Chitungwiza, based on the continuous monitoring program of the city of Harare during 1996. It has shown considerable pollution of the River, as a result of the overloading of the existing waste stabilization ponds system and urban runoff from the satellite town of Chitungwiza, which has a land use pattern similar to high-density urban developments in Harare. The data obtained by the monitoring program served as an alarm for the managing authorities and the upgrading of the existing facilities was undertaken. The new BNR plant was commissioned in 2002.

4 MONITORING PROGRAMS AND REGULATORY PRACTICE

4.1 *Institutions and technical background*

Diffuse pollution monitoring and control should not be viewed as a task on its own, but should be incorporated in the context and objectives of the water quality management of the whole catchment basin. This section presents the status of the water quality monitoring practice in the Chivero basin during the period 1995-2000. Two institutions were responsible for this task – the city of Harare municipality and the Department of Water Resources (DWR), Ministry of Land and Agriculture. Later on the DWR was incorporated into the Zimbabwe National Water Authority (ZINWA). Both institutions have their own water quality monitoring programs, with corresponding laboratory facilities. The monitoring program of DWR has the support of a well equipped and contemporary water quality laboratory, while the water quality laboratory of the municipality is outdated and has limited resources in terms of equipment, space and financial support.

The data storage and retrieval process at DWR was relatively well organized, containing the monitoring data in the form of protocols. An upgrading of this system into a computerized database was under construction. However, the city of Harare did not have a well-organized data record system, with most of the measurements recorded as part of the operational process of the treatment plants in books. These findings show the need to improve the data storage and record process, together with the reporting procedures, in order to make use of the information obtained, and implement it during the management process. If such a practice is not implemented, it could jeopardize the efforts and costs involved during the process of samples collection and analysis.

4.2 *Regulating water quality*

The Zimbabwean regulations (WWEDR 2000) are discharge-oriented and are aiming at enforcing in practice the "polluter pays" principle. They focus on the effluent discharges to surface water, effluent and sludge disposal on land, and solid waste disposal sites. Runoff quality is included in the different classifications only for specific cases as sludge, effluent and solid waste disposal sites, and is treated as a point pollution source. Possible pollution from agricultural or urban runoff is not envisaged, most probably because of the difficulty to identify these types of sources of pollution and corresponding polluters. The regulations provide for a permit classification based on effluent (runoff) characteristics, sludge (effluent) application rates and the protection level of solid waste disposal sites. They include a detailed list of water quality criteria, including blacklisted compounds, which prescribe limit concentrations and qualifies the discharges based on environmental hazard risks (ER) into four major groups: safe, low, medium and high hazard. A partial consideration of the water quality orientated approach has been introduced by identifying environmentally sensitive river catchments, where more stringent criteria are applied. Considering the fact that a large number of the receiving water bodies in the country are ephemeral, the safe prescribed limit of the different water quality criteria could be regarded as water quality objectives for the specified zones. However, the document does not clarify this point. A positive improvement, which has been achieved in this regulatory instrument, is the provision for unified standard methods for water quality assessments and a prescription for sampling procedures in order to provide for a common basis of comparison of the monitoring data. The document also prescribes administrative procedures for practical implementation and provision of penalties.

Another positive development with respect to the regulatory aspects of water quality is the incorporation of this document into the Environmental Management Act (EMA 2002), which allows for the protection of soils and air as well. In the past, the government has been reluctant to impose land use controls on private property, as the sanctity of private property is a vital part of freedoms in society, and is a strong emotional issue. However, the EMA (2002) section 114 now addresses this weakness. An order may be served to the owner, occupier or user of any land on controlling water, including storm water, removing and disposing of litter or other refuse or waste from any land or premises among other pollution sources. This necessitates the use of different land management practices; changes that are not popular with landowners.

Considering the international trends in the development of regulatory instruments, and the specifics of diffuse pollution control, it could be mentioned that there is one aspect of the present regulatory documents in Zimbabwe, which needs improvement. In order to help in the clear understanding and application of the regulations, consideration of the random variability of water quality parameters should be made and the document should define the basis for monitoring frequency and statistical data sets analysis, which would determine the actual value of the constituents observed, and which could be compared to the prescribed limits.

4.3 *The monitoring programs*

The City of Harare monitoring program is very wide in terms of a number of sampling points and diversity of types of water monitored. It consists of several sub-programs with different objectives and strategies to achieve them. The following sub-programs could be differentiated:

- Water works control – this monitoring program aims to support the water treatment process and the plant operation, as well as to provide for the quality assurance of the water produced. The program includes basic parameters and is implemented on a regular basis by the water laboratory of the city of Harare's water treatment works.
- Industrial effluent control – the monitoring program aims at the control of industrial effluents discharged into the City's sewer system and is executed by the Waste Trade Inspectorate – a unit especially designated to control such discharges through permits and to monitor their implementation. This monitoring program includes only selected locations and in terms of resources – technical, financial and human – does not have the capacity to achieve its objectives. This results in a situation where the sewerage system is receiving unauthorized discharges of different origin, which influences adversely the wastewater treatment and disposal process.

- Sewage works control – this program aims at controlling the sewage treatment process at all treatment plants in the city. The program varies in terms of parameters tested and locations, based on the type of the treatment process. More attention is given to the BNR plants operation, while the waste stabilization ponds are very scarcely tested, mainly in terms of effluent quality.
- Potable water from the city's consumer points – this program has the aim to control the potable water at the consumer end and to ensure that no signs of pollution are present in the distribution network. It is executed on a regular basis at selected locations along the distribution network and tests basic parameters.
- Urban drainage runoff (storm water drainage channels) – this program aims at the evaluation of urban drainage water quality at two main discharge channels in the city. The frequency of sampling is relatively low and is not event orientated. More details regarding this program are given in point 5.
- Surface water pollution control – this program aims at determining the natural water quality at locations related to effluent and storm water discharges. It has the purpose to control the impact of such discharges independently from DWR, and includes about 36 sampling stations. In terms of number and location (network design), the sampling stations cover the basin well, within the city of Harare urban areas. The frequency of sampling is not high and the data collected could be used as indicative information regarding the status of water quality, rather than reliable information for supporting the decision-making process.

In general, the city's monitoring sub-programs are covering all aspects of the water quality spectrum. In terms of network design, they are well developed, except for industrial effluent discharges, where the capacity is not enough in order to control all major polluters. The parameters observed cover basic pollutant constituents only. There is an urgent need to upgrade the city's laboratory, in order to include in the monitoring process the testing of toxic elements as well as blacklisted parameters.

The DWR monitoring program, with respect to the Lake Chivero basin concentrates on surface water quality only, and forms part of the national grid for water quality assessment. It contains about 20 sampling points at characteristic locations, the vast majority of which coincide with sampling point locations of the municipal program. It should be noted that the scope of the DWR monitoring program provides information regarding the general status of the water quality of the basin. These objectives need to be reformulated in the light of the requirements of the recently adopted strategy of pollution control – the "polluter pays" principle. It is advisable to update the existing program, so that it provides room for the implementation of auditing functions of DWR, with respect to permit provision and control of its implementation. It would require a much wider scope in terms of sampling point locations, corresponding to the number of permits issued, but the frequency of sampling could be lower, depending on each specific case. During the year 2000, the program included a rather large number of parameters tested, which might not be so relevant to the data analysis process. At the same time, ammonia was not included in this list. The importance of this parameter could not be overemphasized in view of the eutrophic conditions in Lake Chivero and the massive fish death reported by Moyo (1997). This fact shows the need of the regular overview and revision of existing programs in order to have an optimum solution at an economically effective and sustainable cost.

A common drawback of both programs is the lack of reference points for each separate river, which could also be used for the periodic evaluation of background (natural) water quality. In addition, both programs do not envisage and control sources of diffuse pollution from agricultural activities.

The analytical methods applied to determine specific water quality characteristics should be validated with respect to accuracy, precision, range, sensitivity and selectivity (UN/ECE Task Force 1996). It is advisable for all laboratories in the country to apply the Zimbabwe standard test procedures (SAZ), as prescribed in the regulations, and correspondingly, they should be equipped with the necessary resources to implement this requirement. In many cases, this is not applied in practice, which reflects on the level of enforcement of the regulatory documents. A central water quality laboratory, as the city of Harare's one, is required to execute a very wide and diverse scope of water quality monitoring, with a minimum of technical instrumentation, which is old and worn out. In order to provide a reliable data, there is an urgent need for upgrading this laboratory with new technical equipment, which should correspond to the requirements and objectives of the monitoring programs performed.

The necessity to validate the methods applied and to achieve comparability of results obtained in different laboratories is of utmost importance in terms of the optimal use of the resources available and the correct and meaningful data interpretation. In order to emphasize the importance of this point, a comparison between results obtained on the same site location and at approximately the same time by the monitoring programs of the city of Harare and DWR was made. Measurements of pH give a deviation of about 0.5 to 1 unit, which could be attributed to a systematic technical fault of the pH meter. The results of the municipal program are more consistent compared to those of the DWR one. Big differences were found regarding DO measurements, but a more detailed validation of results is necessary in order to determine the specific reason for that. It should be noted that DO concentrations could vary due to water temperature changes and due to the conditions at the specific sampling site (depth and flow velocity). It is also possible that these differences might be due to the improper sampling procedure or malfunctioning of the equipment used for the analysis (DO-sensors). The phosphate measurements are comparable in some cases only, while the results of nitrate concentrations show a very systematic and well-defined difference, which most probably is due to a discrepancy in the methods and analytical procedures applied. The comparison shows that the reasons for fault data might be different and not only due to the analytical methods applied. Problems with the quality assurance of test data regarding water quality are common for almost all the countries in the region. Therefore, specific attention should be given to the following factors:

- The analytical procedures applied and equipment used should be well defined and described and an attempt should be made to use comparable methods of testing in all water quality laboratories.
- An experienced officer should make a detailed validation of the data obtained, before it is stored in the record archive.
- A regular and systematic quality assurance program of the sampling procedure and the analytical methods applied should form an indispensable part of any monitoring program. It should include internal laboratory procedures as: instrument condition, maintenance activities, reference sampling, blanks and spiked samples, as well as external control methods, like inter-laboratory tests.

5 NUTRIENTS VARIATIONS IN THE URBAN STORM WATER OF THE CITY OF HARARE

Considering the advanced stage of eutrophication of Lake Chivero, nutrients contributions to the Lake itself and its tributaries is an important issue, which needs priority attention in the water resources management practice. In this context, the investigation of nutrients concentrations, and their seasonal variation in the urban storm water, should not be underestimated. This section evaluates the status and temporal variations of nutrients in the urban drainage, based on data from the water quality monitoring program of the city of Harare during the period 1995-2000. It discusses the results in a broader context, considering the impacts of the present regulatory instruments and the implementation of the monitoring process and pollution abatement measures, in order to provide environmentally sustainable development.

5.1 *Methodology and study area*

The monitoring program of the city of Harare, regarding urban drainage water quality, includes two major drainage channels. The Coventry Road Drainage Channel is an earth ditch 2.5 m wide and 1 m deep, which discharges into the Marimba River and is referred in this section as DR1. The Central Business District Channel (CBD) is referred as DR2. It is a concrete pipeline discharging into Mukuvisi River through a culvert box with a combined cross section, the bottom part is a rectangular channel of 2m width and 0.35 m depth, the top part is a trapezoidal section with side slopes 1:1 and a depth of 3.2 m. DR1 collects runoff from an area of about 11 km^2 of a mixed land use pattern, including a medium density residential areas, commercial and institutional buildings (schools, hotels, offices), and an industrial area. The former comprises of light industrial enterprises, including meat processing, dairy production, washing materials, automotive enterprises and a variety of small to medium businesses, offices and storage facilities. About 20 % of the whole drained area, in the downstream part of the channel, consists of open undeveloped areas and informal agriculture plots.

DR2 collects the runoff from the town center (CBD-approximate area of 3.3 km^2), where the land use is predominantly commercial, including high story buildings and medium-density residential areas. It is characterized by a high percentage of impermeable surfaces and intensive traffic. About 20% of the drained area consists of developed open spaces as parks and a golf course. The location of the drainage areas of the two channels is shown in Fig.3.4. The urban drainage system of Harare has numerous discharge points, but only these two channels have been included in the monitoring program of the city because of the relatively high value of the drained area, compared to other discharge points, and the specific type of urban land use practice.

The parameters, describing nutrients variations, included in the city's monitoring program are phosphates, ammonia and nitrate. In addition, variations of pH have been included. The planned frequency of sampling was once per month, but at the end of the study period it has been reduced drastically. During 1999 and 2000 only few measurements for the whole year have been executed and for this reason, the available data has been averaged for both years. Analytical methods used are based on the Standard methods of examination of water and wastewater (1989). Phosphates were directly determined by the Vanadomolybdophosphoric acid colorimetric method and represent the total reactive phosphorous. Ammonia was determined by a direct Nesslerization method, followed by a spectrophotometric measurement at 425 nm wavelength. Nitrates were determined by the Cadmium reduction method, followed by spectrophotometric measurement at 540 nm wavelength.

Statistical calculations (descriptive statistics and one-tiled Student's t-test – unequal variance, at 90% of confidence) were performed using the Microsoft EXCEL statistical package. Throughout the text of section 5, significant difference means that $p< 0.1$ for mean's comparison.

5.2 *Results and discussion*

Results for DR1 and DR2, as the annual mean and median values, are shown in Fig. 3.5 and Fig.3.6 respectively. Additional statistical data is presented in Table 3.5. Results are discussed and compared to the blue classification of effluents discharged to surface water (WWEDR 2000), which could be regarded as the safe natural water quality limit and is referred in the text as the "prescribed" limit. The

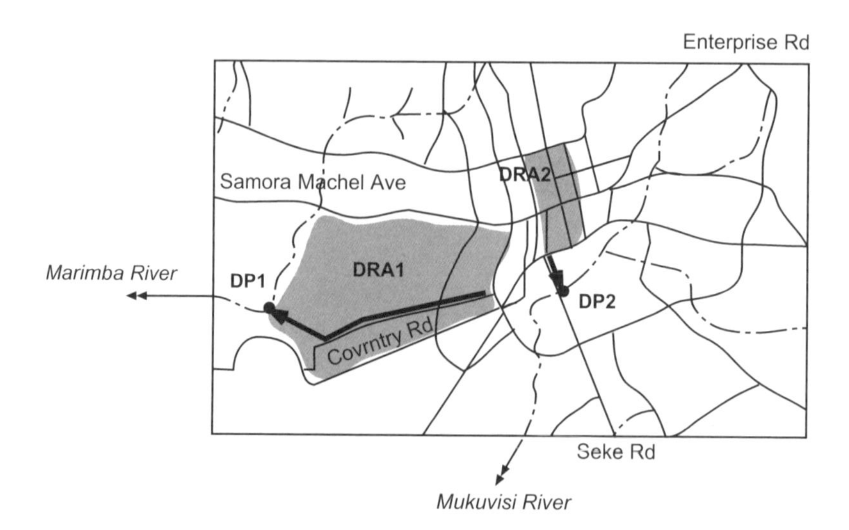

Figure 3.4. Schematic representation of the drainage areas and discharge points of DR1 and DR2, DRA-drainage areas; DP – discharge points.

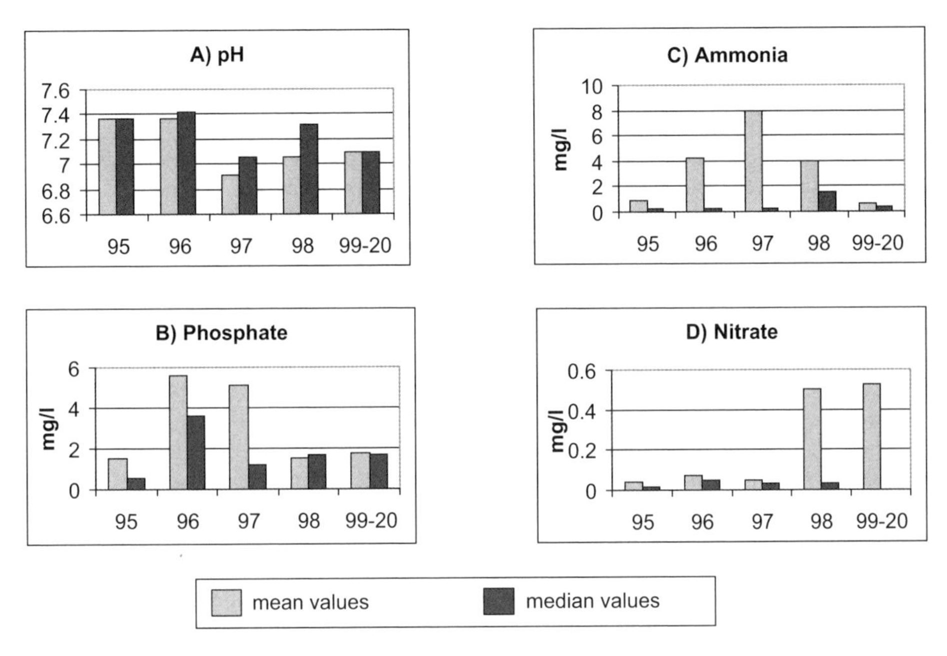

Figure 3.5. Temporal variations of nutrients in the storm water of DR1.

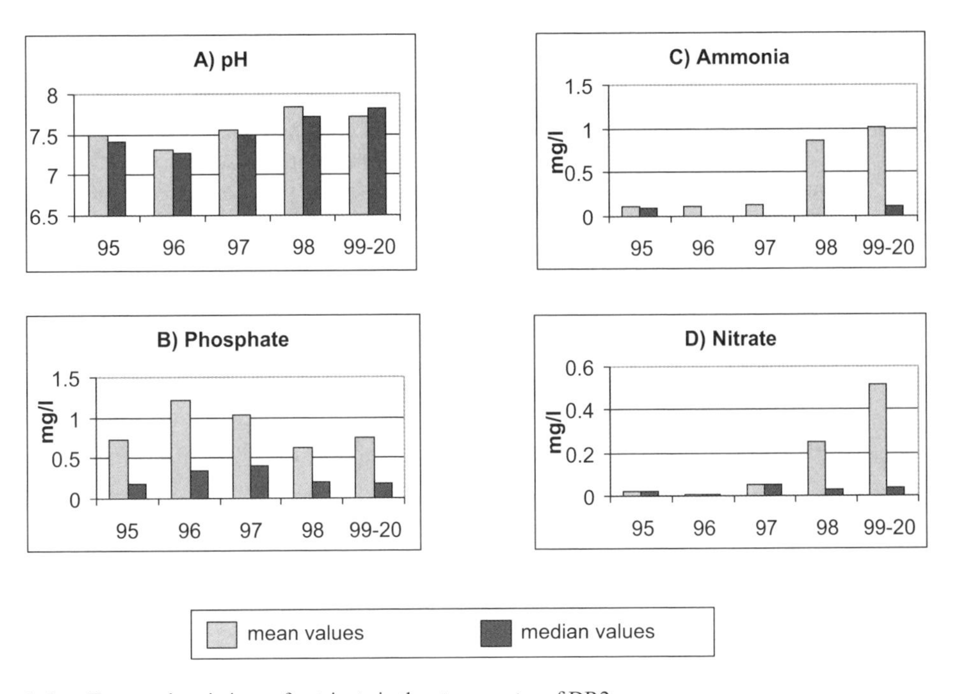

Figure 3.6. Temporal variations of nutrients in the storm water of DR2.

values, included in the city of Harare's data set was of very high variability, with large differences between mean and median values, except for pH. For this reason, results are presented as mean and median annual values, and it is expected that median values of the corresponding data sets reflect more closely the true characteristic of the water quality in both channels.

Despite the distinct seasonal rainfall pattern in the region, where no rainfall or extremely rare rainfall occasions are recorded during the months of April to November, water was flowing and samples were collected throughout the year at both sampling locations. This indicates the presence of informal discharges into the drainage system or drainage of infiltrated ground water. The most probable source of perennial flow in both channels could be associated with informal discharges because the groundwater table in both areas is relatively low during the dry season. Thus, a clarification needs to be done, that the term "storm water quality", used in this section as well as in Chapter 4, refers to the flow in man-made urban drainage channels, which are not always associated with direct surface runoff, but could convey wastewater from informal discharges as well.

Temporal variations of the nutrients concentrations in urban storm water show a general trend of increase during the period. The ammonia concentrations at DR1 (Fig. 3.6) show very high mean values for the period 1996 to 1998, but the median values are within or near the prescribed limit, except for 1998, when the median value exceeded it 3 times. Background pollution values, which are 0.05 mg/l of ammonia for wet season conditions and 0.2 mg/l for dry season conditions (Hranova et al. 2002), have been exceeded by the median values of DR1 throughout the study, which indicates pollution regarding this parameter. Results regarding phosphate variations at DR1 with respect to both-mean and median values exceed the prescribed limit of 0.5 mg/l, with peak mean values during 1996 and 1997. However, it should be noted that the variability of the data set for these two years is very high, resulting in high standard error values (Table 3.5). Phosphates variation (median values) during the study period was between 0.6 and 1.7 mg/l, except for the year 1996, when a peak of 3.6 mg/l was found. Background pollution values for the city of Harare area, regarding phosphorous, was reported to be ranging between 0.1 and 0.4 mg/l with no seasonal variation (Hranova et al. 2002).

The annual median values for phosphate, observed during this study, exceed both background pollution values and prescribed limits up to several times. Similar results regarding high phosphate concentrations at this drainage channel have been reported as well by Jarawaza (1997).

At DR2 (Fig. 3.6) ammonia variations show low values regarding this parameter for the period 1995-1997. During 1998 to 2000, occasional high values were measured but the median values are within

Table 3.5. Statistical characteristic of the data sets.

	pH		Ammonia		Nitrate		Phosphate	
	DR1	DR2	DR1	DR2	DR1	DR2	DR1	DR2
				1995				
Standard Error	0.09	0.067	0.353	0.033	0.02	0.004	0.644	0.535
Count	12	12	12	12	12	12	12	12
				1996				
Standard Error	0.191	0.050	3.851	0.069	0.023	0.004	2.333	0.788
Count	9	8	9	8	9	8	9	8
				1997				
Standard Error	0.118	0.097	7.729	0.073	0.019	0.013	3.254	0.472
Count	12	13	12	13	12	13	12	13
				1998				
Standard Error	0.153	0.271	1.955	0.814	0.022	0.145	0.242	0.230
Count	10	11	10	11	10	11	10	11
				1999-2000				
Standard Error	0.271	0.130	0.284	0.931	0.521	0.448	0.574	0.369
Count	5	7	5	7	5	7	5	7

the prescribed limit. The median phosphate values are below or equal to the background pollution and are much lower than the reported river water quality upstream the discharge point as given in Table 3.4, Point 3. Mean values exceed up to two times the prescribed values throughout the study period with peak mean and median values in 1996 and 1997. Nitrate variations show a similar pattern in both channels and are relatively low, with occasional high values measured during the period 1998-2000.

Spatial variations were analyzed by comparison of the means of the two data sets for each specific year, regarding pH and phosphates only. Nitrogen compounds were not tested because of the high irregularity of the data sets. The test results show no significant difference for the annual mean pH values at DR1 and DR2. The test results regarding phosphates show a significant difference between DR1 and DR2 annual mean values only for the years 1996, 1998 and 1999-2000. DR1 shows a much higher annual mean and median values of phosphates and there is a well-defined trend of increase of these concentrations after 1995. The lack of a significant difference between the annual values in 1997 could be attributed to the irregularity of the data set (SE = 3.24) but not to reduced pollution at DR1.

During 1998-2000 the median values at DR1 are closer to the mean values, suggesting that the process of pollution has a constant nature. However, the reduced frequency of sampling during 1999-2000 might have resulted in data, which does not reflect the real status of the water quality. The significant difference in the level of phosphate concentrations between DR1 and DR2 could be associated with the difference in the land use pattern of the drained area. DR1 drains three different types of areas – residential, industrial and open undeveloped grounds. The residential part consists of medium density one story housing units, where the roof runoff is discharged and infiltrated to gardens. Therefore, its contribution in terms of the quantity and quality of the total runoff could not be significant. The industrial area has a much higher level of impermeable surfaces, thus in terms of runoff quantity its contribution is expected to be much higher than the contribution from the rest of the area. A considerable portion of the open undeveloped areas is used for informal agricultural practice, most often for maize production. Possible sources of phosphate pollution to DR1 could be associated with:

- Illegal discharges from the industrial enterprises;
- Polluted runoff from the industrial areas;
- The excess use of fertilizers by informal small-scale farmers in the open ground area.

It has been mentioned that the water flow in the channel is perennial, thus the most likely source of pollution could be associated with illegal discharges from the industrial areas. The suggestion that phosphorous pollution could be associated with the practice of informal agriculture is not likely, because these are small-scale activities practiced by low-income residents as a form to complement their income.

6 CONCLUSIONS

Concerning the specific case study, diffuse pollution from urban drainage in the Lake Chivero basin has been identified as a considerable source of pollution since the 1970's. It is enhanced by the accelerated population growth, and overpopulated high-density residential areas. However, it should be emphasized that the city's urban planning scheme incorporates a considerable amount of developed and natural open ground spaces, which act as a buffer and most probably, retain a considerable amount of the runoff pollution. Also, it has a relatively well-developed infrastructure, which needs to be maintained and operated in good condition. In general, the city has the potential for the implementation of a proper diffuse pollution abatement program.

The existing regulatory arrangements provide the basis for an effective water quality management but need additional improvements in order to control diffuse pollution sources in a more effective way. However, the level of the monitoring programs, which would back-up the enforcement of the regulations, is below standard, and do not provide an efficient evaluation and detection of water quality. The control of diffuse pollution sources would require a significant alteration of the existing practice, together with the corresponding upgrading of the technical and human resources background.

Considering the specific catchment characteristics and the city's location, the major causes for pollution of the Lake Chivero could be associated with polluted runoff from the city's drainage systems, as well as with informal discharges from point sources of pollution. This fact has been substantiated by the analysis of the monitoring data for the period 1995-2000, which, despite the relatively inconsistent data set, shows a sustained trend of increased nutrients concentrations, especially regarding phosphorous and ammonia. DR1 has been identified as the contributor of the higher pollution concentrations, compared to DR2, which could be associated with illegal discharges from the industrial area drained. This finding points out the need of controlling not only industrial discharges to the sewer system, but also to the drainage system, which should be executed by local authorities.

Diffuse pollution in an urban environment, specific for the countries of the region, is mainly associated with rapid population growth, which enhances the generation of diffuse pollution in different ways – through the overloading of existing treatment facilities, leading to unauthorized discharges of partially treated sewage, formation of informal settlements and a general overcrowding of low-income residential areas. Diffuse pollution pose a threat not only to the surface water quality, but also to ground water quality. However, ground water quality has not been monitored at all, and in the vast majority of cases, is outside the scope of the water resources management practice. The following chapters deal with specific sources of diffuse pollution, which are typical for the countries in the region, and illustrate them by detailed studies with respect to the evaluation and assessment of their impacts on surface and ground water quality.

Acknowledgements – The data presented in this paper was collected during the project "Integrated water and pollution management of the Chivero basin" funded by WARFSA. Thanks to the sponsors, the City of Harare authorities for providing information and support, and to the undergraduate research assistant A. Danha for the help during the process of data collection and handling.

REFERENCES

AQUASTAT (2003). FAO's information System on Water and Agriculture at http://www.fao.org/ag/AGL/aglw/aquastat/countries/zimbabwe/index.stm

Balinger, B.R. & Thornton, J.A. 1982. The hydrology of the lake McIlwaine catchment. In Thornton, J.A. & Nduku, W.K. (ed), *Lake McIlwaine – the Eutrophication and Recovery of a Tropical African Man-made Lake,* 34-43, London: Dr W. Junk Publishers.

Drakakis-Smith, D. & Kirell, P. 1990. Urban food distribution and house hold consumption, a study in Harare. In Paddison R. & Dawson, J.A. (Eds), *Retailing environment in developing countries,* 156-180, London: Routledge

Environmental Management Act (EMA), 2002, Chapter 20:27, Republic of Zimbabwe Government.

Hranova, R., Gumbo, B., Klein, J. & van der Zaag, P. 2002. Aspects of the water resources management practice with emphasis on nutrients control in the Chivero basin, Zimbabwe. *Physics and Chemistry of the Earth,* 27, 875-885.

Jarawaza, M. 1997. Water Quality Monitoring in Harare. In Moyo (ed), *Lake Chivero: A Polluted Lake,* 27-34, Harare: University of Zimbabwe Publications.

Marshal, B.E. 1982. The bentic fauna of lake McIlwaine. In Thornton, J.A. & Nduku, W.K. (ed), *Lake McIlwaine – the Eutrophication and Recovery of a Tropical African Man-made Lake,* 156-188, London: Dr W. Junk Publishers.

Marshal, B.E. & Falconer, A.C. 1973. Physico-chemical aspects of Lake McIlwaine (Rhodesia), a eutrophic tropical impoundment. *Hydrobiology,* 43(1/2), 45-62.

McKendrick, J. 1982 Water supply and sewage treatment in relation to water quality in lake McIlwaine. In Thornton, J.A. & Nduku, W.K. (ed), *Lake McIlwaine – the Eutrophication and Recovery of a Tropical African Man-made Lake,* 201-217, London: Dr W. Junk Publishers.

Moyo, N.A.G. 1997. Causes of massive fish deaths in Lake Chivero. In Moyo (ed), *Lake Chivero: A Polluted Lake,* 98-104, Harare: University of Zimbabwe Publications.

Muthuthu, A.S., Zaranyika, M.F. & Jonnalagadda, S.B. 1993. Monitoring of water quality in the Upper Mukuvisi River in Harare, Zimbabwe. *Environment International,* 19, 51-61.

Muthuthu, A.S., Zaranyika, M.F. and Jonnalagadda, S.B 1995, Water quality assessment by monitoring physical and chemical parameters and heavy metal pollution in the Upper Mukuvisi River in Zimbabwe. In Abdulla, M., Vohora, S.B. & Attar, M (ed.), *Trace and Toxic elements in Nutrition and Health,* India: Jamia Hamdard and Wiley Eastern Limited.

Nhapi I., Hoko Z., Siebel M., Gijzen H. (2001) Assessment of the major water and nutrient flows in the Chivero catchment area, Zimbabwe. *In Proc. 2nd WARFSA/WaterNet Symposium: "Integrated Water resources Management: Theory, Practice, Cases", Cape Town, 30-31st October 2001.*

Nduku, W.K. 1976. The distribution of phosphorous, nitrogen and organic carbon in the sediments of Lake McIlwaine, Rhodesia. *Trans. Rhod. Scient. Ass.*, 57(6), 45-60.

Potts, D. & Mutambirwa, C. 1991 High density housing in Zimbabwe: Commodification & overcrowding. *Third world planning review*, 13(1): 1-25.

Rakodi, C. 1995. *Harare-inheriting a settler-colonial city: Change or continuity*, London: John Wiley and Sons

Robarts, R.D. & Ward, P.R.B. 1978 Vertical diffusion and nutrient transport in a tropical lake (Lake McIlwaine, Rhodesia). *Hydrobiology*, 59(3), 213-221.

Standard methods for the examination of water and wastewater 1989. 17th edition, *American Public Health Association/ American Water Works Association/Water Environment federation*, Washington DC, USA.

Thornton, J.A. 1980 A review of limnology in Zimbabwe: 1959-1979. *National Water Quality. Survey, Report No1*, Salisbury: Municipal Library.

Thornton, J.A. & Nduku, W.K. 1982a. Sediment Chemistry. In Thornton, J.A. & Nduku, W.K. (ed), *Lake McIlwaine – the Eutrophication and Recovery of a Tropical African Man-made Lake*, 59-65, London: Dr W. Junk Publishers.

Thornton, J.A. & Nduku, W.K. 1982b. The aqueous phase: nutrients in run-off from small catchments. In Thornton, J.A. & Nduku, W.K. (ed), *Lake McIlwaine – the Eutrophication and Recovery of a Tropical African Man-made Lake*, 71-76, London: Dr W. Junk Publishers.

Thornton, J.A. &, Nduku W.K. 1982c. Water chemistry and nutrient budget. In Thornton, J.A. & Nduku, W.K. (ed), *Lake McIlwaine – the Eutrophication and Recovery of a Tropical African Man-made Lake*, 43-58, London: Dr W. Junk Publishers.

UN/ECE Task Force on Monitoring & Assessment. 1996. *State of the art on monitoring and assessment of rivers* Volume 5, RIZA report No 95.068, 60-71.

UN Population Division 1997. *World's population prospects-the 1997 revision*, Department of Economic and Social Affair, New York: United Nations.

World Bank. 1997. *Structural Adjustment and poverty: A conceptual empirical & policy framework.* Washington D.C.

WWEDR 2000. *Water (Waste and Effluent Disposal) Regulations*, Statutory Instrument 274 of 2000, Republic of Zimbabwe.

Zinyama L.M. 1993 The evolution of the spatial structure of the greater Harare: 1890-1990. In: Zinyama, L.M., Tevera, D.S. and Cumming, S.D. (eds.) *Harare: The Growth and Problems of the City.* Harare: University of Zimbabwe Publications.

CHAPTER 4

Assessing and managing urban storm water quality

R. Hranova & M. Magombeyi

ABSTRACT: Basic principles and approaches to assess diffuse pollution from urban storm water have been presented. As an example of a pollution evaluation procedure, a case study on the storm water of two major drainage channels in the City of Harare has been done during the period 2002-2003, and results were compared to previous measurements. Results show pollution with respect to TP, TSS, TDS, ammonia, COD, Cd and Pb, and a relatively good correlation has been found between COD and Zn concentrations as dependant variables of TSS concentrations. Spatial variations, as a result of specific land use patterns, did not show a statistically significant difference, except for the contribution from an industrial area. A single storm event has been studied and results show a considerable difference between mean values and EMCs with respect to parameters associated with particulate material only. The development of an event-orientated monitoring program, which would allow more accurate determination of site-specific pollution concentrations, and the need to adopt uniform methodology of pollution loads estimation has been recommended, as well as the integration of storm water management into the process of water resources management at catchment level.

1 INTRODUCTION

Urban development can have a significant impact on catchments by increasing runoff and pollutant loading. When forests and farmlands are converted to industrial and residential developments, this change in the land use practice leads to increased runoff volumes and a rise of downstream flood stages, thus increasing the risks of flooding. In addition, the ground water recharge is reduced due to the increase of impermeable covers of urbanized areas. Water quality problems have intensified through stages in response to the increased growth and contraction of population and industrial centres. Often the problems have been viewed as inevitable consequences of community development and are sometimes even accepted as evidence of affluence and progress. The recent trends in urbanization, as discussed in Chapter 3, show that the scenario of increased urbanization rates will continue during the coming decades, therefore, it is desirable to face the difficulties and to adopt management strategies, which could lead to the alleviation of the negative effects of urbanization.

The increase in paved surfaces has been spurred on not only by urban and suburban development, but also by a steady increase in the use of automobiles, the primary mode of daily transportation. According to a 1999 study, motor vehicle infrastructure, such as roads and parking lots, accounts for close to half of the land area in U.S urban cities (GAO 2001). There have been numerous studies completed since 1994 on the issue of watershed protection techniques, examining the relationship between stream quality and watershed urbanization. In general, these studies point to a deterioration of steam quality with increasing urbanization, e.g. it has been estimated (Paul & Meyer 2002) that a threshold exists at about 10% impervious cover, after which sensitive elements are lost from the aquatic system. As presented in the previous chapter, a well-defined trend of increased nutrient loading with the increased level of urbanization was established in the Lake Chivero basin as well. The increase in impervious surfaces alters the hydrology and geomorphology of streams and results in predictable changes in stream habitat. Common effects are: decline in biodiversity, reduced DO concentrations and increase in invasive species (Paul & Meyer 2002, Gromaire-Mertz et al. 1999). Below is a summary of the most common impacts of impervious covers.

- Hydrologic impacts
  - Increased runoff;
  - Increased peak flow, due to increased velocities in hydraulically more efficient drainage channels;
  - Increased bank full flow;
  - Decreased base flow;
- Physical impacts
  - Changes in stream geometry (channel enlargement);
  - Increased stream meandering;
  - Alteration of stream channel networks (through channel modification) and barriers to fish migration;
- Biological impacts
  - Effects on aquatic insect diversity, wetland diversity and fresh water aquatic ecosystem diversity as a whole.

In many countries of the region, the storm drainage system is treated like the commons, free for anyone to pollute with little risk of being caught and little awareness by the general public that they may be polluting. Niemczynowicz (1999) points out that the long-term cumulative effect of storm water runoff on our fragile urban water environment and the catchment water resources is severe and cannot be ignored. In order to provide for a safe beneficial use (potable, recreational, agriculture, etc), a proper management approach of the catchment area, based on the characterization and quantification of water needs and pollution fluxes, is necessary. Pollution prevention, impact minimization, assessment of assimilation capacity and management of the symptoms of pollution are required for catchment storm water quality planning. The possible effects of increased storm water pollution should be considered at the planning stage of urban development and appropriate steps taken to reduce the impact of urbanization. The only way to avert ecological disasters is to ensure that high quality effluent is discharged into receiving water bodies. Activities in one area of the watershed are correlated to the environmental imbalance of the other and such relationships are independent of administrative and political boundaries. Considering all these factors, the scope and aims of this chapter is to:

- Present background information with respect to diffuse pollution aspects in formal urban areas;
- Present an example of urban storm water quality evaluation and pollution loads estimation, and discuss different aspects of this procedure in the light of the existing conditions in the region. A case study in Harare, Zimbabwe has been used for this purpose. The study was limited in terms of time, manpower and support facilities, thus the results obtained reflect a relatively short time period and should be regarded as a preliminary survey, which indicates the present storm water quality status and could serve as a basis for a further research of storm water quality and its impact in the area.
- Discuss implications of polluted urban drainage on the management practice of local authorities and put forward recommendations for future developments in this direction.

2 DIFFUSE POLLUTION FROM URBAN STORM WATER RUNOFF

2.1 *Sources of pollution*

Non point source pollution and its evaluation and abatement are probably the most pervasive and ubiquitous water quality problems. It is not the land itself or land use per se that causes pollution. Various boundary inputs and polluting processes and activities that occur on the land could cause pollution. The most common inputs and processes that are the cause of urban diffuse pollution are:

- Accumulation of dry atmospheric deposits (dust) and street dirt and subsequent wash-off from impervious surfaces (roofs, parking slots, streets, roads, etc.). The sources of the accumulated pollutants are:
  - Dry atmospheric deposition;
  - Streets refuse accumulation, including litter, street dust and dirt and organic residues from vegetables and animal population;
  - Traffic emission;
  - Erosion of pervious lands, impervious poorly maintained covers and construction related activities;
- Discharge of pollutants, such as oil, paint, detergents and other household/commercial solvents and chemicals, into the drainage systems;
- Application, storage and wash off of chemicals, raw materials and other pollution containing materials from industrial enterprises;
- Leaching of pollutants from septic systems and other sources, such as landfills/solid waste disposal sites, onto surfaces and into ground water and subsequently into storm drainage.
- Application of pesticides and fertilizers in gardens and other cultivated urban lands;
- Illegal discharges of sewage and industrial wastewater designated to be discharged into sanitary sewers
- Cross connection of sewage and/or industrial wastewater from sanitary sewers, failing septic tanks and other sources into the storm water system.
- Pollution contained in precipitation (wet atmospheric deposition).
- Non point sources of pollution from agricultural activities within the municipal areas.

Atmospheric depositions of pollutants washed during the first stages of rainfall have been found to have a significant impact on the quality of surface water runoff. Both industrial (urban and transportation) and agricultural activities may contribute to the pollution content of atmospheric deposits. Rain droplets and snowflakes absorb pollutants from the atmosphere, including acid forming components. In residential areas, the fugitive dust (dry deposition) mostly originates from surrounding soils, construction sites, and refuse disposal sites and from biological sources (pollen, spores and other organic residues). It has been estimated that in larger cities the deposition rate of atmospheric particulates in wet and dry fallout ranges from 7 tonnes/km^2 per month to more than 30 tonnes/km^2 per month (Novotny & Olem 1994). Higher deposition rates occur in congested downtown and industrial zones, and lower rates are typical for residential and other low-density suburban zones. For other measured pollutants (phosphorus, cadmium, chromium, lead and zinc) rainfall contributions were insignificant. It has been found (AWRC 1981) that in terms of pollutant concentrations, surface sources of storm water pollution were much greater than the atmospheric contribution. Exceptions were the atmospheric deposition of nitrate and zinc, which were apparently related to stack outputs from upwind industries The following conclusions were made, based on a precipitation sampling program in Washington DC:

- Pollutants tend to wash out relatively uniformly over the metropolitan area, even though the origins of pollutants may tend to be area specific.
- Atmospheric contamination is washed out during the first stages of a precipitation event. Thus the resulting ground surface loading of atmospheric pollution is largely independent on the magnitude or intensity of precipitation.
- Other than atmospheric conditions prior to or during rainfall, the primary factor affecting pollutant loads from rainfall is the period of time since the previous precipitation event.

It should be noted that in terms of diffuse pollution in the region, atmospheric depositions could be considered a significant pollution source only with respect to heavy industrial sites, which generate a considerable amount of air pollution, and areas of very intensive traffic in urban areas.

Streets refuse deposition is another factor that influences the runoff quality. Particles that are larger in size than dust (> 60 μm) are considered as street refuse or street dirt (Novotny & Olem 1994). In

NURP (National Urban Runoff Program) studies in USA, these deposits were divided into median-sized deposits (street-dirt particles range from 60 μm to 2 mm) and litter (>2 mm). Litter deposits contain items such as cans, broken glass, bottles, pull tabs, papers, building materials, plastic, garbage, parts of vegetation, dead animals and insects, animal excreta, and the like. The sources of street dirt are numerous and often very hard to control. A comprehensive survey of a number of U.S. cities with respect to the accumulation of pollutant on street surfaces (AWRC 1981) points out that runoff from street surfaces is generally highly contaminated. It was estimated that, based upon a complete cleansing of street surfaces by a moderate to heavy storm of one hour duration, the storm runoff water produced during this storm would contribute considerably more pollution load than would the same city's raw sewage during the same period of time. This type of urban diffuse pollution is common for many African countries and contributes a substantial pollution load.

Pollution input from vegetation is generated by leaf fallout and other residues as grass clipping. During characteristic seasons, it dominates street refuse composition and is a source of large urban loads of biodegradable organics. Only a portion of this vegetation residue that accumulates on impervious surfaces is a pollution threat to surface waters. Vegetation fallout on soils becomes an integral part of the soil composition, and in most cases may even improve soil permeability and erosion resistance. During defoliage in the fall a mature tree can produce 15 to 25 kg of organic leaf residue (dry weight) that contains 90% organics and 0.04%-0.28% phosphorus (Novotny 2003). With respect to the countries in the region, grassed open ditches are a preferred option for conveyance of road/street runoff. Grass clippings and other vegetation material accumulate in the ditches and could be a considerable source of pollution if not cleaned regularly.

Motor vehicular traffic is directly responsible for the deposition of substantial amounts of pollutants, including toxic hydrocarbons, metals (from exhaust emissions, worn off tyres, clutch and brake linings, lubricants, coolants, rust and decomposing coatings dropped from the underside of mudguards and undercarriages) asbestos and oils. The particulates contributed by traffic are primarily inorganic. Most of the traffic exhaust pipe emissions are dust size (< 60 μm). Only a small portion (< 5%) of the traffic related pollution could be directly traced to vehicle emissions. However, the pollutants that motor vehicles emit are among the most important because of their potential toxicity. In addition to traffic density and vehicle pollutant emissions, pavement conditions and compactions are significant in determining the traffic impact on pollutant loads. Streets, whose conditions are rated as fair to poor, were found to have total solid loadings 2.5 times greater than those rated as good to excellent. Summarized data on the quality of runoff from highways in the USA show average concentrations of lead and zinc of 0.53 mg/l and 0.37 mg/l respectively (Novotny & Olem 1994).

Toxic chemicals could become a diffuse source of pollution from industrial sites, where no provision for safe storage of fuels, raw materials or auxiliary chemicals is made, and spillages or leakages could be washed out by the runoff and reach the drainage system.

The major pollutants associated with agriculture include sediment, nutrients, pesticides and other toxins, bacteria and salts or increased salinity levels. Different types of agricultural land use are more likely to contribute certain pollutants than others from runoff and subsurface water. Typical types of agricultural land use patterns are dry land and cropland, irrigated cropland, pastureland, rangeland, forestland, confined animal feeding operations, orchards and wildlife land. It has been mentioned (Chapter 3) that informal small-scale agricultural activities become an often-met part of the urban landscape of many African cities. Such practice could be regarded as a source of diffuse pollution only in cases of excessive use of fertilizers or pesticides, or when the cultivation is executed along river/stream banks, thus enhancing soil erosion during flood events.

Illicit (informal) discharges to storm sewers often include wastewater from non-storm water sources. Illicit discharges are so named because storm sewers are not designed to accept and convey raw sanitary waste or discharges of wastewater from automobile stations, car washes, or light industrial facilities, etc. The most often met cases of illicit discharges are illegal connections of wastewater from homes or businesses, which are supposed to be directed to the sewerage system, but instead are discharged into the drainage system. Another form of an illicit discharge is the infiltration of wastewater into the drain system from broken/clogged sewer mains, or from inadequately constructed or maintained septic facilities. Also, the case of polluted storm water from a specified area, which should be directed into the sewer system, but actually discharges into the drain system, should be considered as an illicit discharge.

Soil erosion from construction sites or open mining operations could be a major source of pollution through sediment transport and accumulation in close-by waterways. Although a construction site may seem a small part of the river catchment, the accumulative effect of polluted runoff from a number of sites can have a dramatic impact on water quality. The owner and the builder are responsible for controlling soil erosion and preventing sediment from the site to be washed into storm drains. In the USA, heavy fines may be imposed if a person allows soil, earth, mud, clay, concrete washings or similar material to be placed into a position where it is likely to be washed into storm drains (USEPA 1995).

The Southern Africa region is characterized by large differences in the socio-economic status of the population, which is reflected in the urban and rural patterns of development and lifestyle, with corresponding differences in the land use characteristics. Therefore, all of the listed points of diffuse pollution could be present. However, informal or semi-formal settlements, solid waste disposal sites, and runoff from industrial enterprises could be mentioned as common problems of significant impact in the region. Selected case studies are presented in subsequent chapters in order to illustrate the impacts of such sources. Open mining operations and large construction sites should be regarded as problematic areas in terms of diffuse pollution, which need special attention and investigation as well. In general, detailed prioritising of different pollution sources could be achieved by a correct estimation of pollution loads generated, based on continuous monitoring data of water quality and volumes (rates).

2.2 *Runoff water quality*

The water quality field is dynamic. Profound changes in most of its areas occur rapidly during the last decades in response to the public awareness of environmental matters and gains in knowledge about the technology and sociology of water resources development and uses. This is a field that has had the greatest influence on our understanding of key issues and formulation of changes in policies and strategies. The validity of all decisions concerning water quality and its control hinge ultimately on the information complied during the water quality evaluation process (Chapman 1998). It is not intuitively obvious, but the evaluation of water quality must include careful consideration of available quantities as well. Lamb (1985) points out rightfully that many water quality problems and their solutions can be influenced radically by the volumes and flow rates that are available, their variability, rates of renewals and losses, and the quantity of water for intended future use. The variability of runoff water quality is strongly related to the volumes of runoff generated, flow rates and the general pattern of rainfall events. Thus in practice, both water quality and quantity evaluation and assessments should form the basis for managerial decisions and policy formulation.

It has been widely recognized that urban runoff pollution can lead to a degradation of aquatic life and endanger public health. The magnitude and nature of these effects vary by region, depending on the type and concentration of pollutants in storm water, rainfall characteristics, land use practice and other factors. In the USA, it has been found that: "the size of the non-point source pollution, at least equals, if not exceeds, the total pollutant loadings contributed by all point sources" (AFS 2000). Sometimes, urban storm water runoff may contribute potential pollutants in quantities even higher than those of untreated sanitary sewage from the same community, especially the first flush It was reported that storm water contains more suspended solids than raw sanitary sewage, with BOD, phosphates and total nitrogen in the range of 6-11% of the raw sewage values (Lamb 1985). It is thus obvious that the contribution of storm water pollution to receiving waters cannot be ignored.

In African conditions, many sewage treatment facilities have not met the required standards, thus becoming a serious point pollution source. Under such conditions, the pollution loads generated by runoff might be ignored. However, information presented in Chapter 3 shows that pollution from urban drainage could be substantial and should be considered in the management process.

One factor, which characterizes urban runoff water quality, is its high variability. The concentrations of pollutant constituents may vary in magnitude by 10:1 or more in a single storm, from area to area and from storm to storm, and could be of the magnitude of the effluent from a secondary treatment facility

(AWRC 1981). Therefore, there is a need for event orientated monitoring approaches and for the consideration of the seasonal variability with respect to the monitoring and regulation of urban storm water.

2.3 *Quantifying runoff by a simple procedure*

The rainfall excess or net rain, which is of concern for the determination of diffuse pollution loads, depends on many factors, which could be summarized briefly as follows:

- Soil and surface cover characteristics – Other factors being equal, a soil with low perviousness will produce a higher proportion of net rain than a soil that allows water to infiltrate readily into the ground. There is a greater volume of rainfall excess generation in urban areas due to a large percentage of impervious covers. The presence of vegetation decreases the amount of excess runoff by intercepting rainfall and increasing evaporation losses.
- Topography – Flat land areas produce less runoff excess than steep slopes for similar surface conditions and rainfall patterns.
- Rainfall characteristics (precipitation, intensity and duration of the rainfall event) – Rainfall excess is higher with increased precipitation intensity because water accumulates more rapidly on the ground surface, filling depression storage and giving less opportunity for infiltration before flowing into streams. For a given rainfall intensity, the portion of the net rain increases with rainfall duration because the soil structure becomes more saturated as precipitation continues and reduces the infiltration rate further. For the same reason, the amount of surface runoff generated by a given storm often has been observed to depend heavily on elapsed time since preceding precipitation.

The transport of sediments and pollutants from the surface into the runoff depends on the energy of the rainfall, determined by the rainfall intensity and on other factors such as soil characteristics, slope and vegetation cover. Pollutant fluxes vary drastically over time with fluxes during storm runoff events often being several orders of a magnitude greater than those during low flow periods. It is common for 80% or more of annual loads to be delivered during the 10% of time with highest fluxes (USEPA 1995).

A number of flood survey methods have been developed to estimate the rainfall excess and the peak runoff discharges resulting from rainfall events. In this section a simple procedure will be presented, as it could be applied easily in conditions where the technical background and the limited data available do not allow the application of more advanced methods. The procedure (Schueler 1987) calculates annual excess runoff volumes as a product of the annual precipitation volume and runoff coefficients, based on surface covers, determined by different land use practices. The coefficients also account for the integrated effect of interception, infiltration and depression storage. Ellis (1989) considers the method to be sufficiently robust and reliable to make reasonable pollution management decisions at the planning stage. It should be noted, however, that the procedure should be applied cautiously with a corresponding interpretation of the results obtained. The accuracy of the results would depend strongly on the accurate determination of the percentages of the drainage area with a relatively uniform surface cover, as well as, on the correct estimation of the annual average concentrations of pollution constituents in storm water runoff. In this respect, the application of GIS should be very useful. The excess runoff is calculated by equation 4.1.

$$R = P\, P_j\, R_c \tag{4.1}$$

where R = annual runoff (mm); P = annual rainfall (mm); P_j = fraction of annual rainfall events that produces runoff (usually 0.9); R_c = runoff coefficient.

For the determination of the runoff coefficient Schueler (1987) uses the following relationship:

$$R_c = 0.05 + 0.9\, I_a \tag{4.2}$$

where I_a = impervious fraction (%), representing the percentage of impervious covers, as a fraction of the whole drainage area.

The application of the procedure, with respect to a relatively small drainage area contributing to an urban drainage channel, would require the division of the whole area into small sections, with a relatively uniform surface cover and soil characteristics. For each section, the values for P and P_j would be the same, and the variable parameter would be R_c, which could be determined on the basis of existing maps. If a GIS would be available, the determination of the runoff coefficient could be more accurately and quickly determined.

2.4 *Diffuse pollution loads from urban areas*

The pollutant loads from urban areas are strongly affected by the drainage volume. The lowest pollutant loadings are typical of sub-urban areas with a so-called "natural surface drainage", as well as from parks, playgrounds, and other developed open spaces. The highest pollutant loadings are emitted from highly impervious densely populated or heavily used urban centres with separate or combined sewers. A basic classification regarding the link between the land use pattern and the expected level of diffuse pollution is given below:

- Low pollution loads. – This category includes low and medium density residential land uses and dry-process industrial activities.
- Medium pollution loads – typical examples in this category include high-density residential and commercial land use.
- High pollution loads – typical examples include medium and high intensity industrial uses (wet processes) and densely populated urban centers with a high level of impervious surfaces.

Different land use patterns could account for a different magnitude of the "unit load" in respect to pollution constituents. This term could be defined as follows:

⇒ A *unit load*, sometimes referred to as the *export coefficient*, represents the mass of pollution with respect to any water quality characteristic, exported from a unit drainage area per unit time. Usually, it is expressed in kg or ton/ha per annum (season), and is related to a specific land use practice.

To illustrate this definition, we could consider a given drainage area, with a relatively constant land use pattern, during a specified time period, e.g. one year, which discharges its runoff in a storm water channel. Then the annual unit pollution load with respect to TSS would be equal to the annual volume of runoff during the stipulated period, multiplied by an averaged EMC of TSS during this period and the product divided by the area under consideration. The value obtained could be used for planning and simulation purposes.

Unit loads allow for the calculation and prediction of pollution loads from a catchment basin, by adding the contribution of all individual sub catchments with characteristic land use practice and corresponding unit pollution loads. However, it should be emphasized, that when applied at a larger scale, the pollution loads estimated are determined with respect to a natural water course, and then they would represent not only the runoff characteristics from the drainage area, but the conditions in the stream as well. In such cases, the unit loads might include the combined effect of the surface runoff load, as well as the load from the base flow and background pollution. In most cases, when the term "unit load" is applied to perennial water streams, it would refer to the combined effect, while unit loads determined in ephemeral streams reflect the surface runoff pollution load only. It should be noted that unit loads are highly site-specific and depend on numerous demographic, geographic and hydrologic factors.

It has been mentioned that EMCs represent more adequately the actual runoff quality. However, the EMCs from different storms with respect to the same type of land use practice may vary considerably due to the different characteristics of the storm event. A statistical analysis of data collected during a study in USA, found no correlation between unit pollution loads and typical urban land uses (Novotny 2003), which was explained with the high variability of EMCs for different rainfall events. For this reason, when estimating annual or seasonal loads, a site mean concentration (SMC) should be obtained. It is the arithmetic mean value of the EMCs measured at one site. Commonly, a large variation is observed when comparing

SMCs from different sites, with specific land use patterns, and this gives reason for the determination of site-specific unit pollution loads.

Various approaches and methods for pollution loads estimation exist; the differences in most cases are based on the different methods for runoff estimation applied. For larger catchments, it is advisable to apply computer models. For smaller catchment areas and preliminary surveys, the simple procedure, described above, could be applied.

In general, pollution loads are linked to the total runoff volume generated, the storm duration and intensity. Marshall (1997) concluded that the shorter and more intense storms have the largest impact on receiving waters. The frequency of pollution carrying runoff events is greatly increased in developed watersheds with higher imperviousness. On the other hand, diffuse pollution loads from pervious lands (croplands, woodlands, urban lawns and parks) occur only during very large storms with lower frequency. The time since the last storm event is also important. First rains produce the most concentration of pollutants, a phenomenon known as the "first flush" effect and is dependent on the street cleaning practice and the public's attitude to waste disposal. Considering the rainfall pattern in the region, it is highly advisable to convey specific research for the determination of SMSs. It would be expected that a statistically-significant difference could be found between EMCs, determined during the beginning of the rain season, and those determined during typical wet season conditions, characterized by high frequency rainfall events.

2.5 *Pollution abatement measures*

Measures to reduce pollution from diffuse sources in urban areas, referred also as Best Management Practice (BMP), could be classified as structural and non-structural measures. Structural measures usually involve engineering structures, which help to detain and reduce diffuse pollution by different treatment methods before it is discharged into natural water bodies. Measures concerning urban planning development and the reduction of the percentage of impervious surfaces in general, improved practice of street/road design and construction, introduction of environmental buffer zones and others, could also be classified as structural. Some examples were discussed in Chapter 2. In cases of storms with a high erosive potential, the application of catch basin inserts could be a viable structural measure to reduce and alleviate the diffuse pollution associated with particulate material (Lau et.al. 2001), but such practice would require proper maintenance and regular cleaning.

Non-structural measures are related to the "soft" part of the management process and include such measures as the introduction of proper management policies and strategies, improvement of the regulatory basis, involvement of all stakeholders, community participation, etc. A more detailed listing of possible non-structural measures is given below:

- Good housekeeping such as oil collection and recycling, spill response, regular household and hazardous waste collection, pesticide controls, increased frequency of street sweeping.
- Public education programs, aiming at an increased public awareness regarding the impacts of diffuse pollution and environmental issues in general.
- Improved regulatory instruments to control pollution generation at source, which should:
  - Require that developers comply with storm water regulations and incorporate erosion and sediment control structural abatement measures.
  - Require that owners take personal responsibility for proper housekeeping measures and maintenance of drainage structures.
  - To formalize the informal practice of small-scale urban agriculture, by providing temporary permits for such activities at selected locations and prohibit cultivation near stream banks, to avoid erosion.
- Efforts to identify and eliminate illicit discharges to storm water sewers. This is achieved by identifying failing septic systems, review of building plans and conducting site visits (inspections) at commercial and industrial facilities to verify connections to the storm water and sanitary sewers, identify where they drain and to determine whether a permit is required. Inspectors could also check the flow in drains

and other potential discharge pathways for compliance and educate the owner/operator about improper discharges.

3 ASSESSING URBAN STORM WATER QUALITY

3.1 *The study area*

The study area (Fig. 4.1) is similar to the one described in Chapter 3. It includes drainage area (DA) 1 (3.22 km^2), comprising of the city's central business district and corresponds exactly to the drainage area DRA1 as described in Chapter 3, section 5.2. DA 2 (2.32 km^2) and DA3 (8.24 km^2) correspond to DRA2 as described in Chapter 3 and are drained by the Coventry Rd. drainage channel. During this study, two sample points were selected along this channel (SP2 and SP3) in order to compare the results and to evaluate if a statistically significant variation of water quality exists along the channel for portions of the total drained area with different land use characteristics. DA 2 drains the upper reaches of the channel. Its profile is complex comprising of different types of land use practices, including medium density residential, commercial and industrial areas. The industrial enterprises include meat processing, dairy products processing, a power station, smaller businesses and automotive garages. DA 3 is of a mixed pattern too. However, it contains a larger percentage of open spaces and medium size residential areas. The Workington industrial area forms part of DA3 and consists of smaller scale industrial and commercial enterprises with predominantly dry industrial processes. DA 4 (3.22 km^2) comprises of residential and institutional single/double stories buildings. It differs from DA1 because it contains a larger percentage of medium-size residential dwellings and the area in general has a lower percentage of impermeable surfaces compared to the CBD. DA 5 has a similar land use practice to DA 4 and was excluded from the study. Residential areas could be qualified as medium density with plots varying between 500 and 2000 m^2.

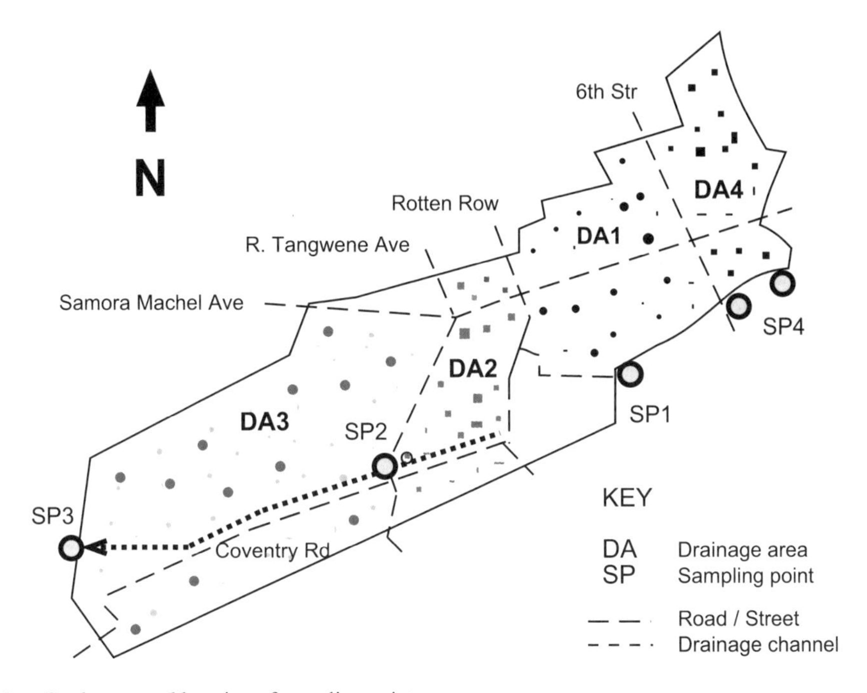

Figure 4.1. Study area and location of sampling points.

3.2 *Methodology*

Grab samples for water quality evaluation were collected at the outlet of drainage channels of the corresponding drainage areas (Fig. 4.1) during and immediately after 7 major rainfall occasions, from November 2002 to March 2003.

SP1 and SP4 collect the storm water from areas DA1and DA4, and discharge into Mukuvisi River. DR 4 has two discharge points. Samples were taken from the two points and mixed proportionally to the corresponding area drained. SP2 is a pipeline, discharging into an open unlined channel, about 4 km long, which collects the drainage from DR3 and discharges into Marimba River.

The variation of storm water quality during a single storm event was investigated at SP1 at one sampling occasion during a storm event of 2.5 h duration, where samples were collected at 30 min. intervals. The conveying structure to SP1 is a trapezoidal concrete lined open channel. Measurements of the water level were taken simultaneously to the sampling procedure and flow rates calculated based on the Manning's equation (Chadwick & Morfett 1993), assuming n = 0.014.

Laboratory analyses of water samples were performed at the water quality laboratory of the Civil Engineering Department of the University of Zimbabwe, within 4 hours of sampling. Laboratory analytical procedures were executed following the Standard methods (1989). COD was determined by the open reflux method. Nitrates (NO_3), pH, and ammonia (NH_4) were measured by selective electrodes (WTW pMX3000/ Ion meter), and standard curves with respect to each parameter have been prepared before the sample collection. TP was determined using the Vanadomolybdophosphoric acid colorimetric method after nitric acid-sulphuric acid digestion, by a Spectronic 21D spectrophotometer (wavelength of 450 nm). The metal concentrations were determined as total metals at the laboratory of the Geology department of the University of Zimbabwe. Samples were digested using the nitric acid-sulfuric acid digestion method. The analyses were performed by a "Varian techtron spectra 50b" atomic absorption spectrometer, employing an air-acetylene fuel.

Statistical data analysis was performed by the use of standard EXCEL statistical tools. The evaluation of a statistically significant difference between two data sets, with respect to each one of the measured parameters for the respective sampling point, was executed by the application of the two-sample t-test, assuming unequal variances at 95% confidence interval. Correlation and regression analyses between TSS and selected parameters were executed by the regression analysis tool at 95% confidence level, based on all measurements for the study period.

3.3 *Water quality assessment and environmental risks*

A comparison between historic data about monthly rainfall volumes for the period 1980 – 2002 and the total monthly rainfall volume during the study period is shown in Table 4.1. MAR value for the study period (840 mm) is close to the historic MAR value of 865 mm, with lower rainfall volumes during the months of December and January and higher values in March. Therefore the results obtained during this study could be assessed as representative for an average wet season.

The evaluation of measured parameters (Fig 4.2) is based on a comparison of the mean values obtained for the different sampling points with the standard effluent discharge regulations in the country (WWEDR 2000). Fig. 4.2a shows variations of TS, TSS, TDS and COD concentrations. COD varies from 100 to 200 mg/l, which presents medium to high environmental risk (ER). The same risk has been evaluated in

Table 4.1. Characteristic monthly rainfall volumes (mm) for the study area.

	Sep	Oct	Nov	Dec	Jan	Feb	March	April	May	June
Av (1980-2002)	2	40	80	180	210	190	120	30	15	0
Max (1980-2002)	80	170	250	420	480	470	300	170	200	10
This study	0	95	90	105	110	190	210	40	0	0

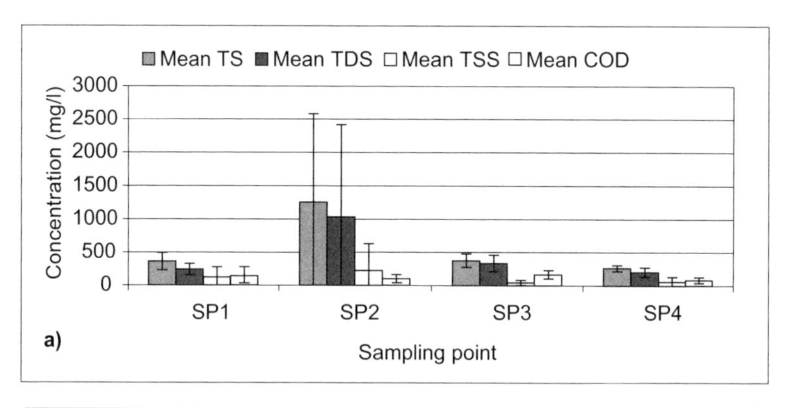

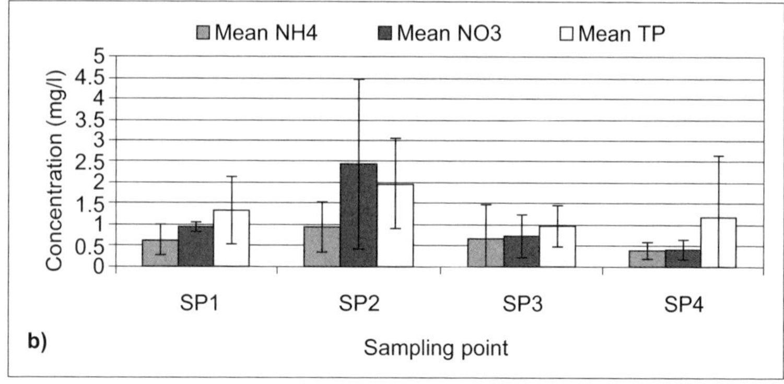

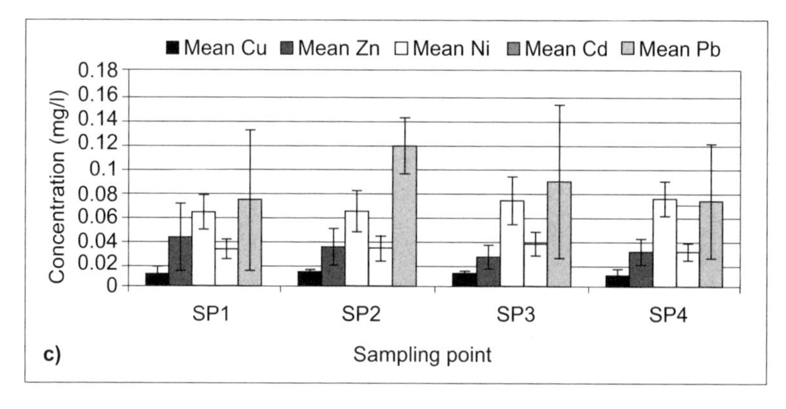

Figure 4.2. Spatial variation of urban drainage water quality (vertical bars indicate standard errors) a) General parameters (n = 7); b) Nutrients (n = 7); c) Metals (n = 5);.

respect to TSS. The concentrations of both parameters are lower than reported EMCs for residential and commercial areas by Choe et al. (2002). TDS concentrations varied between 200 and 1000 mg/l with pronounced high values and high standard error at SP2. TDS concentrations at SP1, SP3 and SP4 are within the safe limit, but at SP2 they present medium ER. PH values were within the safe limits at all points.

Metals variations (Fig. 4.2c) show that Cu, Zn and Ni concentrations are within the safe limit. Cd concentrations present low ER, while Pb concentrations present low to medium ER. Previous research (Mvungi et al. 2003) shows the same trend but Pb concentrations of runoff in high-density suburbs of Harare were two to three times higher than the ones measured during this study. Pb and Cd concentrations evaluated in this study are much higher than the reported EMC of 0.008 mg Cd/l and 0.022 mg Pb/l in California for similar land use practice (Buffleben et al. 2002).

Nutrients variations (Fig. 4.2b) show that nitrates are within the safe prescribed limit, ammonia concentrations present low ER, and TP concentrations present low to medium ER. Comparison with background pollution values in the basin (Hranova et.al. 2002) show that ammonia concentrations in the runoff are 10 to 20 times higher, nitrates and TP – up to 10 times higher than the background values.

In addition, mean concentrations of ammonia and TP in the runoff are comparable and even higher than the same parameters of treated effluents in the two major wastewater treatment facilities of the City (Hranova et.al. 2002). Choe et al. 2002 has reported similar EMC regarding TP concentrations for residential, commercial and industrial areas.

Comparison with previous results from the city of Harare monitoring program (Chapter 3, Figs. 3.5, 3.6) could be done regarding points SP1 and SP3, as these sampling locations coincide with the ones referred in Chapter 3 as DR2 and DR1 respectively. The type of land use practice has not changed considerably since the year 2000. However, the different approaches and methodologies applied to obtain these results, makes such a comparison ambiguous. The results in Chapter 3 are mean values of samples collected throughout the year at different time intervals, and they reflect the flow quality during both – the dry and the wet seasons. Both channels have shown to have perennial flow throughout the year. Correspondingly, these results in Chapter 3 represent also the base flow quality, which might be due to ground water infiltration or to illicit discharges. The results obtained during this study (Fig. 4.2), present the storm water quality during or immediately after rainfall events, correspondingly they could be accepted as representative of runoff quality.

Considering these differences, as well as the different methodologies applied to test the samples, the comparison should reflect the range of variation of the selected parameters, but not the specific values. Considering the CBD (SP1/DR2), the phosphorus variations during this study are comparable and lower than the ones presented in Chapter 3, which could indicate that no substantial increase in these levels have been found. Ammonia concentrations at SP1 vary between 0.5 mg/l and 1 mg/l, which is considerably higher compared to median values observed during the period 1995-2000. With respect to the Coventry Rd. channel (SP3/DR1) phosphorus results are within the same range of variation, while ammonia show a considerable increase. For both locations nitrates measured during this study are about 10 times higher than the monitoring data, which could be associated with a systematic error during the testing procedure.

3.4 *Pollutants transport*

The transport mechanisms of pollutants are complex to evaluate. Numerous factors would influence it, such as: erosive capacity, intensity and duration of the rainfall events; type of cover and imperviousness; the status of streets/roads pavements; the general cleanness of the area drained in terms of adequate refuse collection practice; the intensity of automotive traffic, etc. In this study, an analysis of pollutants association to TSS concentration was done in order to identify which parameters are associated with TSS and at what level. Such an analysis could help to evaluate the effect of the management practices applied, aiming at the reduction of the TSS concentrations in the runoff. The correlation analysis of the TSS data set with respect to the COD, TP, and the metals data sets was performed, and results are shown in Table 4.2. The analysis was executed based on all measurements during the study period for the respective parameters, expressed in mg/l, where for COD and TSS $n = 29$, and for the metals $n = 21$. The regression analysis was executed only regarding COD and Zn as dependant variables of TSS (Fig. 4.3).

Table 4.2. Correlation coefficients of selected data sets with respect to TSS.

	COD	TP	Cd	Cu	Zn	Pb	Ni
TSS	+0.6855	+0.3469	−0.2183	+0.2042	+0.7008	+0.4931	−0.2954

The correlation analysis results show the extent at which the selected parameters are associated to particulate material and the potential of an increase of the pollution loads regarding these pollutants with increased erosivity. Positive correlation coefficients indicate that the increase in TSS concentrations is correlated to the increase in the dependent parameter concentrations, while negative signs show a trend of decrease in the concentrations of the dependent parameter, with increasing TSS concentrations. The TSS and COD data sets show a relatively high correlation, suggesting that a significant portion of the organic pollution in the runoff is present in the form of particulate material. The correlation coefficient between TSS and TP is relatively low, indicating that phosphorus is present in both particulate and dissolved forms in the urban runoff, therefore, the reduction of TSS only, would not affect the pollution loads with respect to this parameter considerably. Considering that the pH values are within normal limits, metals concentrations were expected to show a better correlation to TSS concentrations. However, only Zn results showed well defined correlation with TSS, the Pb data set showed low correlation and for the rest of the metals, correlation was negligible. This suggests that Pb and Zn concentrations might be related to the bulk amount of suspended material, while the rest of the tested metals could be associated with selected particles, which could be randomly found in storm water and their weight with respect to the TSS concentration might be negligible.

The graphical presentation of the results of the regression analysis is shown on Figure 4.3. The linear equations developed could be applied to determine indicative COD and Zn values, when TSS values are known. It should be remembered that the equations are derived, based on a relatively small size of the data set. The regression coefficients are relatively low, and in addition they are representative of the wet season of 2002-2003 only. Additional data will be needed to establish a more reliable relationship.

The data set distribution plot regarding actual measured COD values (Fig. 4.3b) shows that the 95 percentile value is 218 mg/l, the 50-percentile value is 108 mg/l, and the average value of 121 mg/l is close to the 60 percentile of the entire data set. It shows a relatively normal distribution. However, it indicates that if the average values are used for comparison with regulatory instruments, as it is the current practice, it actually means that in 40% of the measured observations the prescribed limit has been exceeded. With respect to Zn, the 95-percentile value is 0.07 mg/l, but the 88-percentile value is 0.04 mg/l, with an average value of the data set equal to 0.035 mg/l, which shows that it represents more accurately the water quality during the period of observations. The need for statistical interpretation of the data sets and the link with corresponding regulatory instruments is discussed in more detail in Chapter 12.

It should be noted that the equations derived do not eliminate the need for a regular water quality monitoring and its importance should not be underestimated. Similar equations, derived for different parameters and sites, could be regarded as valuable tools for planning purposes and the evaluation of the impacts of different scenarios, e.g. if the improved streets pavements reduces the amount of TSS in runoff by 20%, this could lead to a reduction of the COD concentrations by 10%. It is expected that in the future, larger data sets and more reliable monitoring practice would allow for the derivation of more accurate relationships, which could be used with a higher level of confidence during the planning stage and also, could form the basis of the design methods for diffuse pollution control and abatement.

3.5 *Single storm event*

The single storm event characteristics were examined at SP1, representing the most densely populated drainage area with a high percentage of impervious surfaces. Measurements were taken on 26th February 2003, during the peak of the wet season, where intensive storm events were recorded diurnally, thus representing typical wet weather conditions. Results are shown in Figure 4.4

The storm event peak concentrations of TSS, COD and TP (Fig.4.4a, b) were detected one hour after the start of the rainfall, corresponding to the peak flow rates in the channel, while TDS concentrations did not vary considerably. The same applies to nitrates (mean value = 0.61 mg/l) and ammonia (mean value = 0.21 mg/l). This fact confirms that organic pollution and to some extent, phosphorous, are bounded to solid particles, carried with the flow during peak flow conditions, while soluble elements like ammonia, nitrate and TDS have more or less constant concentrations during the whole storm duration. This implies

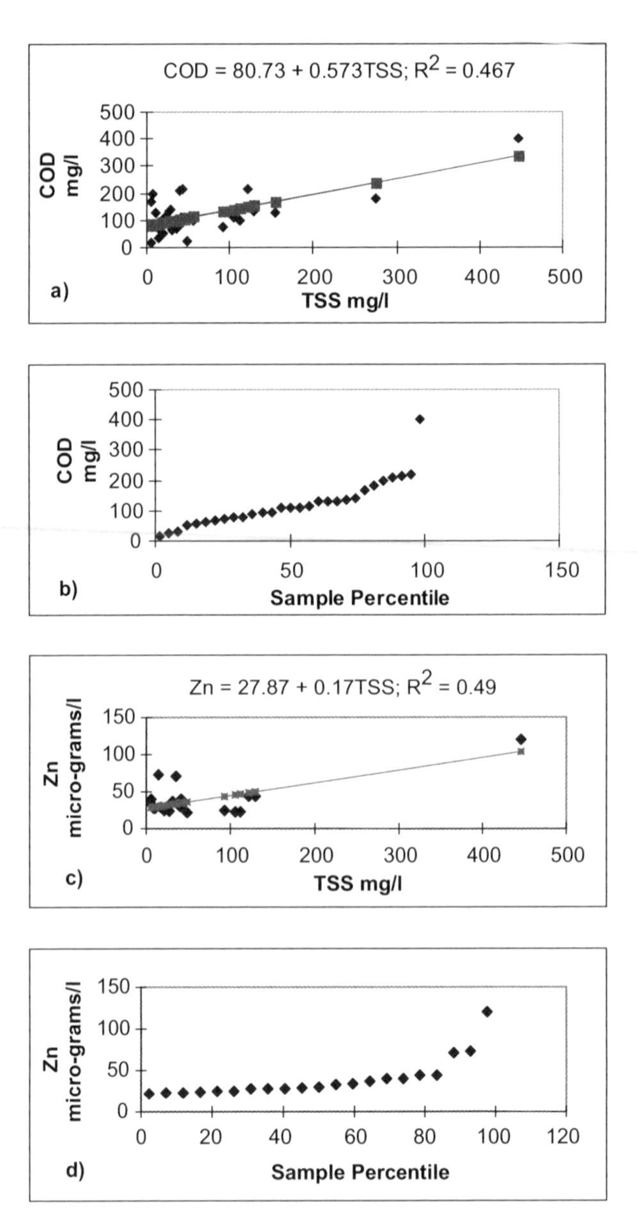

Figure 4.3. COD and Zn as dependant variable of TSS; a) & c) "Best fit" plot of the linear relationship of COD (mg/l) and Zn (μg/l) versus TSS (mg/l), respectively; b) & d) Data set distribution of actual COD and Zn values, respectively.

that the determination of EMCs (equation 4.3) would be of importance with respect to pollutants, which are transported in particulate form and are associated with suspended matter. Also, these findings confirm the trend of correlation found in the previous section. The higher concentrations of COD and TSS at the end of the storm event could be associated with heavier particles and debris, which have not been carried out with the flow, but remain in the channel after the storm wave subsides.

The characteristic parameters of this storm event are presented in Table 4.3. EMC was calculated by equation 4.3, and the flushing effect of the storm was calculated based on cumulative load curves (Choe

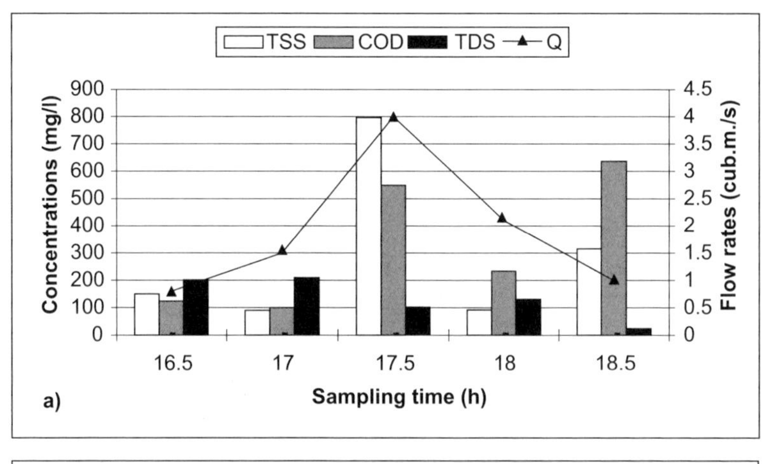

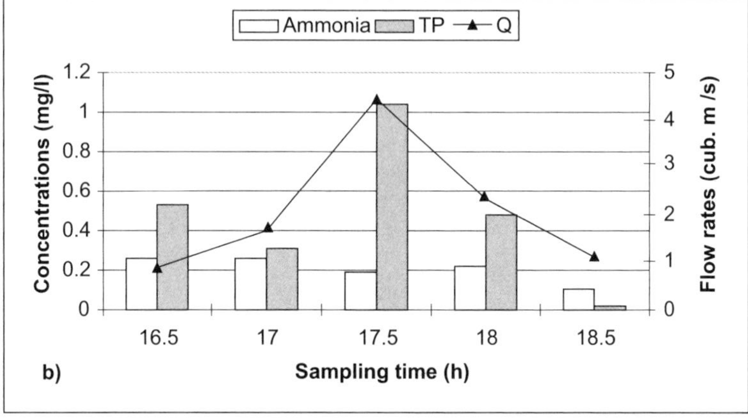

Figure 4.4. Urban drainage water quality variations during a single storm event at SP1.

et al. 2002, Sansalone et al. 1997). For this specific study, the flushing coefficients were calculated by equation (4.4), based on the total duration of the storm event and correspondingly $\sum (Q_i/\sum Q_i) = 1$. The flushing coefficients, which have a value > 1, indicate the presence of flushing effect with respect to the studied pollution constituent. Actual pollution loads represent the sum of the specific load for each time interval, calculated based on the corresponding contaminants concentrations and flow rates. For the purpose of comparison, pollution loads were also calculated, based on the total runoff volume discharged during the storm event and pollutant constituents represented by average concentrations and EMCs.

$$\text{EMC} = \sum (C_i Q_i) / \sum Q_i \tag{4.3}$$

$$\text{Flushing coefficient} = \sum (P_i/\sum P_i)/\sum (Q_i/\sum Q_i) \tag{4.4}$$

where C_i, P_i and Q_i are the contaminant's concentration, pollution load and flow rate, respectively, for each time interval.

The results in Table 4.3 show and confirm that the flushing effect is associated mostly with particulate material. The parameters, which have a flushing coefficient higher than 1, also have EMC concentrations, which are considerably higher than the mean concentrations. This is associated with the erosive capacity

Table 4.3. Characteristic parameters of a single storm event (wet weather).

Constituents	COD	TSS	TDS	Ammonia	TP
Mean concentrations (mg/l)	289	327	134	0.21	0.48
EMC (mg/l)	420	379	126	0.20	0.65
Pollution load – EMC (kg)	7106	6413	2271	4	11
Pollution load – average concentrations (kg)	4887	5533	2132	4	8
Flushing coefficient	1.25	1.13	0.68	0.55	1.03

of the rainfall event and the increased rate of pollutant transport during peak flow conditions, due to the increased suspended solids concentrations in runoff and other pollution constituents associated with them. This result has a significant implication regarding the monitoring and abatement strategy of diffuse pollution. It shows the need for the estimation of EMC regarding constituents associated with suspended material, which would influence considerably, the pollution loads estimation to natural water bodies and corresponding effects and impacts. The application of average concentrations with respect to particulate pollutant constituents leads to a considerable underestimation of the pollution loads contributed to receiving water bodies. A more detailed analysis of the implications due to errors in the estimation of pollution loads is presented in section 4. In addition, the results obtained during the study of this specific single storm event, show that the removal of suspended matter from the runoff before it is discharged into the receiving water body could reduce substantially the pollution loads regarding COD, and to some extent regarding TP, but would not influence the pollution loads regarding dissolved constituents as ammonia and nitrates. Therefore, biological treatment methods should be applied as pollution abatement measures.

3.6 *Spatial variation of water quality*

The comparison of the variation of water quality with respect to the different points, based on Figure 4.2, shows that at SP2, the mean values of TSS, TDS, TP, ammonia, nitrates and Pb were higher compared to the rest of the points. This indicates a considerable contribution of pollutants from the industries located in this area. Data sets for this point show high fluctuation of results, which is typical of runoff water quality in general, but at this specific location could also be associated with possible informal discharges from industrial enterprises. Correspondingly, this is reflected in the statistical analysis and the statistical significance of the difference between mean values. However, the suggested contribution with respect to SP2 is supported by the fact that at this point, the dissolved contaminants as nitrate, ammonia and TDS were much higher compared to the other sampling locations. Pollutant concentrations at SP3 are lower than the ones at SP2, which could be attributed to dilution due to the additional and less polluted runoff from DR3.

The spatial variation of runoff characteristics were examined based on the statistical analysis for a significant difference of data sets ("t-test"). All data sets for the individual sampling points were compared individually for a statistically significant difference of their mean values with respect to all parameters. For data sets with high variability, logarithmic transformations were applied, which did not show any difference, compared to the "t-test" results obtained from the actual values. The results, with respect to statistically significant difference of the mean values (+) between the different locations, are presented in Table 4.4. The parameters, which are not included in the table, did not show a statistically significant difference of mean values. There was no statistically significant difference between SP1-SP4, and SP1-SP3 with respect to all parameters tested.

SP1 and SP4 are characterizing the water quality of runoff from Harare's City center. The difference between the land use patterns of both drainage areas is that at SP1 the impermeable cover area forms a higher percentage of the total area. The density of the building construction is higher and includes high story buildings. The SP4 drainage area includes some medium density residential areas in addition to commercial and institutional developments, and the population density is lower. The results obtained did not show a considerable and statistically significant difference in the water quality from both areas, which

Table 4.4. Results of the test for a statistically significant difference of mean water quality characteristics.

Parameter	SP1-SP2	SP3-SP2	SP3-SP4	SP4-SP2
COD			+	+
Ammonia				+
TP		+		
Nitrate	+			
TDS	+			+

indicates that the whole central part of the town could be accepted to be of uniform quality with respect to its runoff. The area drained by SP2, includes industrial complexes and is of a mixed pattern because commercial areas and medium-density housing developments are drained to this point too. However, medium and low-density residential areas are not expected to contribute considerably to the runoff flow rate at this sampling point, as they consist of single story buildings with gardens and the main portion of the roof runoff is infiltrated into the ground before it reaches the collectors. Consequently, the influence from the industrial sites runoff, which has a high percentage of impervious covers, could be substantial, and this has been reflected by the results of this study, as well as in earlier studies (Jarawaza 1997). Also, Heather et al. (1996) reported that some industries in this area, both small and large, do not have on site pre-treatment facilities or do not bother to ensure that the wastewater and storm water generated by them should meet standards set by the Harare City Council.

The Coventry Rd channel, between points SP2 and SP3, collects runoff from residential areas, small-scale enterprises and open spaces. In general, results show that at SP3 (the discharge point of the Coventry Rd drainage channel), the water quality has improved, compared to SP2 and is similar in strength to the one from the CBD. Thus, it is expected that the runoff from this drainage area is diluting the more concentrated flow coming from up-stream. It should be noted that the dilution might be due to groundwater flow, as along this section, between SP2 and SP3, the channel is not lined.

4 ESTIMATING POLLUTION LOADS

The evaluation of pollution loads plays an important role in the process of pollution control and abatement. It allows the estimation of the expected impacts of pollutants on the water environment and the evaluation of the assimilative capacity of receiving water bodies by means of models. Also, comparing loads from different sources allows to rank them and to identify the critical ones, which should have a priority during the process of planning and the implementation of pollution control and abatement measures.

The estimation of pollution loads from point sources, which are related to treatment plants, is a relatively simple process, as, usually, data regarding quantities and qualities discharged are available. It forms part of the treatment plant operation and is collected on a regular basis. In the case of discharge points of storm water channels, as it is the case in this study, the procedure of pollution loads estimation is more complex, because usually data is limited. In addition, the range of variability of the parameters is much higher, compared to discharges from treatment plants, and also the character of the diffuse pollution implies a much higher level of uncertainty in the correct estimation of qualitative parameters. Storm water channels, discharging into natural water bodies could be considered as point discharges, because of the clearly defined locations. However, considering the character of the pollution loads they could be defined as diffuse pollution sources because of the high uncertainty of both qualitative and quantitative parameters involved. The most complex case is the evaluation of pollution loads from agricultural areas, which are complex in both directions – regarding pollution loads estimation and regarding their spatial variation and identification. In the vast majority of such cases the estimation of pollution loads requires the application of models based on the whole catchment area analysis.

Pollution loads, contributed during the study described in section 3, were calculated as the product of the runoff generated and the average pollutant concentration for each specific drainage area. The runoff has been estimated by the simple procedure, based on the total volume of rainfall during this wet season. The corresponding runoff coefficients have been assigned more or less arbitrarily, based on engineering discretion and information from maps. Unit pollution loads with respect to all parameters tested, have been determined for each specific drainage area and corresponding sampling point. Results are presented in Magombeyi et al. (in press). These loads could be regarded as annual total and unit pollution loads from the corresponding areas for this specific wet season. However, these results should be seen as a rough estimation, given the limitations of the method applied and the time limitation of the project.

In order to illustrate the impact of the accuracy of determination of the individual quantitaxtive and qualitative components of the pollution load on the final result, an example has been solved (Fig. 4.5). The basic data, referring to an imaginary unit drainage area is shown in the text box (Fig.4.5a). Figure 4.5b shows the magnitude of variation of the pollution load, as a result of the deviation from the "true value" (expressed in %), in the case when only one parameter is varying and the other is kept constant at its "true value". The pollution load variation is expressed also in terms of the equivalent population, which would generate the same annual load in the form of raw sewage. The same percentage errors for the quantitative and qualitative parameters were selected. The errors with respect to the quantitative parameter (net rainfall) would be generated by error in the determination of the runoff coefficient. With respect to the qualitative parameters, COD was selected to represent a mean concentration (300 mg/l), and the impact of possible errors of determination, causing deviation from the "true" value, was computed. Figure 4.5b shows that the same error percentage, e.g. + 50%, would account for a pollution load equivalent to the one released by 2465 persons, and would be due to the use of a wrong runoff coefficient of 0.9 instead of 0.6. The same error would be implied by a wrongly determined COD mean annual value of 450 mg/l instead of 300 mg/l.

Usually, errors with the estimation of the runoff coefficients are not of such magnitude, while the difference in the estimation of the EMC value and mean concentrations regarding a single storm event (Table 4.3) show that the expected mean concentration could vary significantly from its true value and a considerable amount of continuous data is necessary in order to determine correct mean concentrations (EMCs). In most cases, errors in the estimation of both quantitative and qualitative parameters occur simultaneously and this type of effect is shown on Figure 4.5c, where both parameters have errors of the same magnitude and direction. It should be noted that errors in the negative zone (underestimation of actual quantity and quality parameters) lead to a lower range of variation in the pollution load computed, than the ones in the positive zone (overestimation of actual values).

The simple method allows the use of weighted runoff coefficients, which give a more accurate estimation of the net rain volume. Their determination is important in cases of mixed pattern land use practices, which have different cover characteristics, slopes, vegetation, etc. It requires the subdivision of the total drainage area on several parts with relatively uniform characteristics and the assignment of area-specific runoff coefficient to each one of them. Then, the weighted coefficient will be calculated by equation 4.5.

$$R_W = \Sigma\ (A_i R_i) / \Sigma\ A_i \qquad (4.5)$$

where R_W = weighted runoff coefficient; R_i = area-specific runoff coefficients; A_i = part of the total drainage area with uniform cover characteristics.

The determination of the net runoff by the SNC curve method (Schulze et.al. 1993), described in more detail in Chapter 5, accounts for more factors, which would influence the accuracy of determination, and if the input data with respect to these parameters is available, it could lead to more accurate results. In both cases, the application of GIS could provide very useful data about the drainage area, thus improving the accuracy of the process and could reduce considerably the time required for the data handling and analysis.

The average runoff quality parameters determined during the study and presented in section 3 could be useful for further investigation of this problem, but could not be accepted as a reliable basis for actual pollution loads estimation, especially regarding parameters showing flushing effects.

Data:
Drainage area: 1 km2;
Runoff coefficient ("true value"): 0.6;
Runoff coefficient variation: 0.2 – 1;
COD ("true value"): 300 mg/l;
COD variation: 100 – 500 mg/l
Unit COD per capita load: 100mg COD/cap. day
a)

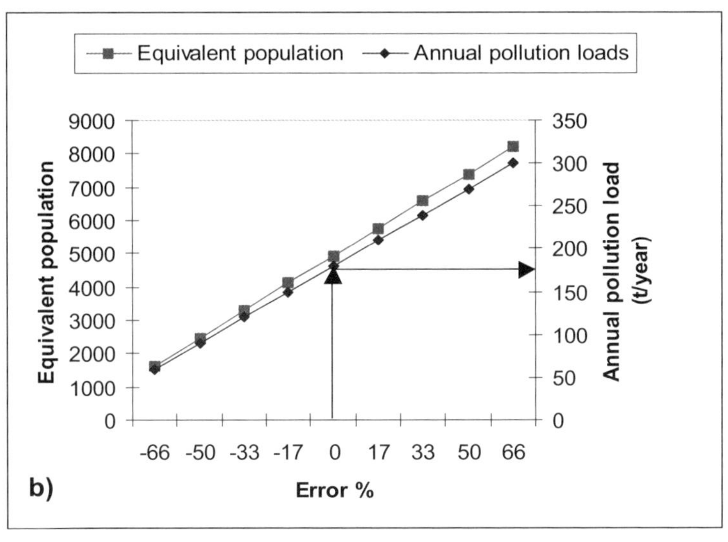

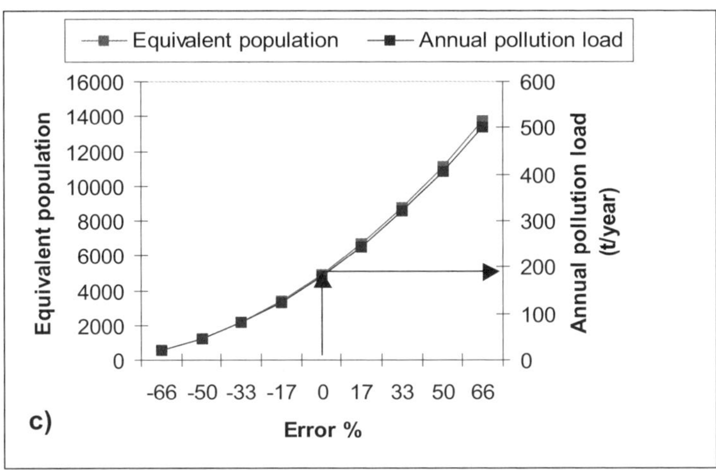

Figure 4.5. An example of the impact of error percentages of quantitative and qualitative parameters on the annual pollution loads a) Initial data; b) one of the parameter is constant at its "true value"; c) both parameters moving in the same direction.

The need for the assessment of flow rate integrated samples and determination of site-specific pollutant constituents' concentrations has been shown in the previous sections. In addition, it would be necessary to determine these qualitative parameters with respect to typical rainfall events. This would require the use of automatic sampling devices or automatic monitoring stations, which could generate a massive database. Unfortunately, automated monitoring stations could not provide information regarding parameters, which are associated with particulate material. Usually, they are able to measure dissolved constituents only, such as conductivity, pH and different types of ions. For this reason, automatic sampling devices at selected locations are an attractive option, while the testing of the necessary parameters could be done at laboratory conditions.

Toxic constituents, which are important in the determination of annual loads and estimation of long term toxicity effects or shot term acute toxicity limits, are of considerable importance as well. During this study, selected metals were tested as representatives of this group. The specific characteristic of such types of parameters is their presence at very low concentrations, usually in the micrograms/liter range. In order to determine the impact of accuracy determination of such parameters on annual loads, the procedure shown on Figure 4.5b was applied for a mean annual concentration of Cd – 0.05 mg/l as "true value". Results show that for the same error percentage variation, the annual loads varied from 10 to 50 t/year corresponding to a variation of Cd from 0.0167 mg/l to 0.083 mg/l. During the study period, a similar variation of the data set was measured, however, with other metals the range of variation is much higher. This implies that in the case of micro-pollutants, the accuracy of the testing method applied should be much higher and a larger data set would be required in order to obtain reliable site-specific values.

The determination of EMC for each specific land use pattern would require a considerable effort and cost. For this reason, the monitoring program regarding urban storm water quality could concentrate on the determination of EMC, associated with suspended particles, and for the case of the city of Harare, these are the two channels at SP1 and SP3. The evaluation of EMCs at these locations for a period long enough to obtain statistically reliable data, together with an adequate record of rainfall data corresponding to each measured EMC value, would allow to determine accurate site-specific values and to link them to typical rainfall events with a corresponding return period. Such information would allow the development of hydrographs representing typical storms with different duration and intensity at the points under consideration, and corresponding site-specific EMCs. A typical storm with given intensity, duration and return period could have different EMCs, depending on the seasonal variation, e.g. if it has occurred at the beginning, during, or at the end of the wet season. These variations should also be detected during the monitoring process for each typical hydrograph. It should be noted that this is a demanding investigation procedure and might be too complex and costly to be performed by a regular monitoring program. Therefore, the combination of data from regular monitoring programs, and data obtained by research projects, having the specific objective to complement the monitoring program, could be a viable option.

5 MANAGING URBAN STORM WATER QUALITY

5.1 *The present management practice*

This section illustrates the opinion of the two major institutions responsible for controlling storm and natural water quality in Harare's catchment basin-the city's municipality and DWR. A questionnaire was prepared and submitted for completion to these authorities, in order to present their experience and opinion on two questions:

i. What are the main sources of diffuse pollution specific for the area?
ii. What are the major problems associated with the diffuse pollution control and abatement?

A summary of the answers regarding the first question are listed below:

- Discharge from sewage blockages – the questionnaire results show that the local authorities recognize this to be one of the serious causes of diffuse pollution in the city. The number of reported and cleaned sewage blockages during the year 2002, supports this statement, with the data presented as follows: 70-80 blockages per month in high-density areas; 30 blockages per month in the city centre; and less than 5 blockages per month in industrial, medium and low – density areas.
- Oils and grease spillage from food enterprises (irregular cleaning of oil and grease traps);

- Illegal discharges of process industrial water into rivers/storm sewers by some industries;
- Leachate from dumpsites/landfill sites;
- Illegal dump of paint, oil, antifreeze and other used chemicals into storm drains;
- Oil and chemicals from garages-detergents from car washing, paints, anti-skid compounds are washed out from the surface into the storm water, especially in cases of informal open-space, low cost garage services;
- Car exhausts and wearing of vehicle parts;
- Erosion of roof material;
- Storm water from industrial enterprises with poor yard hygiene;
- Spills from haulage companies of the goods that they transport.
- Religious/church gatherings in open spaces, where there is no provision for toilet facilities and refuse collection.

The summarized answers of the second question are as follows:

- Control of point sources of pollution – Despite the fact that the emphasis of the pollution control measures has been placed on sewerage treatment, there are treatment plants, which either by-pass substandard flows directly to natural water bodies, or discharge a poor effluent quality that does not meet the Zimbabwe standards for safe discharge into receiving waters. This is due to the fact that some of the treatment facilities are overloaded and are forced to discharge part of the influent directly to the river, in order to protect the treatment plant, especially during or after high intensity rainfall events, leading to a high infiltration rate of ground water to the sewer system. Also, during power cuts, all inflow is directed to the river, as plants usually do not have emergency power supply facilities. In addition, deteriorating economic conditions have an adverse impact on the planned maintenance activities for regular supply and repair of mechanical/electrical equipment, resulting in decreased plant capacity due to breakdowns of equipment. All these reasons lead to a significant contribution of unaccounted pollution from sewage treatment plants.
- Irregular collection of solid waste – The shortage of fuel and vehicle spare parts recently has crippled the management and timely collection of waste in the City of Harare. For instance, in the City centre the refuse collection has become inconsistent, resulting in the piling of litter around refuse containers. Collection of refuse in residential areas at times is done once per fortnight.
- Regulatory instruments – The existing regulatory instruments to control storm water pollution are satisfactory, but there are difficulties with their enforcement by the municipal authorities. This is due to an insufficient staff establishment and in some cases to political interference.
- The attitude of people towards litter – Cases of littering and polluting streets and open spaces are common practice, indicating the fact that people are not aware or are disregarding the causes of such type of behaviour. In the city center, street vending has become a prevalent practice due to economic hardships, resulting in littering and deterioration of the urban hygiene. The public's environmental consciousness is relatively low. In some cases, bus operators wash/repair their buses outside garage bays due to shortage of space for the fleet in garages or due to lack of knowledge that their activities will pollute their drinking water supply source. Illegal and causal dumping of waste at night is also prevalent.
- Public opinion polls – Results of polls to measure public attitudes continue to show wide support for pollution control measures implementation, but strong economic arguments are decreasing the political will to impose the cost of pollution control and abatement. Some industries were willing to implement pollution abatement measures, e.g. as the construction of particle separators or oil traps but they did not have the expertise to do so and the market does not offer ready products for this purpose.
- Financial back up – At present, the council does not have special funds designated for storm water management, and the control of diffuse pollution is executed as part of the activities for the pollution control program, but it is considering the implementation of the "polluter-pays" principle to fund pollution control measures. Government has been reluctant to impose land use controls on private property as the sanctity of private property is a vital part of freedoms in society and is a strong emotional issue. However, the recently introduced Environmental Management Act (2002) section 114 addresses this

weakness. An order may be served to the owner, occupier or user of any land, in cases of water pollution, including storm water, removing and disposing of litter, refuse or waste from any land or premises. The introduction of such regulatory instrument necessitates the use of better land use management practices and would require changes, which often, are not popular with landowners.

5.2 *Management decisions and water quality status*

Under the integrated approach to the water resources management practice and pollution control, a wide spectrum of activities should and could be envisaged as part of any pollution abatement program, but they should be viewed in their inter-relation and mutual interdependence. In general, we could sub-divide them in two major groups: prevention (non-structural) or pro-active measures; and pollution abatement by detention and treatment, also called reactive (structural) measures. As emphasized before, prevention measures should have precedence and priority with respect to diffuse pollution control. The information, presented in the previous section, shows that the managing authorities understand the need for such an approach and are aware of the major sources and consequences of diffuse pollution. They also identified the need for a more detailed survey of diffuse pollution from high-density urban areas, landfills and informal sector, with respect to the urban surface and ground water quality. Results of specific case studies, investigating such types of diffuse pollution sources are presented in subsequent chapters. However, it should be mentioned that the need for acquiring reliable information on the water quality status of natural and man-made bodies, in order to support the decision making process, is of high priority.

Water resources management is an important issue for any country, but in the Southern African region it is even of greatest importance considering the scarcity of this resource and the underdeveloped nature of significant parts of this geographic region. For this reason, the need to make informed decisions regarding the optimal use of available resources is essential. Diffuse pollution regulation and abatement requires a water quality objective orientated approach of the legislative structure, defining the main qualitative objectives to be achieved with respect to the beneficial use of each specific water body (or section of it), based on a catchment principle. The formulation of the water quality objectives is closely related to the quantitative aspect (available water resources, demand estimation and allocation for different purposes and to different consumers). Also, it is related to the type of beneficial use, e.g. a water body, which is designated for the purpose of irrigation only, does not need to have restrictions regarding nutrient concentrations. In addition, surface water bodies in the specific cases of large rivers, lakes or dams, have a considerable natural capacity to self-purify and to restore their water quality. They could assimilate a given pollution load in addition to the one present naturally. Correspondingly, the correct formulation of the water quality objectives and corresponding beneficial use of water resources would depend on the correct estimation of the pollution loads, supplied to the corresponding water body and the assessment of its assimilative capacity. These are site-specific parameters, depending on different factors (geographical, social, economical, etc), which could not be acquired easily from literature sources or adapted from similar case studies. Therefore, it is in the interest of each nation to have its own data sets, representing quantitative and qualitative parameters characterizing the status of water resources. Such data should be obtained based on regular and continuous monitoring programs, with specific objectives and quality assurance tools. Unfortunately, the need for reliable water quality data is usually underestimated in the countries of the region and often is not considered during the process of water resources management. An example is the survey, reported in the previous section, where the lack of reliable water quality data has not been mentioned as a problematic issue by the city of Harare authorities, with respect to diffuse pollution control. As a result of such underestimation, supply orientated management decisions could lead to expensive solutions, such as the construction of new technical structures-dams, long water transportation systems, treatment facilities, etc. In other cases, the lack of qualitative data regarding the status of water resources could cause severe public health or environmental problems associated with the polluted water resources.

It was mentioned in Chapter 3 that the monitoring program regarding storm water quality in Harare has been reduced drastically since 1997 and after the year 2000 is almost non-existent, due to the political, economic and financial difficulties experienced during this period. In such conditions, the correct estimation of pollution loads and fluxes to the Lake Chivero is very difficult and unreliable. The problem is aggravated by the lack of uniform methodology with respect to the estimation of quantitative and qualitative data. As a result, the contribution of pollution loads from storm water runoff in Harare has been estimated at very different scales in different studies (Nhapi et al. 2001, Magombeyi et. al. in press).

6 CONCLUSIONS

The storm water quality from two major channels in Harare could be classified as diffuse source of pollution to the watercourses with respect to TP, TSS, TDS, ammonia, COD, Cd and Pb. No statistically significant spatial variation of pollutants was found regarding SP1, SP3 and SP4. The highest pollution loads were associated with the industrial area (SP2), where the risk to the environment could be evaluated as medium to high hazard, while the rest of the urban areas present a low hazard risk, according to the WWEDR (2000).

Variation of constituents in a single storm event show that the first flush effect was associated with contaminants bounded to particulate material and the EMC concentrations regarding these parameters differ substantially from the mean concentrations.

The specifics of the control and management of diffuse pollution sources require an event-orientated water quality monitoring program. The determination of typical storm events, representative of the case of diffuse pollution transport, with corresponding hydrographs and event mean concentrations, reflecting seasonal variations could help significantly the accurate determination of pollution loads and could reduce the need of monitoring frequency. It could be recommended that such data could be collected by well-defined and planned research projects, to complement a regular and continuous monitoring program. The use of automatic samplers at selected locations could improve the quality and reliability of the results and the cost of the monitoring process.

The lack of a uniform methodology for pollution loads estimation could lead to serious errors in the process of assessment of diffuse pollution and prioritizing abatement measures. It could be recommended that:

- A uniform methodology for pollution loads estimation should be developed and implemented in order to evaluate them more accurately, to compare results from different spatial and temporal data sets and to support the decision – making process.
- Diffuse pollution should be managed in the context of the whole catchment area;
- A reliable water quality monitoring practice should form the basis of the diffuse pollution management process.

Acknowledgements – The Belgian Technical Cooperation provided a scholarship to M. Magombeyi during this study. The cooperation of the city of Harare and the ZINWA is acknowledged with thanks, as well as the help and support provided by the research assistants and technical staff of the Civil Engineering Department, University of Zimbabwe.

REFERENCES

American Fisheries Society (AFS) 2000. AFS Policy Statement no 3: Non-point source pollution. In: www.fisheries

Australian Water Resources Council (AWRC) 1981. Characterization of pollution in urban storm water runoff. *Technical paper No.60*. Department of Natural Development and Energy, Australia.

Buffleben, M.S., Zayeed, K., Kimbrough, D., Stenstrom M.K. & Suffet, I.H. 2002. Evaluation of urban non-point source runoff of hazardous metals entering Santa Monica Bay, California. *Wat. Sci. Tech.* 45 (9), 263 – 268

Chadwick, A. & Morfett J. 1993. *Hydraulics in Civil and Environmental Engineering*, UK: Spon Press, Chapman Hall.

Chapman, D 1998 *Water Quality assessments. A guide to use of Biota, Sediments and Water in Environmental Monitoring*. 2nd ed. London: Spon Press.

Choe, J.S., Bang, K.W. & Lee, J.H. 2002. Characterization of surface runoff in urban Areas. *Wat. Sci. Tech.* 45 (9), 249-254

Ellis K.V. 1989. *Surface water pollution and its control.* London: MacMillan Press
Environmental Management Act (EMA), 2002, Chapter 20:27, Republic of Zimbabwe Government.
GAO (2001). Report to Congressional Requesters: Water Quality: Better Data and Evaluation of urban runoff programs needed to access effectiveness. www.gao.gov/new.iterms/do1679pdf
Gromaire-Mertz, M, C., Garnauds, S., Gonzalez F, & Chebbo, G. 1999. Characterization of urban runoff pollution. *Wat. Sci. Tech.* 39 (2), 1-8.
Heather, B., Kajese, T & Koro.E. 1996 Position paper: Report on pollution issues of Lake Chivero and catchment. Harare, Environment 2000.
Hranova, R., Gumbo, B., Klein, J. & van der Zaag, P. 2002. Aspects of the water resources management practice with emphasis on nutrients control in the Chivero basin, Zimbabwe. Physics *and Chemistry of the Earth,* 27, 875-885
Jarawaza, M. 1997. Water quality monitoring in Harare, a review. In*:* Moyo, N.A.G. (Ed), *Lake Chivero: a polluted lake.*
Harare: University of Zimbabwe Publications.
Lamb, J.C. 1985 *Water Quality and its control.* New York: John Wiley & Sons.
Lau, S.L, Khan, E. & Stenstrom, M.K. 2001. Catch basin inserts to reduce pollution from storm water diffuse pollution *Wat. Sci. Tech.* 44 (7), 23-34.
Magombeyi M., Love D., Hranova R. & Hoko Z. (in press) Storm water quality in the City of Harare – implications for managing storm water pollutants in the urban water cycle In: *Water Management for Sustainable Economic Development – Managing the Urban Water Cycle,* Proc. IWA Specialist Group Conference on Water and Wastewater Management for Developing Countries – Victoria Falls, 28-30 July 2004
Marshall, B.E. 1997. Lake Chivero after forty years: impact of eutrophication. In: Moyo, N.A.G. (Ed). *Lake Chivero: a polluted lake.* Harare: University of Zimbabwe Publisher.
Mvungi, A., Hranova, R. & Love, D. 2003. Impact of home industries on water quality in a tributary of the Marimba River, Harare: implications for urban water management. *J. Physics and Chemistry of the Earth* 28, 1131 – 1137
Nhapi, I., Siebel, M.A. & Gijzen, H.G. 2001. Dry season inflow and export of nutrients from Lake Chivero in year 2000. *Proceedings of the Zimbabwe Institution of Engineers*, **2**, 33-41.
Niemczynowicz, J.1999. Urban hydrology and water management-Present and future challenges. *Urban water* 1, 1-14.
Novotny, V. & Olem, H. 1994 *Water Quality: Prevention, identification and Management of Diffuse Pollution.* New York: Van Nostrand Reinhold.
Novotny, V. 2003. *Water Quality: diffuse pollution and watershed management.* New Jersey: John Willey & Sons.
Paul, M.J. & Meyer, J.L. 2002. *Streams in the Urban Landscape.* Athens (USA): University of Georgia press.
Sansalone, J.J. & Buchberger, S.G. 1997.. Partitioning and first flush of metals in urban roadway storm water. *J. of Environ. Eng. Div. ASCE,* 123 (2), 134-143.
Schueler, T. 1987. *Controlling runoff: A practical Manual for Planning and Designing Urban Best Management Practices.* Washington, DC : MWCOC.
Schulze, R.E.; Schmidt, E.J., & Smithers, J.C. 1993. SCS-SA User Manual: PC – Based SCS design flood estimates for small catchments in Southern Africa. In *ACRU Report No 40*, University of Natal, South Africa.
Standard methods for the examination of water and wastewater 1989. 17th Ed. American Public Health Association/ American Water Works Association/Water Environment Federation, Washington DC.
United States Environmental Protection Agency (USEPA) 1995. Guidelines for preparation of the state water quality assessments. *Report 305b, EPA 84, B-95-001.* Washington DC
WWEDR 2000. *Water (Waste and Effluent Disposal) Regulations.* Statutory Instrument 274 of 2000, Republic of Zimbabwe.

CHAPTER 5

Diffuse pollution in high-density (low-income) urban areas

R. Hranova

ABSTRACT: Diffuse pollution problems typical for low-income residential areas have been discussed in terms of sources, monitoring, regulation and control. The problem has been illustrated by a case study, where the water quality of a natural stream was assessed to evaluate the impact of runoff from storm water generated in low-income residential areas in Harare, Zimbabwe during the 2001-2002 wet season. Results show considerable pollution load with respect to TP, ammonia, Fe and Pb. An evaluation of pollution fluxes along the stream, generated by different types of land use patterns within the study area, showed that the major flux is associated with the dumping of solid waste along the streambed. The analysis of appropriate practices for monitoring, control and abatement of diffuse pollution in such areas is presented and emphasis is made on the need to implement source control and pollution prevention measures, and also to incorporate community education and public involvement activities as part of pollution abatement programs.

1 INTRODUCTION

The characteristics of low-income urban developments in African countries have followed some common patterns, connected to the general economic and social environment. The development pattern, described with respect to Harare, could be accepted as similar to other countries in the region as well, although specific flavor and national characteristics and conditions would lead to differences in each specific case. The link between the economic and social conditions and the pattern of urban development should not be underestimated. In a number of cases, problems, associated with the economic and social status of the society will be reflected in the landscape of urban development schemes and would be closely associated with problems in water resources management, as well as, undesirable public health and environmental impacts.

In Harare, during the 1980's, low-income residential areas were established with uniform houses erected. The standard plot size was 324 m^2, though some earlier schemes utilised smaller plots (Rakodi 1995). However, pressure on space and utilities in these areas, have led to a situation, where many more people often occupy the houses that were originally designed for one family. In 1990, the number of illegal buildings had increased, with the average number of informal buildings per plot variously estimated between three and six. In Highfield for instance, which is the second low-income township to be built, local plan survey revealed a total of 15,230 residential structures of which nearly half have been built without permission (Rakodi 1995). The overwhelming concentration of new development to the southwest and west of the city has reinforced the existing pattern of overpopulation. The growth of these high-density urban environments has remained steady over the years. One major issue has been the emergence of the informal sector (IS) of the economy of the country, which includes activities such as informal agriculture, hawking and petty commodity production. (Drakakis-Smith & Kirell 1990). Population growth and distribution, as noted above, have significant roles to play in the sustainability of the world's vast resources. The number of people, but also the life style, consumption patterns, and the region people inhabit and utilize, directly affect the environment. When we look at the impact of human activities, the situation is more complicated due to the diversity in consumption patterns and societal habits worldwide.

Unlike the notions of the contemporary monopolistic and competitive firm or market, the concept of the IS or informal enterprises lack a similar compelling theoretical grounding. We know the IS when we see it, but its conceptualization for economic analysis and policy purposes has had a number of pit-falls. A generally accepted IS definition is that provided by the International Labor Organization (ILO), which defines the term as "very small scale units producing and distributing goods and services, and consisting largely of independent, self-employed producers in urban areas of developing countries, some of whom also employ family labor and/or a few hired workers or apprentices; which operate with very little capital, or none at all; which utilize a low level of technology and skills; which therefore operate at a low level of productivity; and which generally provide very low irregular incomes and highly unstable employment to those who work in it" (ILO 1996).

IS producers and workers are "generally unorganized.... and in most cases beyond the scope of action of trade unions and employers, ... they generally live and work in appalling, often dangerous and unhealthy conditions, even without basic sanitary facilities, in the densely populated towns of urban areas (ILO 1996). For instance, in Zimbabwe, the urban informal sector consists of establishments that primarily entail self-employment with the addition of one or two helpers who are often family members. The urban informal sector in Zimbabwe has historically been relatively small, both in terms of size and in terms of its role and status in the economy, especially compared with its role and status in West Africa (Mhone 1995). In 1980, when the independence of the country was proclaimed, the IS absorbed about 10% of the labor force. However, with the post-independence growth in the labor force and in the face of the stagnating formal sector, the IS absorbed more than 25% of the labor force by 1991, with a rapidly increasing trend (ILO 1996). The continued growth of IS in Zimbabwe was further propelled by secular economic stagnation that has afflicted the formal sector from the late half of the 1980's up to the present. In Zimbabwe, as in many developing countries with ailing economies, induced urban informal sector activities have proliferated with the intensification and persistence of the economic crisis.

The previous chapter emphasizes the need to study the urban runoff quality and has discussed data and management practices with respect to the CBD, medium – density residential areas, commercial and industrial sectors of the city of Harare. These types of land use practices reflect a well-developed and maintained urban landscape, which generates the runoff. This chapter focuses on high-density areas, which are predominantly populated by the low-income part of the city's population. This type of land use pattern consists of a residential development scheme, with corresponding basic infrastructure (road, electricity, water reticulation and sewerage). However, the accelerated pattern of urban growth has led to a considerable overpopulation of the existing housing developments together with the construction of illegal buildings, which resulted in a significant overload of the existing urban infrastructure. It is most pronounced in the following directions:

- The maintenance of road surfaces is unsatisfactory, with numerous potholes, which increases the erosive capacity of runoff and the transport of suspended material into the drainage system;
- Blockages and overflow of the sewer system are common practice, as reported in Chapter 4;
- Poor solid waste management practice, resulting in the spreading of litter over public places and depositing it into drainage channels, thus enhancing the risk of diffuse pollutants transport;

The drainage system consists of open ditches along the roads, usually with grass cover, which discharge at convenient locations to existing streams and natural depressions. There are no large drainage channels to collect and accumulate the runoff; correspondingly, the pollution generated is of diffuse pattern in terms of both-spatial and temporal variation.

The study focuses on the evaluation of stream water quality, which receives the runoff from low-cost and high-density urban residential areas. For this purpose, a study area in Kuwadzana, Harare, has been investigated. An attempt has been made to evaluate the influence of different types of land use patterns, within the study area, including residential housing, plots designated for home industries (as a representative example of IS economic activities), open spaces, informal agriculture plots. Specific attention is given to the management approaches and corresponding abatement measures that could be applied to control and reduce diffuse pollution in such types of urban developments, which could be regarded as typical for many developing countries.

2 REGULATION AND ASSESSMENT

2.1 *The informal sector of the economy*

The emergence of the home industries activities which forms part of the IS was among other factors caused by the location of about 20 ha of land, designated for low-income housing development, established abut 2.5 km southwest of the urban center. However, the development of municipal housing in the designated areas did not keep pace with the population growth within the municipality as a whole. Major low income housing schemes include Glen View (the development started shortly before independence), Warren Park and Kuwadzana located at 8 and 12 km respectively in the direction to the west of the city center. Others include Budiriro to the southwest between Glen Norah and Hatcliffe, about 15 km to the North of the city center. The expansion of this type of urban development was not linked to a corresponding expansion in the public service and the economic sector in general, and resulted in a failure to secure employment for the increased population of the city (Zinyama 1993). The increasing unemployment rate and the diminishing prospect for formal sector jobs lead to marginal employment activities such as those described as home industries, as the only survival strategy for a large number of the population.

GERMIN (1991) reported that the activities included in the IS and forming the basis of the home industries, consisted of textiles, food, beverages, tobacco, leather production, wood and wood processing, construction, furniture and other manufacturing and micro enterprises. The vast majority of the activities were carried out by women, and consisted of knitting, crocheting and sewing. Males dominate the more complex and relatively larger activities such as carpentry, garages, metal works, brick-making and construction. The medium and small enterprises could roughly be divided into three broad groups. The first group accounts for 70% of all people involved in the urban IS and home industries, and is engaged mainly in grass cane bamboo processing and beer brewing. It includes the bulk of under-employed people, whose enterprises provide a temporary income, thus generating a safety net in respect to live survival. The second group, consisting in about 20% and is mainly involved in construction, retail trade of grocery and welding. The third group accounts for about 10%, enjoys a relatively substantial annual profit, and includes activities such as printing, furniture repair, bottle stores, flour mills, general trade, retail hardware and garments, auto works such as garages and other repairs.

Considering the IS as a possible diffuse pollution source, it could be mentioned that the vast majority of the activities do not generate a considerable amount of waste associated with the manufacturing process, but could be regarded as diffuse pollution source due to improper storage of materials and spreading of residues over the area. Specific attention with respect to pollution abatement measures should be given to garages, leather processing and brick-making activities.

2.2 *The regulatory practice*

Regulatory instruments, such as standards, which are enforceable in the legal environment, usually are designed to address the status of water resources at a national scale, and as such do not include special comments or requirements with regard to low-income areas. Typically, such areas form part of the urban environment and the regulation of diffuse pollution from such areas should be incorporated in the local authorities by-laws and similar regulatory instruments. It is strongly recommended that diffuse pollution control and abatement in low-income areas should be included as one integrated and indispensable part of storm water management and environmental protection programs at local level. Central and regional authorities could support such activities by helping with expertise, and in some cases, with financial support. However, the local community and its management structure should be the developer, owner and executor of such types of programs.

It has been highlighted that diffuse pollution through runoff from these areas is dispersed in terms of both, spatial and temporal variation; therefore, its monitoring and control should be orientated towards the water quality of the streams, rivers or other natural water bodies, which are the interceptor of runoff from such areas. Such type of approach requires a corresponding regulatory basis, orientated towards the water quality of the receiving body. The so-called "clean" water, or in other words, the acceptable standard of water quality, varies depending on the use of the water. The water quality objectives or guidelines, with respect to a natural water body, could be defined as: "the desirable level of water quality to be attained and maintained in receiving waters". They should be formulated, considering the different beneficial uses of the water body, such as municipal and industrial water supply, agriculture, recreation, aesthetics and the propagation of wildlife. The purpose of all pollution control and abatement measures should be aiming at the attainment of these objectives, together with the protection of the water body assimilative capacity, shell and fin fish, wildlife, and the preservation and/or restoration of the aesthetic and recreational value of natural waters.

The regulation of diffuse pollution in the USA is based on the established non-point source management programs under section 319 of the Clean Water Act (USEPA 1995). This approach helps the states to address non point sources, and runoff pollution by identifying water affected by such pollution and adopting and implementing management programs to control it. Such programs recommend where and how to use the best management practices (BMPs) in order to prevent runoff from becoming polluted. In the case of identified pollution levels, the programs recommend specific measures to reduce the amount of pollution that reaches the surface waters. Some of the BMPs include:

- Proper planning to store and dispose materials;
- Erosion prevention by planting fast growing annual and perennial grasses, to shield and bind the soil;
- Specific requirements for urban planning and development, encouraging the introduction of pervious pavements and environmental buffer zones;
- Installation of detention ponds and sand filters and the use of temporary check ditches to divert runoff away from the storm drains.

In addition, a receiving water quality approach has been recommended, which links the specified receiving water quality objectives to the recommended beneficial water use and accounts for the assimilative capacity of the receiving water body. A permissible pollution load (Total Maximum Daily Load or TMDL), which includes point and non-point pollution sources is evaluated and allocated to the water bodies, which are controlled and protected. The consideration of the loads from all point source permits in the catchment of this specific water body, together with all non-point pollution sources, should not exceed the TMDL. It must be emphasized, that such an approach requires legal intervention at national level. Although storm water and environmental management programs are created and executed at local level, national regulatory instruments would be the driving force and would place the framework for such activities. However, as previously mentioned in Chapter 2, the practical implementation of this or similar approach is costly and requires an extensive data base, together with high technical background for its implementation. Its applicability for developing countries could be a point of discussion and most probably a long term and phased implementation strategy should be adopted.

In South Africa, a similar trend has been followed, and the regulatory instrument applied is the National Environmental Management Act (NEMA) (DWAF 1996). It recognizes in one of its principles on water resources that the quantity, quality and reliability of water required for beneficial use, together with the ecological function of water bodies must be recognized and attained. Moreover, after the introduction of the "White paper" in 2000, the integrated water resources management policy was clearly stated. The policy, among other things, aims towards the regulation of pollution and a proper management of the receiving environment. However, at the level of practical application, most of the regulations are in the process of development.

Zimbabwean pollution control regulations are discharge orientated. The latest revisions consider pollution from runoff and suggest that both the owner and the user of an artificial storm drain require a permit for discharge into any surface water body "whether directly or through drainage" (WWEDR 2000). This implies that the main responsibility for the reduction of diffuse pollution lies with the local urban authority and at the same time does not make provision to identify pollution of diffuse origin, as in the case of low-income urban areas, or agricultural pollution, where no specific discharge point could be identified.

The application of the water quality objective approach in the regulatory practice would rely heavily on the site conditions, and the specific geographic, climatic, economic and social factors. For example, in many African countries, the vast majority of the smaller rivers and streams are ephemeral and during the dry season, the flow, which could be found in them, is due to point effluent discharges. Under such circumstances, they could be treated as natural channels for the conveyance of effluent at a different level of treatment. It is clear, that in such cases, we cannot rely on the natural assimilative capacity of the body, and that an effluent orientated approach would be feasible. In cases of larger surface water bodies, a water quality orientated approach for protection and pollution prevention, together with assessments of their assimilative capacity would be feasible. Regulatory documents with respect to ground water resources should account for the fact that their assimilative capacity is very limited and should be orientated towards minimum or zero pollution loads.

Many of the small-scale ephemeral streams and rivers, which are located downstream of urban population centers or industrial enterprises, are used by rural population for bathing, drinking and washing activities, directly and without any pretreatment. This scenario could be met very often in many African countries and examples are given in Chapters 8 and 11. In such cases, the effluent discharge water requirement should meet the WHO criteria. This requirement makes the effluent treatment cost very high and could influence the development of the whole region. In such conditions a balanced solution with respect to regulatory instruments, involving compromises from all interested parties, together with a wide public and all stakeholders' involvement is mandatory.

2.3 *River water quality monitoring and data collection*

Monitoring data regarding runoff water quality from high-density housing developments is very limited in the region. Considering the low economic status of such types of areas, and the lack of large runoff collection channels, makes such a monitoring exercise unfeasible and costly. However, the potential for diffuse pollution from such areas is high. Therefore, monitoring programs should be orientated towards the examination of the streams and rivers, which collect the runoff from such areas and compare the results with control points, which have water quality characteristics close to background pollution values. In order to minimize the monitoring efforts, the location of such points should be chosen carefully and should concentrate on major streams, collecting runoff from larger areas. In numerous cases, the monitoring stations, which represent large drainage areas, might include not only the impact of low-income areas but also the contribution of other types of land use practice and point source discharges as well. Consequently, the specific contribution of runoff from low-income areas might not be possible to be identified and separated from the whole load.

The previous chapter discussed the need to introduce the concept of EMC and site-specific pollutant concentrations with respect to typical storms. In cases of drainage channels, conveying storm water only, the EMC represents an averaged value for a selected storm event, and as such defines clearly the contribution of pollution from storm water regarding a selected parameter. However, when the estimation of EMC of rivers and streams for a selected storm event is needed, the determined EMC value at a given location would represent not only the contribution from runoff, but also, the contribution from:

- Natural sources (background pollution). It would have two components – natural pollution of base flow from ground water recharge and natural pollution from materials present in the river bed, which have been contributed from the surrounding environment or grow naturally;
- Materials dumped into river beds by human actions;
- Effluent discharges from point sources up stream the sampling location.

During the dry season (low flow conditions), EMC would represent only the input from background pollution and point sources upstream the sampling location, but the term "event" needs to be well understood. In this case the "event" does not refer to a rainfall event, but to low-flow events with a corresponding return period e.g. one year return period would present results regarding the quality of low flows, which could be expected to occur each year, while 10 years low flow could be expected to occur once per 10 years. The duration and frequency of the sampling procedure and the return period should be chosen and planned accordingly to obtain representative results. In most cases it could be expected that dry weather EMC would be higher than the wet weather EMC at the same location, with respect to the vast majority of pollutant constituents, except for suspended solids. These values should be used for comparison with prescribed limits and regulations or for evaluation of toxicity effects.

The EMC multiplied by the corresponding average flow rate for the event would represent a pollution flux for this specific event. An accurate estimation of the annual pollution load contributed by a river or stream, would be the sum of all fluxes calculated on a daily basis. Such a procedure for the estimation of annual pollution loads would require extensive data sets regarding flow rates and corresponding EMC, which could be collected only if specific monitoring stations are available and equipped with automatic measurement devices, which could provide data in respect to flow rates and the corresponding pollution constituents. The former could be measured in-situ or by laboratory tests. This is an expensive option, especially in the case of pollution reduction and prevention in low-income areas.

Another, less expensive option in terms of monitoring efforts, which requires a continuous flow rate measurement and ordinary approach to water quality monitoring, could be applied with some compromise in terms of accuracy. It would consider the sum of seasonal pollution loads, where the different seasons would be determined based on the determined seasonal variation of EMC. In the case of the South African region, a significant difference in seasonal EMC variations during the dry and wet season could be expected. Within the wet season, differentiation could be made with respect to average values of EMC during the starting period of rains, usually in October and November, and during typical wet conditions, with respect to typical annual rainfall events. An example of possible seasonal differentiation could be done as follows:

- Dry season – from May to September with an average seasonal EMC equal to EMC_{dry} and a total flow volume = Vol_{dry}
- Start of the wet season – October – November – with an average seasonal EMC equal to $EMC_{s.\,wet}$ and a total flow volume = $Vol_{s.\,wet}$;
- Wet season – December to April-with an average seasonal EMC equal to EMC_{wet} and a total flow volume = Vol_{wet};

Under such an approach, which could be referred as "seasonal" approach for annual pollution load (PL_{annual}) estimation, the load could be calculated as:

$$PL_{annual} = Vol_{dry}\ EMC_{dry} + Vol_{s.\,wet}\ EMC_{s.\,wet} + Vol_{wet}\ EMC_{wet} \tag{5.1}$$

Obtaining data for the application of the seasonal approach would require the design of a monitoring program based on a monthly frequency of the sampling occasions. The EMC_{dry} should be determined as the averaged value of the results during the dry months. The determination of each one of the monthly EMC values during the wet season should be determined during and/or immediately after rainfall events, catching the variation of concentrations along the hydrograph. The most convenient approach would be the collection of flow – integrated samples during the whole duration of the hydrograph, where the results of the laboratory tests would represent the corresponding EMC. During the start of the rain season, the frequency of the sampling occasions could be increased to achieve a representative average value. If results after several years of monitoring, show no statistically significant difference between $EMC_{s.\,wet}$ and EMC_{wet}, the monitoring program could be re-designed and only two seasonal values could be used for the estimation of PL_{annual}.

The seasonal approach is relatively simple and less expensive. However, the operation and maintenance of flow rate devices and a regular monitoring program might not be economically feasible for low-income areas. It would be advisable, to establish and properly operate such monitoring stations at important sections of rivers and streams, after major pollution sources or at confluence points to lakes/reservoirs with high water quality requirements, based on their beneficial use. Under such circumstances, the monitoring program should include additional water quality monitoring stations, after major discharges of effluent or storm water channels. A control point, reflecting background water quality in the upper reaches of the stream, which is relatively unaffected by human activities is essential. For this specific location, the frequency of sampling could be lower, as no high variation of the water quality is expected. The availability of background pollution data, in addition to data regarding the quality and quantity of point discharges, would allow the estimation of the contribution of the combined diffuse pollution load from the area under consideration. Also, it would allow:

- Evaluating trends and variations in the status of the river water quality;
- Differentiating between the contributions of point and diffuse pollution sources and;
- Estimating the positive results of any water quality management program in the River's catchment, with respect to point or diffuse pollution sources.

The data record and storage of such type of monitoring programs should include meteorological data, related to the determined values of EMC, which would complement and help the data analysis process and the results interpretation.

For specific stretches of urban rivers and streams flowing through low-income areas, the application of bio-monitoring methods (Chapter 12) would contribute to the evaluation of the environmental status of the water body, and could compliment the data from monitoring programs reflecting the physical, chemical and microbiological status of the stream.

3 WATER QUALITY ASSESSMENT OF AN URBAN STREAM

3.1 *The study area*

Part of the Kuwadzana high-density suburb was chosen as the study area for this investigation because most of the home industries are located in a centralised, especially designated plot. Thus, the area accessibility and the location of the residential areas and the home industries plot close to a natural stream of water, made the study area enough representative and easier for investigation and sample collection. The study was performed during the wet season 2001-2002.

The Kuwadzana high-density suburb is located in the southwest part of the city of Harare. The area comprises of about 13 km^2. The high-density suburb, which was first occupied in 1984, consists of housing density approximated to 11.2 dwelling units per hectare, with a total of 110 persons per hectare. The residents include mostly a low-income generating group, living in a moderately dense populated area. The people engage themselves in informal sector activities such as the home industry establishments and it was approximately estimated that about 30% of all residents rely on such activities as their only source of income. The majority of the home industries enterprises were set up in the late 1980's and especially in the 1990's when ESAP (Economic Structural Adjustment Program) was introduced. Table 5.1 shows the general growth trend of the home industries in Kuwadzana since the 1980's, based on Kuwadzana District Council records, which were studied during 2002.

The major activities, which are incorporated into the home industry area are: automotive service garages, furniture making, panel beating, painting, milling machine operation (food processing), brick making and

Table 5.1. Home industries growth rate in Kuwadzana.

Year	Number of home industries
Before 1980	0
1981-1985	9
1986-1989	11
1990-1993	25
1994-1997	48
1998-2001	82

leather craft. In addition, street vending and small-scale urban informal agricultural activities are a widely spread practice.

The study area forms part of the Marimba River catchment, which has a total of 189 km^2 and is a sub catchment of the Manyame River, as presented in Chapter 3. The geological formation of the study area mainly comprises four major rock formations. The eastern part of the area comprises course to medium grained, quartz-rich clastics with preserved or remnant sedimentary textures. The western part that contains the home industry comprises Mashonaland dolerite. The northern part consists of Metadacite of Passfor formations (Baldock et al. 1994). Based on the classification system proposed by Nyamapfene (1991), the major soil types in Harare are kaolinitic. They are moderately to strongly leached soils. Clay fractions are mainly inert, together with appreciable amounts of free sesquioxides of iron and aluminium. The soils are either classified as fersiallitic (mixed clay) or paraferrallitic (inert clay) or orthoferrallitic (very inert clay) soils. The important consideration is that all these formations are rich in ferro-magnesium minerals and this results in clayey soils that are red, reddish brown to yellowish red in the drained areas.

In order to obtain information regarding the specific patterns of land use practice, the storm water drainage system and the general condition of the infrastructure, two reconnaissance visits to this specific area and other possible alternative options were done. It was found that most of the high-density suburbs have similar physical set up with the location of the home industries along the streams nearby. The types of the informal sector activities were very much alike in the different suburbs visited and correspond to the description made above. During the second visit, the natural drainage pattern was identified by the use of topographic series maps of Harare TR8027 and TR8025. The major drainage channels from different land uses such as residential only, residential and part of home industries, residential and runoff from informal agriculture were identified, with corresponding drainage areas. Also, it was observed that the adjacent stream to Kuwadzana 1 home industry area was flowing during typical dry season conditions (after 6 months of a dry spell), indicating the presence of perennial flow. This suggests that the presence of base flow is due to groundwater recharge. An approximate estimation of the base flow rate was done by the application of the velocity-area method (Mays 2001). Physical investigation and the geology of the area confirmed the perennial nature of the stream due to base flow. In addition, it was observed that due to improper solid waste collection practice, the stream was used as a dumpsite for solid waste of predominantly inorganic character. Old utensils as cooking pots, paper, plastics and rubber waste, old clothes and a variety of other waste material could be found along the stream banks and within the stream bed and the flood plain. An illustration of the stream is shown in Figure 5.1.

As result of the recognisance visit the following types of land use patterns were identified:

- Formal Housing – This is the original land use pattern established by the municipality. It consists of low cost houses that are provided with a central water supply system and sewerage. However, houses formally designed for up to six people now accommodate more than twelve people. Due to the existence of an acute shortage of housing and the high cost of renting in the formal housing developments, some landlords have illegally constructed small adjacent houses also known as "shacks" for leasing. These informal housing establishments, within the formally developed areas, do not have adequate water and sewerage collection systems and this fact, in most cases, results in illegal discharges of raw sewage to the existing drainage system and nearby water bodies. An illustration of a formal low cost house together with informal "shacks" is shown in Figure 5.2.
- Commercial Establishments – These represent the markets and any trading activities within the residential suburb. In most cases, the market places consist of unregulated shades, and are the source of considerable amount of solid waste, which is not collected and is spreading along the area and the drainage channels, which are the favourite dumping sites.
- Schools – These usually have a reticulated water supply and proper sewerage systems. Solid water management is again a problematic issue but it is handled at a higher level in schools and other educational facilities. In Zimbabwe, most schools comprise large open grounds, with corresponding areas for recreation, gardens and sports facilities, which, however, are susceptible to pollution originating from the improper disposal of litters.

Figure 5.1. The Kuwadzana 1 streambed near the "Home industry" plot.

- Urban informal agriculture – This practice has spread widely during the last twenty years and consists of the informal and unauthorized use of open spaces within the municipal area, for the purpose of small-scale farming activities, predominantly maize growth, which would complement the family budget. The farming depends entirely on rainfall. The need of water for irrigation purposes makes the stream banks preferred places for informal agriculture and at suitable locations they are occupied first. The preferred crops consist of a variety of traditional vegetables, maize and potatoes.

Figure 5.2. A low-cost house with informal additional "shacks" in Kuwadzana 1.

Figure 5.3. Part of the "Home industries" plot in Kuwadzana 1.

- Home Industry Establishments – These comprise specially designated plots for small enterprises, established formally by the local authorities and located close to the residential areas. Such plots are usually supplied with several taps for clean water, and do not have connections to the sewerage, thus providing wide options for the pollution of the adjacent streams. For this specific study area, the survey identified the major enterprises established in the home industry plot at Kuwadzana 1. Data with respect to the different types of enterprises (Table 5.2) was collected during the study, based on personal communication with the owners and the counting of the different small enterprises within this specific home industry plot. A typical illustration of such a type of land use practice is shown in Figure 5.3.

3.2 *Methodology*

The location of the study area is shown schematically in Figure 5.4, including the different types of land use patterns. The selection of the sampling points location was done, based on the different types of areas drained. Point F represents the contribution from the informal agriculture. Point C is representative of the

Table 5.2. Type and number of home industries.

Type of home industry	Number
Garages	21
Motor spares & electrical related	9
Panel beating	7
Brick making	8
Maize milling	15
Painting	6
Salon	15
Leather craft	8

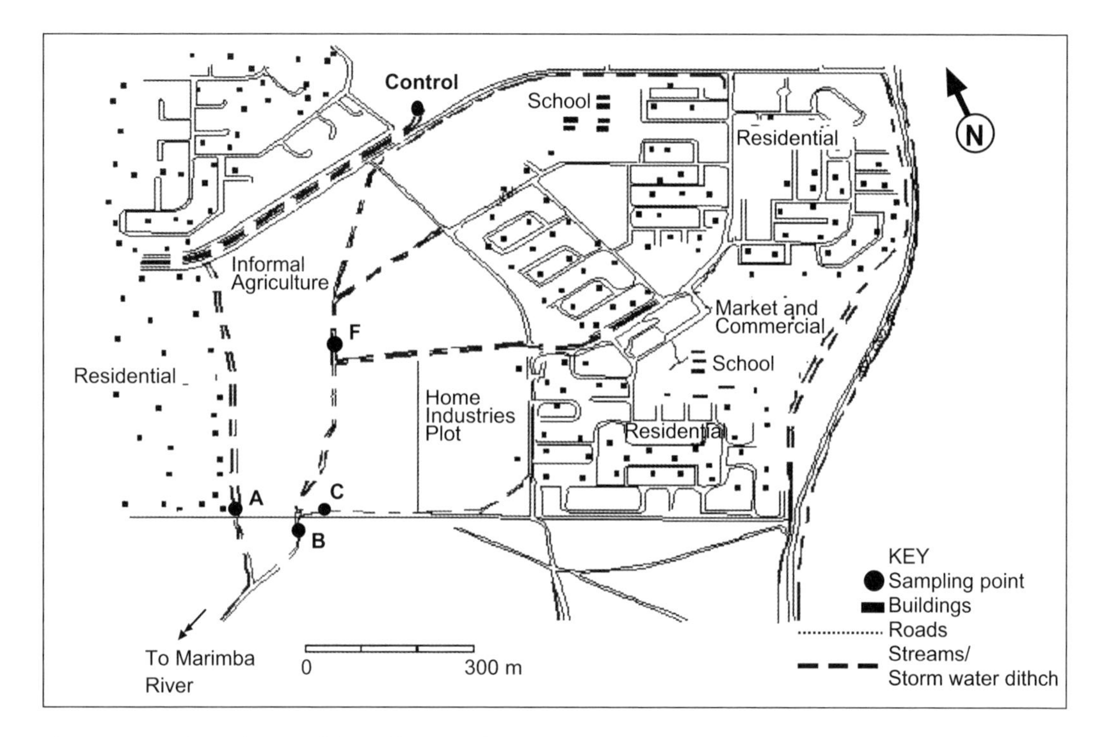

Figure 5.4. Schematic map of the study area and sampling points.

runoff quality from a mixed pattern area, including residential and commercial patterns, and part of home industries plot. Point A, on a tributary stream, reflects the contribution of a purely residential area; and point B represents the combined contribution of all different types of land use patterns.

The control point was selected at a ground water seepage point, which is the source of the main stream (points Control, F and B). Sampling was done on five occasions for points A, B, and F, and on three occasions on point C (when a runoff flow was observed), and at the control point, during the period November 2001 to February 2002. The parameters tested were EC, pH, COD, TKN, Ammonia, Nitrate, TP, ortho-P, Fe, Ni, Cd, Pb, Zn, Cu and FC. The last parameter was determined as numbers per 100 ml of sample according to the Standard methods (1989). Preservations for TKN (addition of H_2SO_4) and for TP (addition of $HgCl_2$) were executed according to the same standard methods. Nitrate and ammonia were measured immediately after arrival at the laboratory by selective electrodes (WTW pMX3000/Ion meter); standard curves have been prepared before the sample collection. For TKN, the Macro-Kjeldhal method was applied. The rest of the parameters were tested by the same methodology as described in Chapter 4.

3.3 *Land use patterns and impacts on water quality*

The results regarding stream water quality at points C, F and control are presented in Table 5.3, results regarding points A and B have been presented and discussed in Mvungi et.al. (2003). The median values are calculated based on the data sets with a minimum count of 4. The stream water quality at F presents the impact of informal agriculture, while point C represents runoff quality from a mixed land use pattern in a low income area, which includes residential development, commercial area and part of the home industries plot.

The water quality at the control point reflects the background pollution for Harare area (Hranova et.al. 2002), for all parameters except for nitrate, which at this point was about 10 times higher. The stream water quality at F shows a distinct increase in pollutant concentrations with respect to all tested

Table 5.3. Water quality characteristics at points C, F and control.

Parameter	Point C			Point F			Control		
	Mean	Median	Std. Dev.	Mean	Median	Std. Dev.	Mean	Median	Std. Dev.
pH	8.7	9.3	1.11	6.7	7.0	0.75	7.2	7.6	0.89
Conductivity (μS/cm)	538	-	184	206	200	19	165	-	3
COD (mg/l)	64	45	54.5	39	40	11.4	7	5	2.9
TKN (mg/l)	2.31	1.12	2.713	1.23	0.56	0.964	0.26	0.37	0.191
Ammonia (mg/l)	2.00	0.62	2.693	0.53	0.65	0.332	0.12	0.08	0.138
Nitrate (mg/l)	3.20	2.40	1.931	2.30	1.82	1.633	1.22	1.54	0.642
TSS (mg/l)	51	38	53.7	39	40	89.1	17	4	24.3
TP (mg/l)	1.90	2.00	1.212	1.8	2.00	0.871	0.05	0.00	0.084
Ortho-P (mg/l)	1.10	1.20	0.963	0.74	0.53	0.751	0.04	-	0.075
Fe (mg/l)	5.70	4.71	8.012	5.73	3.22	7.244	0.041	-	0.053
Pb (mg/l)	0.060	-	0.0711	0.355	-	0.3353	0.050	-	0
Ni (mg/l)	0.215	-	0.1863	0.280	0.311	0.1674	0.013	-	0.0051
Zn (mg/l)	0.520	-	0.7252	0.201	0.179	0.1792	0.012	-	0.0023
Cu (mg/l)	0.043	-	0.0581	0.148	0.192	0.0981	0.005	-	0.0007
Number FC/100 ml	0-100	-	-	0-80	-	-	0	-	-

parameters, except for pH. This reflects the impact of the informal agriculture area together with the impact of solid waste dumped on the streambed and stream banks. It is impossible to distinguish between the two possible causes of pollution, but it is more likely that it is due to the solid waste depositions, because of the very low contribution of runoff during this specific season, as discussed in the following section. The data set regarding this point shows a good distribution with a relatively low difference between mean and median values. The comparison with the acting regulations with respect to the safe limit (WWEDR 2000) shows that the mean values measured during this study exceed the safe limits in respect to ammonia (< 0.5 mg/l), TP (< 0.5 mg/l), Fe (< 1 mg/l) and Pb (< 0.05 mg/l). The excessive mean concentration of Fe could be due to the metal wastes dumped into the river, rather than agricultural activities or background pollution. This suggestion is in agreement with the relatively high variability of the data set, where individual sample concentrations varied from 0.1 mg/l to 11 mg/l, which could be explained with washed metal particles in some of the samples.

The statistical significance test of the mean values between the control point and point F showed positive results regarding COD, TKN, Cu, TP, Ni and ammonia. The rest of the parameters were not significantly different due to the high standard deviations of the data sets. The comparison of the water quality at point F, representing a natural stream collecting runoff from low-income areas, and the runoff quality described in Chapter 4, shows a lower level of pollution with respect to COD, the TP concentrations are comparable, and much higher metal concentrations, which is an indication of a potential source of toxic pollution from the drained areas, especially with respect to Pb.

Point C represents runoff quality from a drainage ditch in low-income areas with mixed land use patterns. On certain occasions, the ditch collects overflows from blocked sewer manholes. One of these occasions was witnessed during the recognisance visits. The comparison with control point water quality, and with point F water quality, shows higher pollutants concentrations with respect to pH, conductivity, COD, TKN, ammonia and nitrates, which reflects a contamination typical for sewer discharges. The relatively high values of pH are due to two specific sampling occasions, and could be associated with the discharge of some waste products from the home industry area. The safe limit concentrations (WWEDR 2000) are exceeded with respect to ammonia, TP, Fe and Zn. FC were found at both points in selected samples, indicating the presence of faecal pollution, most probably due to sewer blockages. The test for statistically significant difference with the control point gave positive results only with respect to TP. This could be explained by the low number of samples (due to low rainfall events, which generated runoff), and the considerable variation of the measured parameters. The TP data set showed a relatively lower variation, and this trend has been observed with respect to the results presented in Chapter 4 also. At this specific point, the Pb concentrations were lower compared to point F and comparable to the results presented in Chapter 4, but the Zn and Ni values were higher than the ones measured in the runoff of the City's central and industrial areas.

The results of the tests performed at points A and B have been discussed in the light of the presence of the home industry area and its impact on the natural water quality (Mvungi et.al. 2003). In general, they show the same trend in the water quality and identify the same parameters as the main pollutant hazards. The pollutants concentrations and pollutant loads were higher at point B, but no significant differentiation due to the home industry site was identified. Point B reflects the total pollution load, including the areas drained at C and F with a much higher runoff volume. In addition, the pollution loads calculated, could not be refereed as contributed by the runoff only as they reflect stream water quality and include the background pollution loads in terms of concentrations, but exclude the base flow rate in the stream.

In general, the results obtained show that the stream water quality was affected adversely by diffuse pollution sources. The major pollutant constituents of concern are TP, ammonia, Fe, Pb and Zn.

The concentrations of TP observed during this study, exceed considerably observations, reported in literture sources. Finnemore & Lynard (1982) reported TP concentrations of 0.16-0.4 mg/l in urban storm water, while USEPA (1995) presented results regarding the water quality of various streams in the US, where TP ranged from 0.011-0.4 mg/l, which are comparable to the control point concentrations during this study, as well as, to the background water quality concentrations in the Harare area (Hranova et.al. 2002). The results obtained during this study, show values ranging between 1 mg/l and 2 mg/l, which is a point of concern, considering the advanced status of eutrophication of Lake Chivero.

4 ASSESSMENT OF POLLUTION FLUXES ALONG AN URBAN STREAM

4.1 *Pollution loads and pollution fluxes*

Pollution loads contributed by runoff are important characteristics to be determined during the analysis of the status of water quality at catchment level, and the evaluation of potential risks to public health and the environment. It is an important parameter to be considered during the process of the development of pollution mitigation programs. The term is associated mainly with loads from point sources with well-identified quality and quantity characteristics. It could be applied successfully in cases of runoff pollution loads contributed by man-made drainage channels, with defined drainage area and discharge location.

The process of determination of pollution loads is mostly related to the effluent quality approach of pollution control and regulation. We could consider the pollution load of one tributary to a river or the load contributed by a river to a lake as well. By definition, it refers to the evaluation of the combined effect of water quality and quantity characteristics of a given man-made or natural channel, before its point of discharge to other water body.

In cases when the water quality objective approach is applied, it envisages the quality control of the status of natural water bodies – streams, lakes or aquifers. The contribution of diffuse pollution from runoff to the water body could not be isolated from the influence of other factors, such as upstream discharges, background pollution, etc. In such cases the term "pollution load" is not adequate, as we are evaluating the variation of the water quality status of the body itself, but not specifically a load contributed from outside. In other words, we are evaluating the effect of a load, which has already reached the stream. Instead, the term "pollution flux" will be more adequate, as it reflexes the varying nature of the combined effect of the quantitative and qualitative characteristics of the water body at the point of consideration. In the case of the monitoring of a given pollutant constituent, at a given location along a river or a stream, we should bear in mind that the value of this constituent would be influenced by:

- The background quality of river water;
- The contribution from runoff from the drained area, as well as drained areas upstream the location;
- The contribution from point sources upstream the location;
- The contribution from sediments or other residues along the streambed.

The pollution flux at a given location will be equal to the product of the measured flow rate and the corresponding value of the pollutant constituent. During a specific rainfall event, the pollutant constituents, associated with transported suspended solids, will vary during the duration of the hydrograph. Therefore, it is advisable that the pollution fluxes during such events would be calculated based on EMCs, which would reflect not only storm water pollution constituents, but also those contributed from point sources upstream and from riverbed sediments. For a given storm duration, the variation of the pollutant constituent concentration would be due to the runoff quality mainly. The contribution from the other two sources could be accepted as constants. A schematic presentation of the formation of a pollution flux at a given location in a stream or river is shown in Figure 5.5.

The total pollution flux during the rainfall event (PF_e) could be calculated as the product of the total volume of flow for the event duration and the EMC for the corresponding constituent. During dry periods, the total flux variation would be due mainly to the variation of the pollution load contributed from the point sources upstream the location and from those present along the streambed.

An accurate estimation of seasonal and annual pollution fluxes would be equal to the sum of all diurnal fluxes during the anticipated period of time. If a monitoring station is located immediately before the confluence point of a river (stream) into a lake, then the pollution fluxes evaluated would be equivalent to the pollution load contributed to the lake. In cases of automatic monitoring stations, a considerable amount of data might be available for accurate evaluation of actual pollution fluxes. However, in most cases of regular monitoring, measurements are performed once or twice per month, which do not allow an accurate estimation, and assumptions with respect to averaging available data should be made, which might lead to errors in the estimated loads. In this respect the remarks made in Chapter 4 and in section 2.3 of this chapter should be considered.

4.2 *Estimating urban runoff volumes by the SA-SCS method*

The evaluation of pollution loads and fluxes generated by runoff requires quantification of rainfall excess. A simple procedure for such a type of estimation was described in Chapter 4. In this section, a more detailed description of the SA-SCS method (Shulze et al. 1993) has been done. In order to illustrate the different steps and procedures, the methodology applied for the specific case of Kuwadzana, during this specific study will be presented in more details.

The first step of the procedure was the evaluation of several key drainage area characteristics. The first characteristic that was determined was the drainage area size, based on the topographic series maps of Harare TR8025 and TR8027. The other characteristics were the shape of the drainage basin for each

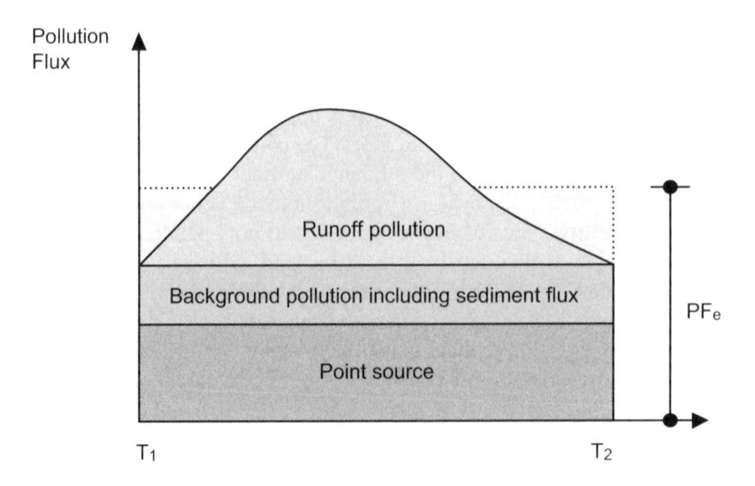

Figure 5.5. Pollution flux at a given location of a stream for a storm with duration $T_2 - T_1$.

separate sampling point and its various slopes. In addition, the type of cover for each specific area was evaluated as a percentage of the total drainage area. These parameters would determine the runoff flow pattern-the steeper the surfaces the faster the runoff can drain to its outlet and the higher the impervious cover – the larger of the volume of the runoff generated by the specific rainfall event.

The SA-SCS method uses curve numbers, where the initial curve number selection is based on the soil and land characteristics only. The second step of the method includes the determination of the different soil characteristics and was executed based on the soils classification in Zimbabwe (Nyamapfene 1991). The study area was subdivided into three major soil types:

- The first soil type is Kalahari sand of Permian to Triassic age together with that of alluvial Holocene age, soils that are relatively youthful. These were to be found mostly in the built up part of the study area. The residential areas consist of this type of geological formation, but the fact that they have been compacted over time for many years, has resulted into a soil type of slow infiltration rates and moderate to high runoff potential. The home industry and commercial areas, gravel street roads also fall under this category. This category is classified as soil group C in the hydrological soil groups.
- The second soil type comprises of the vertisoils. These consist of clayey soils. The area with such characteristics are subject to some degree of water logging and as a result develops structure and cracks which may be confined to the top 30 to 50 cm in most areas. This type is mainly found along the stream area, where informal agriculture takes place and consists of a combination of the soil groups B and C that is classified as B/C.
- The high storm flow potential, low infiltration rates and permeability have resulted in classifying tarred street roads and storm drains in the soil group D.

The results of the third step of the method application are shown in Table 5.4, which summarizes the soil type and the percentage of the area covered as adopted for the purpose of the study. Once the soil type and group were known, the model soil groups, associated with each sampling point, were adopted from SCS hydrological soil groups, and the initial curve numbers for each land use pattern, including the cover and treatment were selected (Table 5.5). Then the weighing factors were applied to account for the areas covered by soil type, cover and land use in the whole catchment area. The sum of these weighted initial curve numbers represents the initial composite curve number for the catchment area.

After the determination of the initial composite curve numbers follows the adjustment of the initial curve numbers, with respect to the soil moisture conditions of the area prior to the rainfall events. The rainfall data for the study area was obtained from the Meteorology Department, Harare, and is presented in Figure 5.6. A significant feature of the rainfall is its unreliability in terms of rainfall height and duration. The onset of the rains, which is critical, is rather unpredictable. The variation from year to year is such that in some years, as little as quarter of the arithmetic mean may fall while in other years 200-300% of mean may fall.

Table 5.4. Adjusted soil type distributions in the study area.

Catchment component	Soil group	% Area covered	Area covered 10^6 m^2	Weighing factor
Pervious				
Rain fed row crop	B/C	22	1.430	0.22
On plot gardens	B	5	0.325	0.05
Open spaces, parks	A	5	0.325	0.05
Impervious				
Roofs	C	43	2.827	0.43
Tarred streets, roads and home industry	D	12	0.761	0.12
Commercial	C	11	0.715	0.12
Gravel street roads	C	1.5	0.0975	0.015

Table 5.5. Initial composite curve numbers.

Sampling Point	% Area covered	Weighing factor	Initial composite CN
A	21.6	0.216	27.2
B	45.8	0.458	29.155
C	19.1	0.191	13.52
F	13.6	0.136	20.17

During the period of study, a few odd showers had occurred at the end of October, but the temperature and evapotranspiration rates were high enough during the day resulting in virtually no effective rainfall. In general, relatively low rainfall events occurred, which represents a low rainfall wet season. For this reason, the "dry-five day antecedent rainfall" soil moisture condition was adopted.

Usually the Median Method and Joint Association Curve Numbers are used to adjust the initial curve numbers regarding the soil moisture conditions into climatic zones according to the Koppen climatic classification criteria (Shulze et. al. 1993). But in this case the study area was too small and was not possible to adopt this method. Instead, the curve numbers (CN) were determined by equations 5.2 and 5.3.

$$CN_d = CN_a / 2.334 - 0.01334CN_a \quad (5.2)$$

$$CN_w = CN_a / 0.4036 - 0.0059CN_a \quad (5.3)$$

Where d= dry; a = average; and w = wet.

From Figure 5.6 it can be seen that there was little or no rainfall in most of the five days prior to the sampling events. This has been due to an acute dry spell that has occurred in almost all parts of the country. In this case the "dry" antecedent soil moisture condition had been prevailing. The "dry" antecedent moisture conditions are considered when the rainfall during such five days period is less than 35 mm.

After the initial curve numbers have been adjusted, the potential maximum soil water retention (S) was determined. It is related to the hydrological soil properties, land cover and land management conditions, and the soil moisture status of the catchment prior to the rainfall event. This potential is related to the dimensionless response index termed "the catchment curve number" (CN). It is the final curve number, adjusted for antecedent moisture conditions. The relation between CN and S is shown in equation (5.4).

$$S = (25400/CN) - 254 \quad (5.4)$$

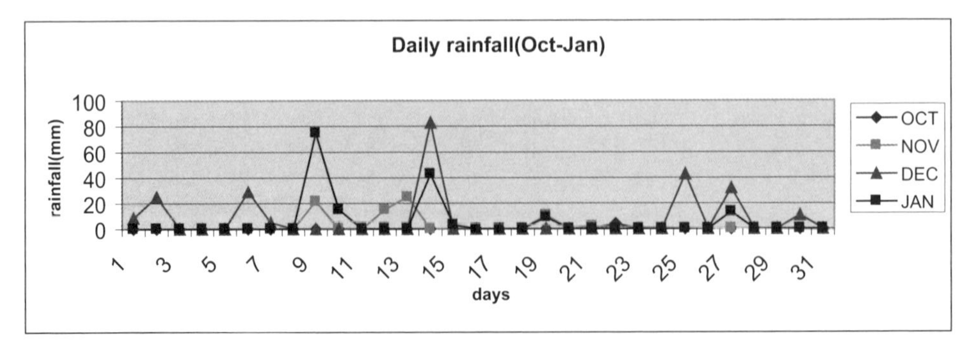

Figure 5.6. Daily rainfall patterns during the period of study.

The storm flow depth was calculated using the SCS storm flow equation (5.5), where Q is defined as the direct runoff response to a given rainfall event, and consists of both surface runoff and subsurface flows, but excludes base flow or a delayed subsurface response.

$$Q = (P - I_a) \,/\, [(P - I_a) + S] \qquad \text{for } P > I_a \qquad (5.5)$$

where Q = storm flow depth (mm), which can be converted to volume by introducing catchment area; P = rainfall event depth (mm); I_a = 0.1 S = initial losses prior to commencement of the storm flow, comprising of storage, interception and initial infiltration (mm); S = potential maximum soil water retention (mm) = index of the wetness of the catchment's soil prior to the rainfall event.

The total net runoff volume for the study period for each specific point was evaluated as the sum of the storm flow volume for each rainfall event and this cumulative value, including runoff from up-stream points was used to estimate the pollution fluxes in the subsequent section.

It is obvious from the procedures described that the manual estimation of net runoff volumes is a tedious and time consuming procedure, therefore for a regular monitoring and control practice, the implementation of GIS systems and their incorporation in a model for pollution loads and fluxes evaluation would lead to a more precise and user-friendly procedure.

4.3 *Pollution fluxes evaluation*

The study area under consideration (Figure 5.4) could be used as an example to illustrate the procedure for pollution fluxes evaluation along a river or a stream. Such evaluation could help to differentiate the contribution of different pollutant constituents, generated by different types of land use practice and to prioritize the corresponding abatement measures. From the previous section, the concentrations of the selected parameters were not evaluated as EMCs but as composite samples taken after the rainfall event, due to financial, transport and manpower restrictions during the study period. This would reflect on the accuracy of the determined pollution fluxes, especially with respect to parameters as COD, TP and metals, which were expected to be associated to sediment particles. The ratio between TP and ortho-P (Table 5.3) shows that 50-65% of the TP was present in dissolved form and this fraction of the TP should not be affected considerably. For the purpose of this example, pollution fluxes with respect to ammonia were determined, because this parameter is usually present in dissolved form and its concentration should not vary considerably during the storm event.

4.3.1 *Methodology*

The fluxes were evaluated based on the flow volume during the study period and the median values of the ammonia concentrations at the different sampling locations. The background flux was calculated as the product of the base flow, measured during the dry period at a location close to point F and the ammonia median concentration at the control point (Table 5.3). The base flow was measured on two occasions during typical dry season conditions by the velocity-area method (Mays 2001).

The fluxes calculated at each sample point location, reflect the background flux plus the flux generated by the storm water of the corresponding drainage area and the upstream areas during the period of study. The location C$^{|}$ reflects the pollution load mainly from storm water, contributed to the stream by drainage area C. The final flux, downstream to Marimba River, includes the contribution from drainage area A, shown down stream of point A$^{|}$ (the discharge point of the stream collecting runoff from area A).The same procedure should be applied to the rest of the parameters measured for a full evaluation of the pollution fluxes and their variation.

4.3.2 *Results*

The results (Fig. 5.7) show that the major pollution flux is generated between the control point and point F. This drainage area is with a low percentage of impervious cover and correspondingly does not contribute

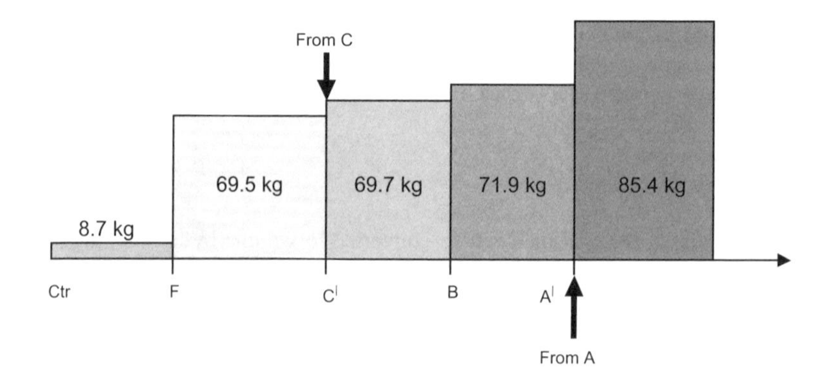

Figure 5.7. Variation of pollution fluxes along the stream with respect to ammonia.

a significant volume of runoff. For the whole study period, the runoff from area F formed only 0.5% of the base flow volume. Therefore, the sharp increase in the pollution flux is due mainly to the increased ammonia con-centration between the two points, which could be explained only with pollution generated by the solid wastes dumped in and around the stream bed.

This trend was sustained for the rest of the parameters as well. It shows the importance of maintaining natural water bodies clean of litter and solid wastes. The cleaning of the streambed could reduce 80-90% of the diffuse pollution load contributed to Marimba River from this tributary, especially during dry season conditions.

5 ABATEMENT AND MANAGEMENT

5.1 *Solid waste management and diffuse pollution*

There is no need to conduct a special research investigation in order to ascertain that unacceptable practice of solid waste collection is a serious threat to public health and a significant source of diffuse pollution. This study has shown the extent, at which the natural stream is polluted due to unacceptable solid waste management, but this extent is site and time specific and the results obtained could not be generalized. The main conclusion, which could be drawn, is that in such types of urban developments and socio-economic conditions, the management activities to reduce the pollution of water resources should be directed to improve the solid waste collection process and the general hygiene of the residential areas. The application of source control measures is imperative, as it could lead to a high percentage of diffuse pollution reduction by the application of relatively low cost measures. In most cases, the general public considers that this is a task, which should be undertaken by the municipality and consequently, it is the only responsibility of municipal authorities. It should be emphasized, that even the best practice of solid waste management could not cope with an unacceptable behavioral practice of littering and polluting the environment by the community members. Thus, the cleaning and maintenance of public spaces and streams is mainly a community problem, which should be addressed at all levels and should be considered an important aspect of pollution abatement programs. It is in the interest of municipal authorities to consider carefully the specifics of the community they are serving, and to apply creative tactics to mobilize, educate and provide incentives to the public in order to achieve their objectives.

The practice of separate solid waste collection and recycling is gaining momentum and becoming part of the daily life of people in developed countries. In contrast, the solid waste collection and recycle in the countries of the South African region is very rare. It has already been mentioned that in the study area, even an ordinary practice of solid waste collection is not performed regularly. One of the reasons pointed out by the municipal authorities, was the insufficient funding for the proper maintenance of the transport park, and also, the limitations in fuel supply at national level. However, even under normal operational conditions, local authorities do not have the understanding of the need to develop and implement broader programs, which could involve innovative management activities, such as public and community participation, education, involvement of the media, organization of special events orientated to the public education and promotion of values and goals, in addition to the ordinary technical and operational activities.

One possible form of encouraging the collection and recycle of waste material would be the implementation of a network of small centers, where people could return waste material, as paper, metal scrap, plastics, tins, etc., and in return be remunerated modestly. Such centers could provide jobs for some community members, who will be responsible for the proper packaging and sorting of the collected material. The implementation would not require a large investment; a simple shade structure would be feasible. The transport of the collected material to a central location could be executed by the local authority or delegated to a private company. Such an arrangement at local level would require a nation wide policy and support on recycling of waste materials and the creation of a conducive environment to encourage and regulate activities on solid waste generation, collection, recycling and treatment.

5.2 *The community and the general public*

5.2.1 Education and information needs

In general, the public is well aware of the health risks associated with an inadequate solid waste management practice and pollution of water resources, especially concerning the link between water borne diseases and polluted water sources. However, the fact that polluted runoff could also be a public health threat is not well understood even at higher levels of management. In most cases, typical for the countries in the region, it is a well-known fact that people living in low-income residential areas and informal settlements experience acute water and sanitation problems and the generated solid waste is spread over the area and in the surroundings. Such conditions are accepted to be part of the development of the country and people living in a better environment try to isolate themselves and to leave the problems to the managing authorities. The vast majority of the public is usually not aware and informed that the polluted runoff and untreated sewage from such areas reach the natural streams and pollute the water sources and recreational areas, which they are using on a daily basis. Also, they are not aware about the increased costs involved to treat a polluted water source. Thus, the pollution generated by low-income residential areas affects the urban environment as a whole, and should not be regarded as an isolated issue, but requires the attention and efforts of the whole urban community in order to solve the problems.

When dealing with educational or public awareness programs, it is important to define who is the public, or to which specific part of the public the program will be orientated, and based on this, to choose suitable educational and information activities. Different sectors or segments of the public will be interested in different issues:

- The low-income community in general will be interested in the regulatory and financial aspects and the following questions might arise:
    - Who is going to support financially a program for diffuse pollution abatement?
    - What could be the possible fees and charges as a result of the implementation of such a program?
    - What are the forms of involvement of community members in the implementation of the program?
- School children need to be educated properly with respect to the health risks associated with pollution and littering. Also, schools are the centers to provide the background towards a proper behavior in respect to littering and protection of water resources.
- The general public will be interested in utilities fees and charges;
- The commercial and industrial sectors would be interested in regulatory aspects and charges with respect to storm water discharges;
- The professionals involved in design and other technical issues, need to be aware of regulatory requirements, by-laws and technical criteria.

The objectives of such types of programs should be well formulated to match the specific information and awareness needs of the different groups involved. Correspondingly, the program should include an appropriate form of public awareness activity in order to achieve the objectives.

There are numerous techniques to educate and inform the public, and to achieve the successful implementation of such programs. Debo & Reese (1995) provide a detailed list, but some examples, appropriate for the conditions in the region, are:

- Media releases and news information;
- Feature stories on specific cases of pollution and possible consequences;
- Pamphlets and brochures;
- Community and neighborhood meetings;
- One-on-one meetings with important stakeholders or decision makers;
- Phone survey research;
- Information workshops;
- Up-dated school programs to include issues on water pollution, including diffuse pollution and solid waste management;
- Extra curriculum school activities to promote positive behavior with respect to littering and clean environment.

One important aspect of public educational programs could envisage the training of community members and their involvement in the process of identification and reporting of illicit discharges. An educational program with such an objective could include the following:

- Provide the training of volunteers;
- Develop educational and informative brochures and guidance;
- Provide methods for public reporting of illegal discharges;
- Coordinate the efforts of individual volunteers for the visual inspection of outfalls and other possible discharge spots;
- Provide the necessary conditions for the collection of commonly dumped waste materials.

5.2.2 Community participation and public involvement

The need to implement preventive and source control measures with respect to diffuse pollution control is imperative and it could not be achieved effectively by structural measures only. Point source pollution abatement also requires source control and preventive measures, but in most cases it is achieved by technical means in treatment facilities. Diffuse pollution from urban storm water discharged by man-made drainage channels, and diffuse pollution from highways could be subject to partial treatment as specified in Chapter 2. However, diffuse pollution of surface water from low-income areas, informal settlements, and different agricultural practices is extremely scattered, thus it is difficult to apply technical solutions, such as treatment facilities, for its control and reduction. For this reason, the best possible practice is to limit the emission of pollutants at the source. Such an approach would require the change of established behavioral practice in the community or established agricultural practices in the region. Thus, such an approach would be difficult to achieve by the efforts of one managing institution only, but would require the support and involvement of all sectors of the community and the public at local and national level.

One very important aspect of any public involvement program is the extent at which the public will be convinced in the need for their contribution and involvement. In many cases, the link between the managing authority and the community is non-existent, and this is more often the case in low-income areas. Therefore, the first step of any efforts to involve the public in the management of water quality in their neighborhood should be the establishment of such a link, where the local authorities should take the lead in this process. Probably, one of the most important objectives of any public involvement program should be to obtain the trust and confidence of the people, who are expected to be involved in the consequent stages of the program.

As in the case of educational and information activities, the forms of public involvement are numerous and differ according to the specifics of each country. As the first step of the process, it would be advisable to identify the major stakeholders, or groups of stakeholders, which should be addressed. These could be:

- The local community:
  - Residential;
  - Churches;
  - Schools;
  - Business representatives;
  - Political representatives;
  - Council wards;
- Local media:
  - Print and broadcast organizations;
- Official authorities:
  - Municipality representatives;
  - City commissioners;
  - Government officials;
- Non-governmental organizations and environmentalists;
- Residential and commercial developers;
- The public.

The preparation and execution of programs to mobilize the public into the management process of diffuse pollution abatement should involve and integrate different techniques. It is important to point that each activity must be clearly understood, within the context of the target group and the objectives of the whole process or program. Some common activities or techniques have been emphasized in Chapter 2. In the case of low-income areas, volunteer programs to maintain the cleanness of the area, to perform simple monitoring functions and to educate the community could be a suitable option. For a successful volunteer program to be implemented the following factors are important:

- A good formulation of the tasks to be performed and instructions on how to perform them;
- A formal assessment of achievements and acknowledgement of the volunteers' contribution and work done;
- Provision of opportunities for personal development and education of volunteers;
- Creation of sense of common goals and shared values between the volunteers and the professionals in the field could help significantly for the implementation of the target.

5.3 *Urban planning solutions and diffuse pollution*

The link between urban planning and diffuse pollution has been emphasized in terms of the different types of land use patterns and corresponding impacts on natural water quality. Appropriate urban development methods could be regarded as a source control of diffuse pollution. In the case of low-income areas, some specific measures could be undertaken during the planning and design stage, which could help alleviate future problems and reduce the cost of pollution abatement measures. These could be:

- Provision of larger buffer zones and environmental corridors surrounding low-income urban developments (such an arrangement is viable in numerous cases in the region, as land is usually available);
- Provision of a larger percentage of impervious covers, specifically at open spaces and commercial areas, in order to reduce the runoff volume;
- Provision for erosion-protective measures and the preservation of natural vegetation during the design stage;

- The plots designated for home industries should be located away from natural water bodies and should be supplied with basic water and sanitation facilities;
- A possible solution of the solid waste management problem could be the provision of protected local decentralized disposal sites, which could be operated and maintained by the communities with support from local authorities.

6 CONCLUSIONS

The study shows a considerable increase of pollutant constituents in a natural perennial stream draining a low-income residential area with respect to all tested parameters, compared to a control point reflecting background water quality. The main pollutant constituents identified were ammonia, TKN, TP, Fe and Pb. At point F, reflecting the impact of informal agriculture practice and illegal dumping of solid wastes and at point C, reflecting storm water quality from a drainage channel, the TP, ammonia and Fe concentrations were similar and exceeded the control point concentrations more than 10 times. The recommended safe environmental limits were also exceeded considerably. The Pb concentration at point C was similar to the control point concentration, but at point F it was exceeded more than 7 times. The pollutant constituents at point F were associated mainly with dumped solid waste along the stream banks.

The different land use practices identified did not show a significant difference in pollutant concentrations. This fact could be due to the relatively low runoff volume during the study period. A comparison with results from Chapter 4 shows the same trend in respect to TP and ammonia, but in this study area, the COD concentrations were lower, while Pb values at F were higher

The method applied to evaluate pollution fluxes in cases when natural streams water qualities are examined, showed that the major impact on stream water quality is due to the sharp increase of pollutant constituents along the stretch from the control point to point F, in the upper reaches of the stream. The impact of runoff from areas located downstream of point F was relatively low due to the comparable values of the pollutant constituents in the runoff and in the stream and the relatively low volume of the runoff during this period.

Based on the results of this study, it could be recommended that the monitoring of water quality in low-income areas should be directed towards the evaluation of the water quality status of natural streams, and should be executed as an integrated part of a larger scale monitoring programs, rather than individual monitoring exercises. Such an approach could allow the concentration of efforts to obtain a full data set of EMCs with respect to major points of interest and the possibility for accurate pollution loads or fluxes evaluation and trend variations. The recommended management options were stressing on a proper solid waste managing practice and the implementation of source control methods for pollution abatement.

Acknowledgements – The authors would like to thank the management of the WREM program for the financial support offered during this study. To the technical staff of the laboratories of the Departments of Civil Engineering and Geology at the University of Zimbabwe – thanks for their assistance during the sampling and laboratory analysis work.

REFERENCES

Baldock, J.W., Styles, M.T., Kalbskopf, S. & Muchemwa, E. 1991. The geology of Harare Greenstone Belt and surrounding Granitic Terrain. *Zimbabwe Geological Survey Bulletin* 94, Harare.

Debo T. N. & Reese, A. J 1995. *Municipal storm water management.* Boca Raton, Florida: CRC Press, Lewis Publishers.

Department of Water Affair and Forestry (DWAF) 1996. *Water quality guidelines*. Vol. 1-7, Pretoria: Government publications.

Drakakis-Smith, D & Kirell, P 1990. Urban food distribution and house hold consumption: a study in Harare. In R. Paddison & J. A Dawson (Eds), *Retailing environment in Developing countries*, 156-180, London: Routledge.

Finnemore, E. J., & Lynard, W. G 1982. Management and control technology of urban storm water pollution, *Journal of Water Pollution Control Federation*; 54, 7, 1097-1111
GERMIN 1991. *Changes in the small-scale enterprise sector form: results of 2nd National-wide survey in Zimbabwe.* GERMIN Technical report no 71, Maryland: Bothesda.
Hranova, R., Gumbo, B., Klein, J. & van der Zaag, P. 2002. Aspects of the water resources management practice with emphasis on nutrients control in the Chivero basin, Zimbabwe. Physics *and Chemistry of the Earth,* 27, 875-885
International Labour Organization (ILO) 1996. The informal sector in Africa. Geneva: ILO.
Mays L. W. 2001. *Water resources engineering.* USA: John Wiley & Sons.
Mhone C. Z 1995. *Impact of Structural Adjustment Program (SAP) on the urban Informal sector in Zimbabwe* Harare: University of Zimbabwe publications
Mvungi, A., Hranova, R. & Love, D. 2003. Impact of home industries on water quality in a tributary of the Marimba River, Harare: implications for urban water management. *J. Physics and Chemistry of the Earth* 28, 1131 – 1137
Nyamapfene, K. 1991. *The soils of Zimbabwe*, Harare: Nehanda publishers.
Rakodi, C 1995. Harare: Inheriting a settler-colonial city: Change or continuity, Sussex: John Wiley & Sons.
Schulze, R.E.; Schmidt, E.J., & Smithers, J.C. 1993. SCS-SA User Manual: PC – Based SCS design flood estimates for small catchments in Southern Africa. In *ACRU Report No 40*, University of Natal, South Africa.
Standard methods for the examination of water and wastewater 1989. 17th Ed. American Public Health Association/ American Water Works Association/Water Environment Federation, Washington DC.
United States Environmental Protection Agency (USEPA). 1995. Guidelines for preparation of the state water quality assessments. *Report 305b, EPA 84, B-95-001.* Washington DC
WWEDR 2000. *Water (Waste and Effluent Disposal) Regulations.* Statutory Instrument 274 of 2000, Republic of Zimbabwe.
Zinyama L.M. 1993. The evolution of the spatial structure of the greater Harare: 1890-1990. In: Zinyama, L.M., Tevera, D.S. & Cumming, S.D. (Eds.), *Harare: The Growth and Problems of the City.* Harare: University of Zimbabwe Publications.

CHAPTER 6

Impacts on groundwater quality and water supply of the Epworth semi-formal settlement, Zimbabwe

D. Love, E. Zingoni, P. Gandidzanwa, C. Magadza & K. Musiwa

ABSTRACT: Many developing countries experience rapid urbanization without rapid economic growth. This encourages rural to urban migration, but governments are rarely in a position to immediately service the burgeoning urban population. These trends have led to the development of informal or semi-formal peri-urban settlements in the municipal areas of large cities in the developing world. This chapter presents an investigation of a semi-formal settlement as a diffuse pollution source. Epworth settlement, southeast of Harare, has characteristics of both formal and informal settlements. The settlement began developing an urban character in the 1970s and has grown and densified substantially during the past twenty years. This growth has not been accompanied by significant investments in water supply and sanitation infrastructure. A survey of water supply and sanitation practices was undertaken, and it was determined that close to half the residents use unprotected wells for water supply and over 90% use pit latrines. The problem is worst in the oldest parts of the settlement. A groundwater quality investigation was therefore undertaken in this ward. Shallow boreholes were drilled, groundwater sampled and chemical and microbiological analyses performed. The results revealed significant levels of contamination, particularly with regard to nitrates and coliform bacteria, and highest down flow in the wetland areas. Elevated levels of these parameters are most likely caused by the abundant pit latrines in Epworth, although urban agriculture could be a secondary factor influencing the nitrate levels. - Preventing further deterioration in groundwater quality requires major investment in piped sanitation. Unless this comes from central government, it will have to be self-financed and residents are unlikely to contribute to costs unless they have security of tenure. The most pragmatic short-term solution for water supply would be for the local authority to extend provision of a limited water supply, via communal taps, to be used for drinking purposes only.

1 INTRODUCTION

1.1 *Informal and semi-formal settlements*

Zimbabwe like most developing countries, is experiencing rapid urbanization: for example the population of Harare increased from 665, 000 in 1982 to 1.4 million in 2002 (CSO 1999, CSO 2003). The urbanization process, coupled with inadequate public or private housing finance, slow economic growth and natural disasters in the rural areas, has led to the mushrooming of informal settlements (Makoni 2001). Such settlements now contain more than half of the households in many developing cities (Pugh 2000). Sometimes informal settlers occupy land alongside, or even within, a formal settlement. The settlement thus develops a semi-formal character, with some households occupying land legally, and subject to some form of settlement planning, and other households (within the same settlement) occupying land illegally and outside of any official planning process. Such illegal land occupation and dwelling construction characteristically shows a higher density of development than in surrounding legal areas and a random orientation of dwellings and land parcels or stands (Zegarac 1999).

Informal settlements are considered to be illegal in Zimbabwe and are often demolished and the residents evicted. Such a reaction is typical of government responses in post-colonial Africa (Obhudo and

Mhlanga 1988). Even when such settlements are tolerated, they are given low priority in local authority and government development or planning programs (Kombe in press). Furthermore, even if a settlement is legal or recognized by government, the development of sanitation infrastructure in particular, tends to be implemented later than the initial urbanization phase in a new settlement (Foster 2001). This is particularly true for peri-urban settlements, with their rapid growth patterns, where local authorities often have ambivalent attitudes towards the development and limited planning capacity (Kyessi, in press). Where peri-urban settlements are administratively separated, under independent local authorities, problems of administrative capacity and resources for infrastructure development and services provision arise (Dahiya 2003).

Urban utilities often fail to provide service to low income customers, particularly those settled on illegal or low-grade land. This is because low-income communities are often perceived by the utility (public or private) to be financially unreliable, transient, difficult to identify and expensive to service (Wright 1997).

Growth in sanitation coverage in low-income areas worldwide has been much slower than that for water supply (Cairncross 1998). Lack of resources is not the only reason for slow progress in sanitation coverage. Low-cost sanitation programs are far more difficult to implement than water supply schemes for several reasons. It is more expensive and time consuming to install infrastructure and services in densely populated illegal or informal settlements after they have developed, as space is constrained and utility servitudes have not been reserved, as there was no planning process (Hardoy and Sattethewaite 1989). It is therefore not unexpected that poor cost recovery has been identified as one of the major causes of failure of post-independence sanitation programs (Manase et al. 2004).

Because of these problems, informal settlement residents are rarely provided with proper water supply, sanitation or waste disposal services. As a result, residents are frequently forced to rely on shallow groundwater for domestic water supply (Butcher 2003, Dahiya 2003' Tsvere et al. 2004). In fact, urban groundwater is thought to supply up to half of the world's urban population (Foster, 2001). This exposes settlement residents to health risks from the contaminated groundwater. Informal settlement communities have become highly prone to water-borne diseases (Chidavaenzi et al. 2000) - although respiratory diseases are also common in overcrowded slums (Gulis et al., 2004). In cases where supply is erratic, the situation is worsened as residents store water in containers or sinks for more than a day (Tsvere et. al. 2004).

Inadequate water supply and sanitation are largely responsible for the high levels of such diseases in Southern Africa (Makoni 2001). Human excreta may contain eggs of helminthes (worms), protozoa, bacteria and viruses. These may be excreted in vast numbers depending on the age and state of the individual. Fecal matter contains on average 10^9 bacteria per gram (not necessarily pathogenic) and the excreta of infected individuals as high as 10^6 viruses per gram (National Academy of Sciences 1987). When feces are deposited in shallow water an easy pathway is provided for pathogens to enter surface and groundwater, and hence to the local population via contaminated drinking water.

1.2 *Impact on groundwater of semi-formal settlements*

In developing countries, the subsurface environment frequently acts as a major sink for domestic waste, leading to significant degradation of groundwater quality (Forster 1999). Rapid urbanization is associated with rapid deterioration in water quality (Ren et al. 2003). Contamination of shallow groundwater from peri-urban semi-formal settlements is especially problematic in cities where the settlement is in a water source area for municipal reservoirs. The semi-formal settlements of Old Naledi, Gaborone, Botswana (Gwebu 2003), and Epworth, Harare (the settlement studied in this chapter) are cases of this kind.

Nitrate, nitrite, phosphate and organic compounds are the main contaminants that frequently percolate to groundwater from informal settlements (Wright 1999). Microbiological contamination is also frequent. The combination of a lack of well-developed service infrastructure and a lack of proper settlement planning means that a variety of activities in informal and semi-formal settlements contribute to groundwater contamination, see Table 6.1.

Probably the most important of these problems are those caused by on-site sanitation – since the latrines are abundant, and spread throughout an informal settlement. On-site sanitation can be of various

Table 6.1. Summary of possible groundwater contamination problems associated with informal settlements.

Activity	Associated Contaminants
On-site sanitation (various methods)	Nitrate, phosphate, coliform bacteria, other pathogens
Food and produce markets	Nitrate, phosphate
Home industries	Transition metals, ammonia, TDS
Urban agriculture	Nitrate, phosphate, potassium

types: standard pit latrines, ventilated improved pit latrines (known in Zimbabwe as "Blair toilets"), pour-flush pit latrines or, occasionally, septic tanks.

Bacteria and viruses may be transported with percolating effluent into the groundwater. Many pathogens, especially coliform bacteria, can survive in shallow groundwater for longer than two months (McCarthy et al. 2004), and these organisms may be ingested causing infection. Whether or not an individual will become infected will depend on the concentration and persistence of the pathogen in groundwater and the infectious dose required to initiate disease (Feachem et al. 1981). Infectious water-borne diseases such as diarrhea-type diseases, dysentery, cholera and hepatitis are almost endemic in many rural and informal urban areas (Manase 2001). Children under the age of five years of age living in settlements with no access to safe water are the most susceptible (Obi et al. 2004). These factors clearly indicate the major problem that poor sanitation in informal settlements presents to water supply for the same settlements.

Home industries, which are also a source of diffuse pollution in informal settlements, are small-scale industries such as panel-beating, brick-making, painting, leather craft and grain milling. They are widespread in both high-density formal residential areas and in informal settlements in many cities in Zimbabwe (including the study area), with no zoning controls in the latter case. The operations of these small industries has been associated with release of a number of metals (Mvungi et al. 2003), and their impact on surface water has been discussed in Chapter 5.

Produce markets are found to occur at irregular intervals in most informal settlements, either as small "commercial centers" or as aggregations of two or three traders at bus-stops, cross-roads or other places where people congregate (Kyessi, in press). They are associated with nutrient contamination-and also coliform bacteria at markets large enough for public urination to become a nuisance.

Urban agriculture is widespread in informal and semi-formal settlements, as residents grow staple crops or vegetables to augment their diet and/or income (Kyessi, in press, Lynch et al. 2001). This can be either on the land they own or occupy for residential purpose, or distant from their dwellings, in open spaces-either unusable for housing development or simply not yet occupied. Figure 6.1 illustrates the urban agriculture practice and in this case it is practiced within a residential dwelling. Note the monitoring borehole in the foreground. The pit latrine on the right was constructed after the borehole was drilled.

Some residents also raise small or large livestock, as medium term investments, or for milk production (Losada et al. 2000). Agriculture can contaminate shallow groundwater with potassium, nitrate, nitrite, phosphate, dissolved organic carbon and salinity (Conrad et al. 1999), although most of these parameters only become elevated when fertilizers or manure are applied.

1.3 *Epworth Settlement – its growth and urban structure*

There are three main informal or semi-formal settlements in the Harare area: Hatcliffe Extension, Epworth and Snake Park (also known as White Cliff). Epworth is the largest and oldest of these settlements, and is located to the southeast of the city of Harare, outside the city borders, but within the Harare Metropolitan Province (see Fig. 6.2). This is a water source area – the headwaters of a tributary of the Manyame River. It has been well established that contamination of shallow groundwater leads to contamination of streams that either arise from or flow through the area of contaminated groundwater, although the extent of

Figure 6.1. Urban agriculture within a residential land, informal settlement in Ward 1, Epworth.

groundwater contamination and nature of seepage to surface water controls levels of stream contamination (Winde & van der Walt 2004). Therefore a study of contamination of shallow groundwater in Epworth is strategic to diffuse pollution management in the Harare area.

The Epworth settlement grew up around the Epworth Mission, established by the Methodist Church in Zimbabwe in 1900. For the first forty years, Epworth was a church farm with an established population of around 240 people (Clarge 1999). Settlement of larger numbers of people near the mission began in 1950, starting with 500 (formally settled) families. During the struggle for independence, many refugees deserted their rural homes and settled at the mission farm, especially during the 1970s. They were allocated pieces of land to construct houses and to grow crops. Many people remained at Epworth after Zimbabwe's independence in 1980 and the settlement continued to grow, becoming a clearly urban settlement, although agriculture was still (and is still) practiced (Rakodi 1995). In 1983 the Epworth Mission transferred ownership of the farmland to the Ministry of Local Government (Clarge 1999) – excluding a small portion upon which the Mission, a theological college and a children's home are built. Subsequently, the population continued to rise, reaching around 35,000 people by 1987, most of whom were informally settled (Rakodi 1995) and rising further to 113,884 recorded in the 2002 national census (CSO 2003). The settlement is continuing to grow rapidly towards the south and east. Epworth has always been and remains to be one of the lowest income areas in Harare Metropolitan Province (Rakodi 1995). Density of settlement is now very high, with as many as 40 people per stand in some areas by the early 1990s (Butcher 1993).

Epworth is an unusual case, since it is the only informal settlement in the Harare area that the post-independence government has been prepared to upgrade (in terms of settlement planning and infrastructure development), rather than demolish. Although the decision to upgrade was taken in 1983, work only commenced in the 1990s (Butcher 1993). The process of providing the Epworth area with mains water supply, sewage reticulation, an electricity grid and rubbish collection services has been extremely slow. By 1988, 75% of residents were still using wells for water supply and over 80% were using pit latrines (Butcher 1993). Newer areas to the north and west are essentially un-supplied. Households without piped water rely on wells. Epworth is still not provided with sewerage or electricity reticulation-or with rubbish collection. The lack of piped sanitation continues to be a major health problem for the households relying on groundwater for domestic water use. Gastrointestinal diseases are endemic, and particularly pronounced in the rainy season.

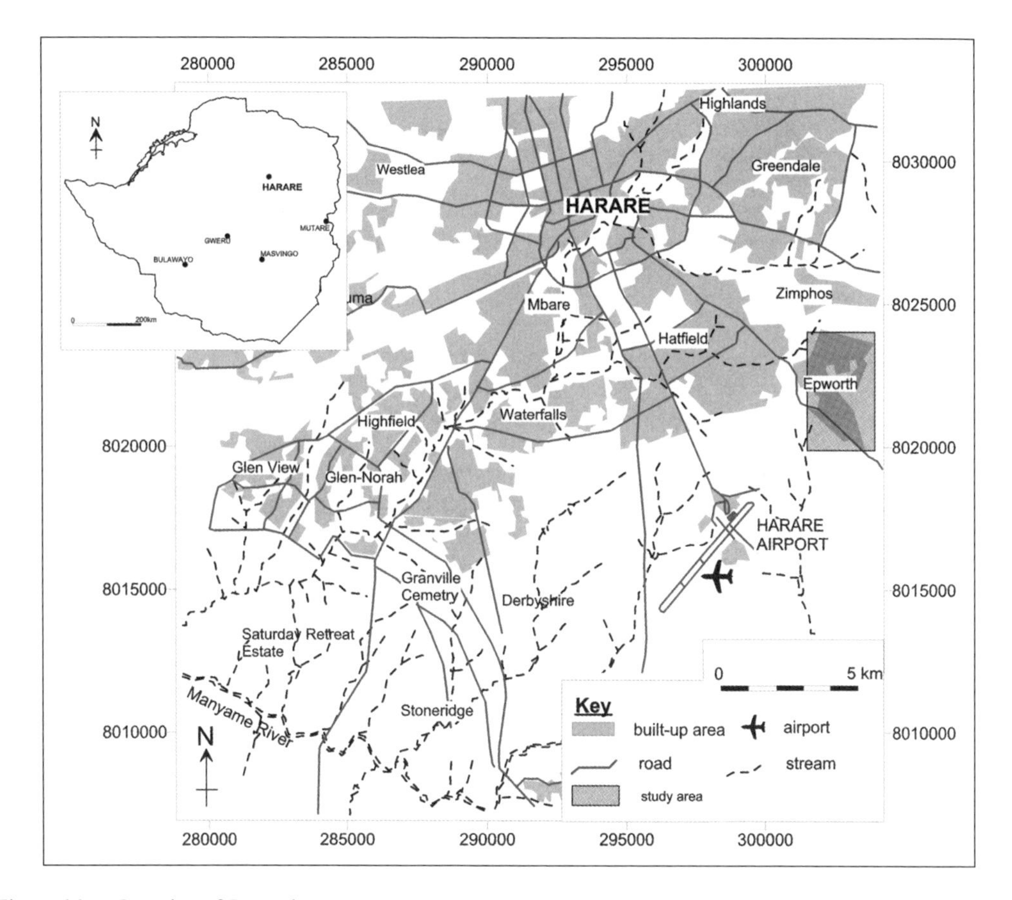

Figure 6.2. Location of Epworth.

An elected Local Board was established under the Urban Councils Act (1992) to manage the affairs of the settlement. The Epworth Local Board gets grants from the central government for infrastructure development and it also collects rates from residents in the serviced areas. The settlement now comprises five administrative sections of Maguta, Makomo, Zinyengerere, Chinamano and Overspill villages (Fig. 6.3). Each of these sections has an extension village.

Settlement patterns vary across different villages, with some exhibiting more regular patterns and others being characterized by haphazard patterns. Each village has both formal settlements and informal "gada" settlements. Gada residents occupy untenured, unserviced and low-lying or otherwise unattractive land. The "gadas" are considered to be informal settlements by the Epworth Local Board. The overall character of Epworth is thus a semi-formal or mixed formal/informal settlement. An illustration of such type of residential development is shown on Figure 6.4. Note the small dwelling sizes, pit latrine outhouses and the dambo (wetland) in the foreground.

1.4 *Objectives of this study*

This study was undertaken to determine the impact that the Epworth semi-formal settlement has on groundwater quality, and the implications for the settlement's water supply. It was carried out via two activities: a water supply and sanitation survey (to determine how residents obtain domestic water and how sewage is disposed of) and a groundwater quality investigation (to determine levels of key parameters in the shallow groundwater in Epworth).

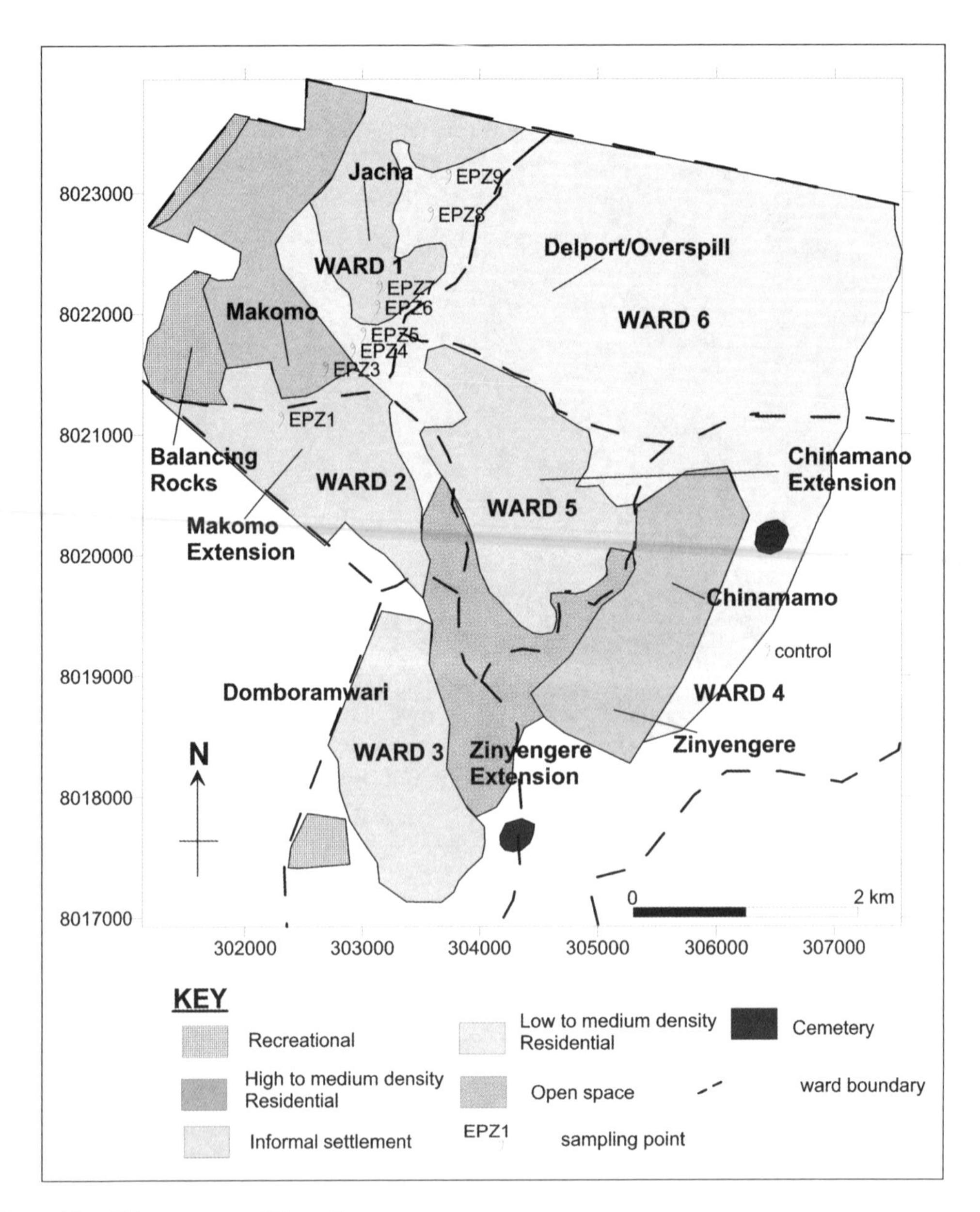

Figure 6.3. Urban structure of Epworth.

2 METHODOLOGY

2.1 *Water Supply and Sanitation*

A questionnaire was designed to collect qualitative and quantitative primary data from Epworth residents, aimed at getting information on water and sanitation services and practices. It was orally administered to 384 households and addressed a series of socioeconomic and demographic parameters. The households to be selected were spread across the six local board wards of Epworth, using a population-based weighting, which in turn determined the number of clusters required and the sample sizes (Table 6.2).

Figure 6.4. The informal "gada" settlement of Jacha, Ward 1, Epworth.

Within each cluster, systematic random sampling was used. Every fourth residential stand was chosen and a total of ten households randomly chosen were interviewed. Having determined which households to visit, the questionnaires were administered to present heads of household or a spouse, child or other relative or lodger regardless of whether they owned or rented their residencies. That ensured a mixture of people with varying land tenure status and socioeconomic characteristics.

2.2 *Groundwater Quality*

Reconnaissance sampling across the whole of Epworth settlement established problematically high levels of coliforms and nitrate (Zingoni et al. 2004). Detailed groundwater sampling was restricted to Ward 1, which was selected as the area where a high proportion of residents rely on shallow groundwater for water supply and on pit toilets for sewage disposal (based upon the water supply and sanitation survey, see below).

Nine boreholes were drilled. A control borehole was sited outside the Epworth built-up area, but within the same geological unit, which is the Harare Granite (Baldock et al. 1991). The boreholes were drilled

Table 6.2. Sample size and clusters per ward for household questionnaire, Epworth.

Ward	Population Size (CSO, 2003)	Weight	Sample size	Clusters required
1	22,976	0.202	77	6
2	13,101	0.115	44	3
3	10,865	0.095	37	3
4	20,253	0.178	68	5
5	17,174	0.151	58	5
6	29,515	0.259	100	8
Total	113 884	1.000	384	30

using a Vonder rig and cased with 6-inch plastic PVC casing. Concrete was used to secure the PVC casing and a lid made of PVC was used to cover the boreholes.

A sampling campaign was conducted: water samples were collected aseptically from the nine boreholes, refrigerated with ordinary ice in transit and analyzed within 24 hours.

Coliform bacteria were determined by the membrane filtration method: a measured volume of water was filtered through a membrane composed of cellulose esters. The pore size was such that the organisms to be enumerated are retained on or near the surface of the membrane, which was placed, face upwards, on a differential medium selective for the indicator organism sought. Volumes were chosen so that the number of colonies to be counted on the membrane would lie between 10 and 100. Membranes were incubated for 14 hours and at 37°C to determine total coliforms and separate membranes were incubated for 4 hours at 30°C, and then for 14 hours at 44.5°C. Shiny yellow colonies were counted. The following formula was used: (Coliform colonies counted/ml of sample filtered) $\times 100$ and expressed as colony forming units (Clesceri et al. 1989).

Electrical conductivity, pH, temperature, chloride concentration and phosphate concentration were measured in the field using a Horiba U 20 multi-meter. The determination of cations was done at the University of Zimbabwe Geology Department by atomic absorption spectrophotometer (AAS) (VARIAN SPECTRAA 200HT). A 21D spectrophotometer was used for nitrate determination by the ultraviolet spectrophotometric screening method at wavelengths of 220 nm and 275 nm on filtered samples.

3 RESULTS AND DISCUSSION

3.1 *Water Supply and Sanitation Survey*

The results of the survey are given in Tables 6.3 and 6.4.

Table 6.3. Water supply by ward, Epworth; sample size as per Table 6.2.

Ward	Indoor Tap	Outdoor Tap	Unprotected Well	Communal Tap	River/ Dam
1	6%	46%	48%	0%	0%
2	3%	66%	31%	0%	0%
3	0%	0%	43%	56%	1%
4	10%	52%	38%	0%	0%
5	2%	20%	78%	0%	0%
6	4%	77%	19%	0%	0%

Table 6.4. Sanitation type by ward, Epworth; sample size as per Table 6.2.

Ward	Flush Toilet	Ventilated improved pit latrine (Blair toilet)	Standard pit latrine
1	5%	53%	42%
2	0%	35%	65%
3	3%	56%	41%
4	13%	45%	42%
5	0%	38%	62%
6	25%	32%	43%

Almost half of all respondents use unprotected wells for water supply and over 90% use "Blair" ventilated improved pit toilets or ordinary pit toilets. In Ward 1, the oldest part of Epworth, only a small minority have either indoor taps or flush toilets. Most of the residents depend on shallow groundwater for both domestic water supply (via shallow wells) and sanitation (via pit latrines) - this often leads to a dangerous proximity of latrines to wells, as illustrated in Figure 6.5. The water is also used for urban agriculture, and for livestock watering by richer households.

In Ward 2 (Makomo Extension) all residents use Blair toilets or pit toilets, though only around 30% use unprotected wells for water supply. In Ward 3 (Dombowamwari) the majority are supplied with water by communal taps but over 90% use Blair toilets or pit toilets. Ward 4 (Chinamano) has a slightly lower pit toilet usage (over 80%) and lower usage of unprotected wells (less than 40%). Ward 5 (Chinamano Extension) has the worst situation with almost 80% of respondents supplied with water from unprotected wells and all respondents using Blair toilets or pit toilets. Ward 6 (Overspill) has the best water supply and sanitation picture with less than 20% of respondents supplied with water from unprotected wells and 25% using flush toilets rather than Blair toilets or pit toilets.

3.2 *Groundwater Quality*

The results of the groundwater quality study are given in Table 6.5.

There is a substantial variation in nitrate, phosphate and coliform distribution. All these parameters peak around boreholes EPZ4 and 5 (see Figs. 6.6 and 6.7) - which are located close to a "dambo" (seasonal wetland). Shallow aquifers collect water from rainfall events and slowly release excess water into dambos (Mharapara 1995). Therefore, local groundwater flow should be towards the dambo, accordingly, accumulation of nitrate, phosphate and coliforms is to be expected in such a downflow site.

Nitrate concentration is below risk level of 10 mg/l (WRC 1998) or 6 mg/l (DWAF 1996) except for boreholes EPZ4 and 5 (Fig. 6.6), which are located in the dambo. The elevated nitrate levels could be related to urban agriculture or to pit latrines, or to a combination of these.

Drinking water standards (WHO 1993) require no fecal coliforms in samples. However, water in boreholes EPZ3, 4, 5, 6 and 8 contained fecal coliforms (Fig. 6.7).

Figure 6.5. Shallow well (on left) in close proximity to a pit latrine (on right).

Table. 6.5. Groundwater quality in Ward 1, Epworth.

Sample	pH	Na (mg/l)	Fe (mg/l)	Zn (mg/l)	Ni (mg/l)	Co (mg/l)	Cu (mg/l)	Cl (mg/l)	NO_3 (N mg/l)	P_2O_5 (mg/l)	Total coli-forms (cfu)	Faecal coliforms (cfu)
Control	7.55	17.67	0.07	0.02	0.34	0.46	0.01	0.00	1.00	0.00	0	0
EPZ1	7.42	53.96	2.87	1.22	0.27	0.54	0.02	0.40	7.00	1.01	18,500	14,400
EPZ3	7.93	51.85	0.00	0.24	0.26	0.69	0.02	0.27	6.00	2.03	27,000	20,000
EPZ4	7.78	71.18	17.90	0.18	0.32	0.59	0.02	0.13	14.00	5.01	52,800	13,000
EPZ5	8.12	73.60	3.05	0.17	0.29	0.79	0.02	0.31	12.00	27.2	33,200	9,500
EPZ6	7.61	92.09	9.52	4.32	2.46	2.29	0.61	0.35	1.00	17.2	0	0
EPZ7	7.60	60.79	47.00	0.16	0.19	0.63	0.03	0.18	0.00	3.01	34,000	10,000
EPZ8	7.13	40.72	3.71	0.37	0.30	0.76	0.01	0.09	1.00	2.02	0	0
EPZ9	7.07	36.48	4.24	0.10	0.22	0.74	0.02	0.04	0.00	5.01	20,500	12,400
Mean	7.58	60.08	11.04	0.85	0.54	0.88	0.09	0.22	5.13	7.81	23,250	9,913
SD	0.37	18.39	15.56	1.45	0.78	0.58	0.21	0.13	5.57	9.38	17,781	6,911
Maximum recommended levels for different water uses												
Domestic (WHO, 1996)	---	---	---	---	0.02	---	2.00	---	50.00	---	0	0
Domestic (DWAF, 1996)	6.0-9.0	100	0.10	3.00	---	---	1.00	100.00	6.00	---	5	0
Livestock (DWAF, 1996)	---	2000	10.00	20.00	1	1	1.00	1500.00	100.00	---	---	200
Crops (DWAF, 1996)	7.5-8.4	70	5.00	1.00	0.20	---	0.20	100.00	5.00	---	---	1

The small proportion of households served by sewerage (about 5%, as per Table 6.2) indicates that high coliform levels in Ward 1 of Epworth are unlikely to be derived from sewer leakages – unlike in better-serviced urban environments, e.g. city of Bulawayo as reported by Mangore & Taigbenu (2004). Accordingly it is suggested that high coliform levels are a result of the widespread use of pit latrines of various types – which is 95% in Ward 1 of Epworth (Table 6.2).

The shallow depth of wells-as well as the incomplete lining of most wells-is liable to lead to very high bacterial contamination drinking water (Conboy 1999). Gastro-intestinal problems and other more serious health risks are likely to occur if the water is consumed and many people can be infected. This makes it risky to use the water as potable water. The water can be used for other purposes such as agriculture, and if groundwater is the only source of drinking water, then it must be boiled, since chlorinating would be an expensive alternative.

Metal levels are only problematic in the cases of iron and nickel. Nickel levels are unsafe throughout the area, including the control, suggesting that this is a natural regional problem. Iron levels are much higher in the settlement than at the control. This could be related to excavation and rusting of metal fixtures during construction activities, as more dwellings are built – both in terms of densification in the formal areas and in terms of further informal settlement. Levels of zinc are also much higher in the settlement than at the control, although they are within safe limits.

It is clear that the major shallow groundwater contamination problems associated with Epworth semi-formal settlement are increased levels of coliform bacteria and nitrates. High levels of these parameters are most likely to be caused by the abundant pit latrines – although nitrate levels could also have been influenced by urban agricultural practices. It is therefore suggested that efforts to mitigate the impact of semi-formal settlements on groundwater should focus firstly on improving sanitation – which also would improve livelihoods.

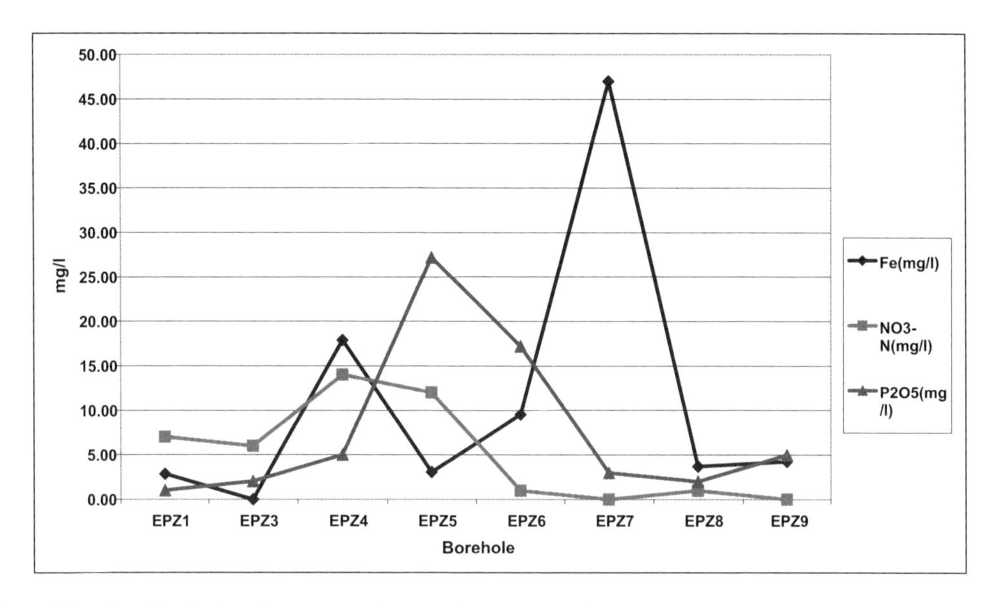

Figure 6.6. Levels of selected parameters in groundwater, Epworth.

It can be noted from this study that groundwater contamination is higher closer to dambos and these could be zoned as aquifer areas unsuitable as water sources for drinking purposes.

The only significant problems encountered with metals in the shallow groundwater at Epworth are the cases or iron and nickel. The latter is a natural regional problem and the former could be related to ongoing construction activities. Zinc distribution in the groundwater follows that of iron, and could be related to the same cause. However, the levels are sufficiently low to render zinc unproblematic as an environmental health issue currently.

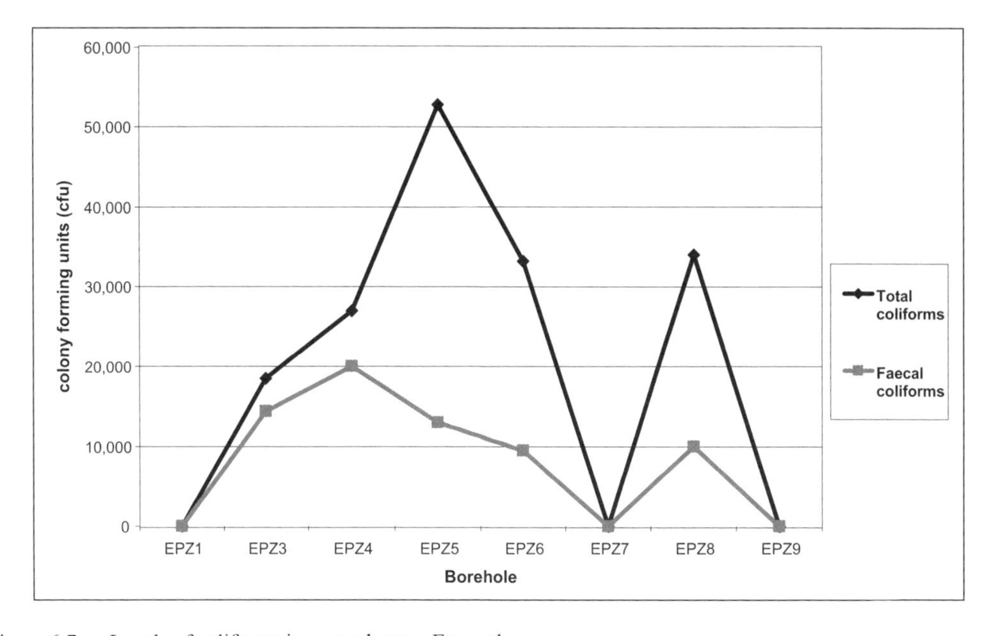

Figure 6.7. Levels of coliforms in groundwater, Epworth.

In the last decade a number of African governments have taken up the challenge of upgrading already existing squatter settlements (e.g. Kombe, in press). Upgrading is a locality-based infrastructure improvement strategy, that does not involve resettling people (Yeh 1987). Instead, such schemes implement on-site improvements in areas that are already occupied, but that have inadequate or non-existent basic services. Upgrading thus avoids both the inhumanity (and political consequences) of eviction and clearance and also the high cost expense of providing new housing. Upgrading is thus inexpensive, whilst preserving the existing socio-economic network for the urban poor. The alternative of full-scale slum redevelopment, completely replacing existing structures are re-housing residents, as carried out in Mumbai, India (Mukhija 2002), whilst having more far-reaching social benefits, is generally beyond the scale of funding central governments prefer to provide to peri-urban settlements. Upgrading has taken place in Epworth to a limited extent, but the growth of the settlement has far outstripped the development of infrastructure.

As a cost-effective measure to reduce health risk, the local authority could consider the provision of a limited water supply, via communal taps (already in use in Ward 3 as shown in Figure 6.8), to be used for drinking purposes only. Residents could then continue to extract water from wells for other domestic purposes, such as washing. Such infrastructure provision could be carried out by the state, budget permitting, or through tripartite public-NGO-private partnerships, which have proved successful in some informal settlements, such as in Kenya (Otiso 2003). It is also possible for residents to become private water vendors, whilst providing water of potable quality (Kyessi, in press).

Water supply and sanitation tariffs in low-income settlements have been kept at very low levels by central government controls on local authority rates (Gandidzanwa 2003). This is a challenge to the sustainability of any infrastructure development. However, it has been shown that in many low-income areas, a majority of residents would be prepared to pay higher rates if services were improved (Manase et al. 2004).

Point of use interventions, such as filters, have been suggested as a possible way to reduce the health impact of poor quality domestic water (Moyo et al. 2004). However, for such interventions to be sustainable in low-income areas, including informal settlements, the technologies need to be readily available and at low cost.

Figure 6.8. Residents queuing at a communal tap in Epworth, the "card" required is proof that the resident has paid for water.

Providing security of tenure for gada (informal) residents would also assist the situation. Provision of tenure has helped in the success of some informal settlements rehabilitation or upgrading programs (Choguill 1999, Kombe & Kreibich 2000, Pugh 2000). Not all informal settlement residents are from the lowest income group of society. Some residents occupy land in informal settlements due to lack of adequate sites in formal suburbs or lack of construction or purchase credit (Magatu 1991). Many of such residents are likely to be willing to invest in upgrading their household water supply or sanitation, but are unlikely to do so as long as they hold no tenure. Security of tenure has been shown to readily lead to community and individual investment in infrastructure development-even without financial assistance from central or local government (Winayanti & Lang 2004).

With better water supply and sanitation servicing (either communal or individual), Epworth residents would experience a variety of benefits. The immediate benefits of better health from safer drinking water would also have downstream impacts on nutrition and education. It has also been shown that such upgrading could benefit small- and home-based enterprises (Kigochie 2001, Tipple 2004), thus stimulating local economic growth. If household incomes and standards of living rise, residents are more likely to be willing to invest their own resources in improving their household sanitation. This, in turn, would lead to a gradual drop in pollution loadings from pit latrines.

In conclusion, provision of a limited amount of water supply servicing, together with security of tenure for gada residents, is probably the best medium-term solution to both health and shallow groundwater pollution problems at Epworth. This should be done as part of a holistic urban development process (Pugh 2000). In the case of Epworth, this clearly relates to the eventual incorporation of Epworth into the City of Harare, as required by the Combination Area Master Plan (HCMPPA 1989). This requires a clearly enunciated urban growth management policy (Frenkel 2004) as well as a far more substantial budgetary appropriation for Epworth from central government than has hitherto been provided. In addition to benefits to the Epworth residents, this would serve long term national strategic interests, such as decreasing diffuse pollution of the metropolitan water supplies and slowing urban sprawl and associated loss of agricultural land. It is to be expected that Epworth shall continue to grow, and as the Harare metropolitan area expands, the settlement is likely to grow, gradually developing an increasingly formal character. This is the trend with informal settlements in Africa, as a peri-urban settlement gradually gets built around and becomes urban (Fekade 2000). Accordingly, action is needed sooner rather than later as upgrading infrastructure becomes more difficult and costly with increased densification, and the implications of shallow groundwater pollution become more severe as an increasing number of users wish to draw on the resource.

4 CONCLUSIONS AND RECOMMENDATIONS

- The most significant groundwater quality problems in the Epworth settlement are high levels of nitrates and coliform bacteria.
- Nitrate and coliform bacteria levels are highest in the dambo areas.
- The abundant pit latrines in Epworth most likely cause the elevated levels of nitrates and coliform bacteria, although urban agriculture could be a secondary factor influencing the nitrate levels.
- Preventing further deterioration in groundwater quality requires major investment in piped sanitation. Unless this comes from central government, it will have to be self-financed and residents are unlikely to contribute to costs unless they have security of tenure.
- In the meantime, the local authority should increase its efforts to provide clean drinking water via communal taps, which is a cost-effective solution.
- Long term developmental needs of the metropolitan area require that upgrading of infrastructure in Epworth be done sooner, rather than later, as the settlement continues to grow.

Acknowledgements – The data presented and discussed in this paper were collected as part of the "Harare Urban Groundwater Project" funded by WARFSA and incorporate work carried out by Mr. Zingoni and Mrs. Gandidzanwa as part of their M.Sc. studies at the University of Zimbabwe (Zingoni, 2003; Gandidzanwa, 2003). The authors would like to thank the Epworth Local Board for information and support. Acknowledgement is also made to Mr. J. Tom, technical assistant in the Department of Geology, University of Zimbabwe for his help during data collection.

REFERENCES

Baldock, J.W., Styles, M.T., Kalbskopf, S. and Muchemwa, E. 1991. The geology of the Harare Greenstone Belt and surrounding granitic terrain, *Zimbabwe Geological Survey Bulletin* 94.

Butcher, C. 1993. Urban low-income housing: a case study of the Epworth squatter settlement upgrading programme. In: Zinyama, L.M., Tevera, D.S. and Cumming, S.D. (eds.) *Harare: The Growth and Problems of the City*. University of Zimbabwe Publications, Harare, Zimbabwe.

Cairncross, S. 1992. *Sanitation and Water Supply: Practical Lessons from the decade*. World Bank, Washington D.C.

Choguill, C.L. 1999. Community infrastructure for low-income cities: the potential for progressive improvement. *Habitat International*, 23, 289-301.

Chidavaenzi, M., Bradley, M., Jere, M. and Nhandara, C. 2000. Pit latrine effluent infiltration into groundwater. *Schiftenr Ver Wasser Boden Lufthyg*, 105, 171-7.

Clarge, N. 1999. *An assessment of the human excreta disposal practices in the poor urban area of Epworth, Zimbabwe: a review of the institutional, social and technical aspects*. Institute of Irrigation and Development Studies, Southampton University, UK.

Clesceri, L.S., Greenberg, A.E. and Eaton, A.D. (eds.) 1989. *Standard methods for the Examination of Water and Waste Water* (20th Edn). American Public Health Association, Washington D.C.

Conboy, M.J. 1999. Factors affecting bacterial contamination of rural drinking water wells: comparative assessment. In: Fitzgibbon, J.E. (ed.) *Advances in Planning and Management of Watersheds and Wetlands in Eastern and Southern Africa*. Weaver Press, Canada.

Conrad, J.E., Colvin, C., Sililo, O., Görgens, A., Weaver, J. and Reinhardt, C. 1999. Assessment of the impact of agricultural practices on the quality of groundwater resources in South Africa. *Water Research Commission Report*, 641/1/99.

CSO (Central Statistics Office). 1999. *Demographic and Healthy Survey 1994*. Government Printer, Harare.

CSO (Central Statistics Office). 2003. *National Population Census*. Government Printer, Harare.

Dahiya, B. 2003. Peri-urban environments and community driven development: Chennai, India. *Cities*, 20, 341-352.

DWAF (Department of Water Affairs and Forestry) 1996. *South African Water Quality Guidelines, 8: Field Guide*. Government Printer, Pretoria.

Feachem, R.G., Bradley, D.J., Garelich, H. and Mara, D.D. 1981. *Health Aspects of Excreta and Wastewater Management*. Johns Hopkins University Press, Baltimore.

Fekade, W. 2000. with informal settlements Deficits of formal urban land management and informal responses under rapid urban growth, an international perspective. *Habitat International*, 24, 127-150.

Foster, S.S.D. 1999. The interdependence of groundwater and urbanisation in rapidly developing countries. *Urban Water*, 3, 185-192.

Frenkel, A. 2004. The potential effect of national growth-management policy on urban sprawl and the depletion of open spaces and farmland. *Land Use Policy*, 21, 357-369.

Gandidzanwa, P. 2003. *Attitudes and practices among small urban communities in relation to water services in an area-beyond-network: case of Epworth*. Unpublished MSc thesis, Environmental Policy and Planning Programme, University of Zimbabwe.

Gullis, G., Mulumba, J.A.A., Juma, O. and Kakosova, B. 2004. Health status of people of slums in Nairobi, Kenya. *Environmental Research*, 96, 219-227.

Gwebu, T.D. 2003. Environmental problems among low income urban residents: an empirical analysis of old Naledi-Gaborone, Botswana. *Habitat International*, 27, 407-427.

Hardoy, J. and Sattethewaite, D. 1989. *Squatter Citizen – Life in the Third World*. Earthscan, London.

HCMPPA (Harare Combination Master Plan Preparation Authority). 1989. *Harare Combination Master Plan: Report of Study*. HCMPPA, Harare.

Kigochie, P.W. 2001. Squatter rehabilitation projects that support home-based enterprises create jobs and housing. *Cities*, 18, 223-233.

Kombe, W.J. *in press*. Land use dynamics in peri-urban areas and their implications on the urban growth and form: the case of Dar es Salaam, Tanzania. *Habitat International*. 29, 113-135.

Kombe, W.J. and Kreibich, V. 2000. Reconciling informal and formal land management: an agenda for improving tenure security and urban governance in poor countries. *Habitat International*. 24, 231-240.

Kyessi, A.G. *in press*. Community-based urban water management in fringe neighbourhoods: the case of Dar es Salaam, Tanzania. *Habitat International*. 29, 1-25.

Losada, H., Bennett, R., Soriano, R., Vieyra, J. and Cortés, J. 2000. Urban agriculture in Mexico City: functions provided by the use of space for dairy based livelihoods. *Cities*, 17, 419-431.

Lynch, K., Binns, T. and Olofin, E. 2001. Urban agriculture under threat: the land security question in Kano, Nigeria. *Cities*, 18, 159-171.

Magatu, G. 1991. Site and service for low income housing. In: Turner (Ed.) *Review of Rural and Urban Planning in Southern and Eastern Africa*. World Bank, Washington D.C.

Makoni, F.S. 2001. *An assessment of water and sanitation facilities and related diseases among poor urban communities in Zimbabwe*. Unpublished MSc thesis, Public Health Programme, University of the North, South Africa.

Manase, G. 2001. *Willingness to pay for improved sanitation: The case study of poor urban areas in Zimbabwe*. Unpublished Ph.D. Thesis, University of Cape Town.

Manase, G., Ndamba, J. and Fawcett, B. 2004. Financing services for the urban poor: the case of willingness to pay for sanitation at "Growth Points" of Zimbabwe. In: *Proceedings of the IWA Specialist Group Conference on Water and Wastewater Management for Developing Countries*, Victoria Falls, Zimbabwe.

Mangore, E. and Taigbenu, A.E. 2004. Land-use impacts on the quality of groundwater in Bulawayo. *Water SA*, 30, 453-464.

McCarthy, T.S., Gumbricht, T., Stewart, R.G., Bradnt, D., Hancox, P.J., McCarthy, J. and Duse, A.G. 2004. Wastewater disposal at safari lodges in the Okavango Delta, Botswana. *Water SA*, 30, 121-128.

Mharapara, I. 1995. A fundamental approach to dambo utilization. In: Owen, R., Verbeek, K., Jackson, J. and Steenhuis, T. (Eds.) *Dambo Farming in Zimbabwe: Water Management, Cropping and Soil Potentials for Smallholder Farming in the Wetlands*. University of Zimbabwe Publications, Harare.

Moyo, S., Ndamba, J., Gundry, S.W. and Wright, J. 2004. Point of use interventions as a strategy to improve and maintain rural household water quality. In: *Proceedings of the IWA Specialist Group Conference on Water and Wastewater Management for Developing Countries*, Victoria Falls, Zimbabwe.

Mukhija, V. 2002. An analytical framework for urban upgrading: property rights, property values and physical attributes. *Habitat International*, 26, 553-570.

Mvungi, A., Hranova, R.K. and Love, D. 2003. Impact of home industries on water quality in a tributary of the Marimba River, Harare: implications for urban water management. *Physics and Chemistry of the Earth*, 28, 1131-1137.

National Academy of Sciences. 1987. *More Water for Arid Lands-Promising Technologies and Research Opportunities*, National Academy of Sciences Publications, Washington D.C.

Obhudo, R.A and Mhlanga C.C. 1998. The development of slum and squatter settlements as a manifestation of rapid urbanisation in Sub-Saharan Africa. In: Obhudo, R.A. and Mhlanga, C.C. (Eds.) *Slum and Squatter Settlement in Sub-Saharan Africa: Towards a Planning Strategy*. Praecer, New York.

Obi, C.L., Green, E., Bessong, P.O., de Villiers, B., Hoosen, A.A., Igumbor, E.O. and Potgieter, N. 2004. Gene encoding virulence markers among *Escherichia coli isolates* from diarrhoeic stool samples and river sources in rural Venda communities of South Africa. *Water SA*, 30, 37-42.

Otiso, K.M. 2003. State, voluntary and private sector partnerships for slum upgrading and basic service delivery in Nairobi City, Kenya. *Cities*, 20, 221-229.

Pugh, C. 2000. Squatter settlements: their sustainability, architectural contributions, and socio-economic roles. *Cities*, 17, 325-337.

Rakodi, C. 1995. *Harare: Inheriting a Settler-Colonial City: Change or Continuity?* Wiley, Chichester.

Ren, W., Zhong, Y., Meligrana, J., Anderson, B., Watt, W.E., Chen, J. and Leung, H.-L. 2003. Urbanization, land use, and water quality in Shanghai, 1947 – 1996. *Environment International*, 29, 649-659.

Tipple, G. 2004.Settlement upgrading and homebased enterprises: discussions from empirical data. *Cities*, 21, 371-379.

Tsvere, M., Tirivarombo, S. and Nhunzvi, P. 2004. Poverty, conflicts and domestic water resource use in poor areas of Zimbabwe. In: *Proceedings of the IWA Specialist Group Conference on Water and Wastewater Management for Developing Countries*, Victoria Falls, Zimbabwe.

WHO (World Health Organisation) 1996. *Guidelines for Drinking Water Quality, Volume 1: Recommendations* (2nd Edn). World Health Organisation, Geneva.

Winayanti, L. and Lang, H.C. 2004. Provision of urban services in an informal settlement: a case study of Kampung Penas Tanggul, Jakarta. *Habitat International*, 28, 41-65.

Winde, F. and van der Walt, I.J. 2004. Gold tailings as a source of waterborne uranium contamination of streams - The Koekemoerspruit (Klerksdorp goldfield, South Africa) as a case study. Part II of III: Dynamics of groundwater-stream interactions. *Water SA*, 30, 227-232.

WWEDR 2000. *Water (Waste and Effluent Disposal) Regulations*, Statutory Instrument 274 of 2000, Republic of Zimbabwe.

WRC (Water Research Commission) 1998. *Quality of Domestic Water Supplies, Volume 1: Assessment Guide* (2nd Edn.). Water Research Commission, Pretoria.

Wright A. 1997. *Towards a Strategic Sanitation Approach: Improving the Sustainability of Urban Sanitation in Developing Countries*. UNDP-World Bank Water and Sanitation Program, Washington D.C.

Wright, A. 1999. Groundwater contamination as a result of developing urban settlements. *Water Research Commission Report* 514/1/99.

Yeh, S.H.K. 1987. Urban low income settlement in developing countries characteristics and improvement strategies. *Regional Development Dialogue*, 8, 1-29.

Zegarac, Z. 1999. Illegal construction in Belgrade and the prospects for urban development planning. *Cities*, 16, 365-370.

Zingoni, E. 2003. *Microbiological and chemical groundwater contamination from a semiformal settlement, Epworth, Zimbabwe*. Unpublished MSc thesis, Tropical Resource Ecology Programme, University of Zimbabwe.

Zingoni, E., Love, D., Magadza, C., Moyce, W. and Musiwa, K. 2004. Groundwater use zones and water supply options for Epworth semi-formal settlement, Zimbabwe. In: *Proceedings of the 5th WaterNet-WARFSA Symposium*, Windhoek, Namibia.

CHAPTER 7

Impacts of a solid waste disposal site and a cemetery on groundwater quality in Harare, Zimbabwe

W. Moyce, D. Love, K. Musiwa, Z. Nyama, P. Mangeya, S. Ravengai, M. Wuta & E. Zingoni

ABSTRACT: Among the more significant large municipal facilities to impact on the quality of shallow groundwater are landfills and cemeteries. At the same time, there is an increasing small-scale use of shallow groundwater as a water source in rapidly expanding cities. In this chapter, the pollution of shallow groundwater around Harare's largest landfill and its largest cemetery is characterized and recommendations made for mitigating the facilities' impacts. At each site, boreholes were drilled, groundwater sampled and chemical and microbiological analyses undertaken. The investigation at the landfill revealed typically high levels of contamination, particularly with regards to metals, nitrates and coliform bacteria. These can be related to the landfill, although activities in the suburb downflow will also have contributed. The investigation of groundwater at the cemetery established contamination with respect to some metals, nitrates, phosphate and coliform bacteria. These can be related to decomposition of bodies, and to natural hydrolysis of the aquifer rock unit. Grouting and the use of gum trees as a pressure barrier are recommended for the downflow margins of both the landfill and the cemetery. Conditions at the landfill could also be treated with locally available iron and sulphate minerals and by revegetating the top of the dump. The problems experienced at both sites relate to poor planning practices, and will be repeated in the future unless environmental and geotechnical considerations are given priority in site selection for such facilities.

1 INTRODUCTION

1.1 *Objectives of this study*

In the City of Harare, Zimbabwe, there is a widespread small-scale use of shallow groundwater by industrial and private residential establishments. Research in some parts of Harare has shown that such shallow groundwater is polluted (Hranova et al. 2003, Ravengai & Love 2004, Zingoni et al. 2004). At the same time, Harare is rapidly urbanizing and is faced with the effects of urbanization such as population growth, pollution, and demand for land for settlement. The rapid population growth and demand for land has lead to non-ideal residential development, where residential stands are developed in close proximity to pollution sources. This is especially problematic where shallow groundwater is used as a water source.

Among the more significant municipal facilities to impact on groundwater quality are landfills and cemeteries. In this study, the pollution of shallow groundwater around Harare's largest landfill and its largest cemetery is characterized and recommendations made for mitigating the facilities' impact and for medium-term urban planning.

1.2 *Solid waste disposal and its impact*

Due to rapid urbanization groundwater is becoming increasingly vulnerable to pollution from human activities (Aldrick et al. 1999). Some of the most problematic sources of groundwater pollution in the urban environments are municipal landfills (Dutova et al. 1999). Formal municipal landfills are often major sources of contamination – especially where they are unlined, as is common in the developing world. A whole suite of contaminants such as sodium, potassium, ammonia, nitrate, nitrite, chloride and heavy metals such as iron, manganese, lead, mercury and chrome, can leach out of landfills (Fetter 1994, Stone 1996, Zhu et al. 1997), although leachate composition and pollution loadings are also affected by climate (Tatsi & Zouboulis 2002). The strategic impact of a landfill can be considered as a result of three factors: pollution load, permeability and reactivity of the unsaturated zone and the value of the groundwater resource under threat (Parsons & Jolly 1994).

The location of a landfill, like any other potential contaminator of groundwater, is critical. Proximity to faulted or unstable areas and wetlands can be highly problematic (USEPA 1993), as such proximity increases the risk of contamination, or the speed of movement of the contaminant plume, or both. Faulted areas allow easier contamination of groundwater by leachate, unstable areas (such as seismically active zones or sinkhole areas) risk collapse of the landfill and wetlands are ecologically highly sensitive to many forms of contamination.

1.3 *Cemeteries and groundwater quality*

Large-scale human decomposition processes associated with cemeteries could be interpreted as special kind of landfills (Dent & Knight 1998, Dent et al. 2004). In cemeteries, human corpses may cause groundwater pollution not by virtue of any specific toxicity they possess, but by increasing the concentrations of naturally occurring organic and inorganic substances to a level sufficient to render groundwater unusable or unfit for potable purpose (Usisik & Rushbrook 1998).

The nature of pollution emanating from a cemetery site differs from that produced by conventional solid waste-disposal sites, primarily because cemetery leachate pose a greater health hazard. Studies in a graveyard in Wolverhampton found out that the groundwater was contaminated by bacteria from decaying bodies (Trick et al. 2002). Microbiological tests indicated thermo-tolerant coliforms and fecal streptococci at concentration above World Health Organisation standards (Trick & Klinck, 2001). This is despite the fact that some peri-urban communities near cemeteries depend upon shallow groundwater as a water source. There is good evidence that bacteria, which decompose corpses can reach the water table, especially where groundwater flows through fractures, which provide faster and rapid movement (Trick et al. 2002). Studies in Germany reported high concentrations of bacteria, ammonium and nitrate in a contamination plume, which rapidly diminished with distance from graves and a study in the Netherlands reported a very saline (2300 μS/cm) plume of chloride, sulphate and bicarbonate ions beneath graves (Usisik & Ruchbrook 1998).

The body of a 70 kg adult human male contains approximately: 16 kg carbon, 1.8 kg nitrogen, 1.1 kg calcium, 500 g phosphorous, 140 g sulfur, 140 g potassium, 100 g sodium, 95 g chlorine, 19 g magnesium, 4.2 g iron, and water 70–74% by weight (Usisik & Rushbrook 1998). Human corpse undergoes autolysis and putrefaction in the early stages of decomposition, then liquefaction and disintegration and final skeletonization (Dent et al. 2004). The leachate (liquids generated during decay) generally contain contaminants such as ammonia, Ca, Mg, Na, Fe, K, P, Cl and SO_4, as well as a variety of pathogenic organisms, such as protozoa, many algae, fungi, bacteria and viruses (Engelbrecht 1998, Environmental Agency 2002). The role of microbial contaminants in groundwater is in the transmission of waterborne disease. Waterborne diseases of concern such as cholera, typhoid fever, cryptosporidiosis and viral gastroenteritis can originate from protozoa such as *Cryptosporidium parvum*, *Giardia lamblia* and bacteria such as *Legionella* or *Vibrio* species of enteric and other viruses (Centre of Disease Control and Prevention 2000, Environmental Agency 1999). It is difficult or even impossible to detect some pathogens of concern with the current technology; hence indicator bacteria (coliforms) are used as indicators of possible contamination and as an index of water quality deterioration.

1.4 *Landfill study area*

The recently developed suburb of Westlea, Harare is located very close to the municipal Golden Quarry landfill, downwind and down-flow of the waste disposal site. Harare has two active landfills: Golden

Quarry in the Warren Hills area and Teviotdale in the Pomona area. Both are former quarries. Solid waste also used to be dumped close to the Mukuvisi River, immediately south of the main railway station (Zaranyika 1997). The Golden Quarry is the largest municipal landfill in Harare, and is located some 7 km from central Harare, on the edge of Westlea suburb, along the main road to Bulawayo, westwards from the city center (Fig. 7.1.). The Golden Quarry landfill was selected for study because of the availability of sixteen boreholes drilled in the late 1990s under a project sponsored by SIDA-SAREC, and since it took the vast majority of the city's waste (Rakodi 1995).

The landfill has been used for disposal of both domestic waste and industrial waste. Figure 7.2 presents a location map of the study area, and Figures 7.3 and 7.4 illustrate the physical appearance of the disposal site. At the same time, some of the residents depend upon shallow groundwater as a water source.

There are trenches dug on the western side of the dump to capture runoff from the western side of the landfill, and prevent it flowing to the Westlea residential area. Gum trees are planted along the western and southern sides of the landfill. Shallow groundwater flows from a high at the dumpsite, westwards under Westlea suburb, towards a tributary of the Marimba River (Fig. 7.2). The Marimba River flows into Lake Chivero, Harare's main water source. Some Westlea residents extract shallow groundwater, especially for urban agriculture, but also for limited domestic uses, such as washing.

Dumping at Golden Quarry started in 1985 on an abandoned gold mine, located in banded ironstone (Baldock et al. 1991). Operation of the site as been hampered from time to time by problems such as a shortage of compacting equipment (Tevera 1991) and occasional outbreaks of fire. Dumping has now

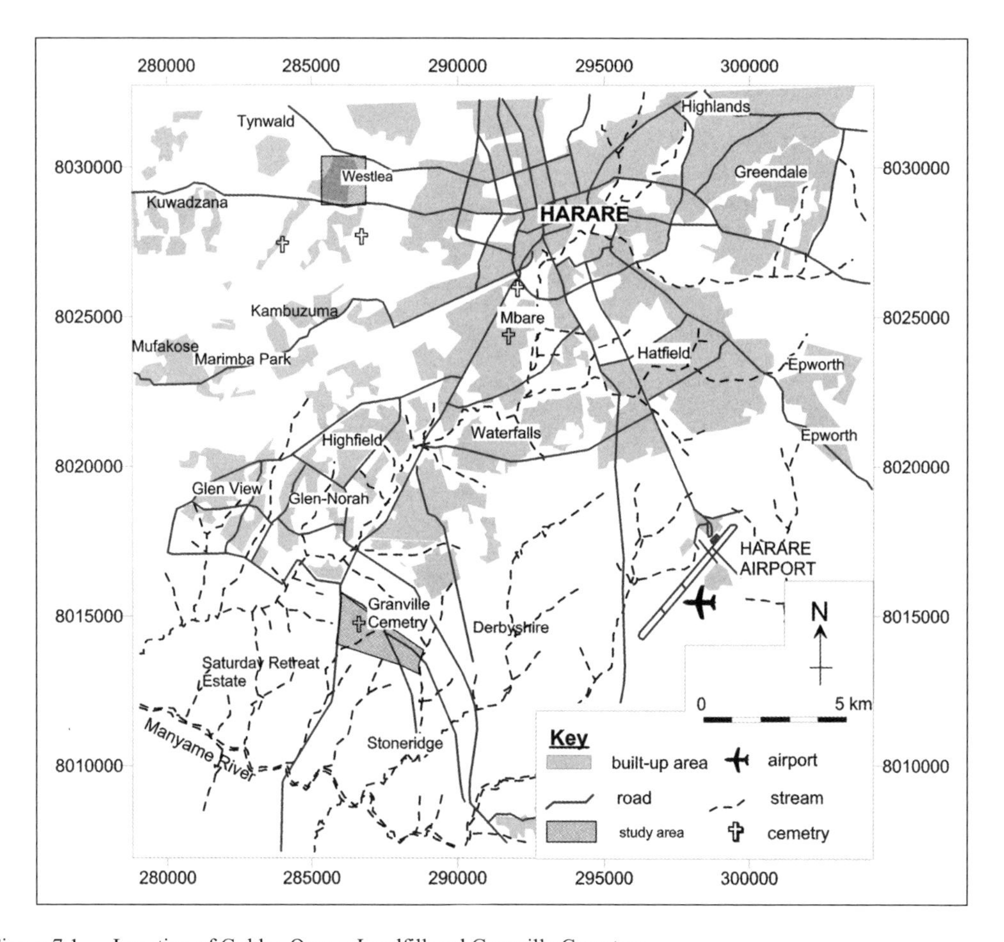

Figure 7.1. Location of Golden Quarry Landfill and Granville Cemetery.

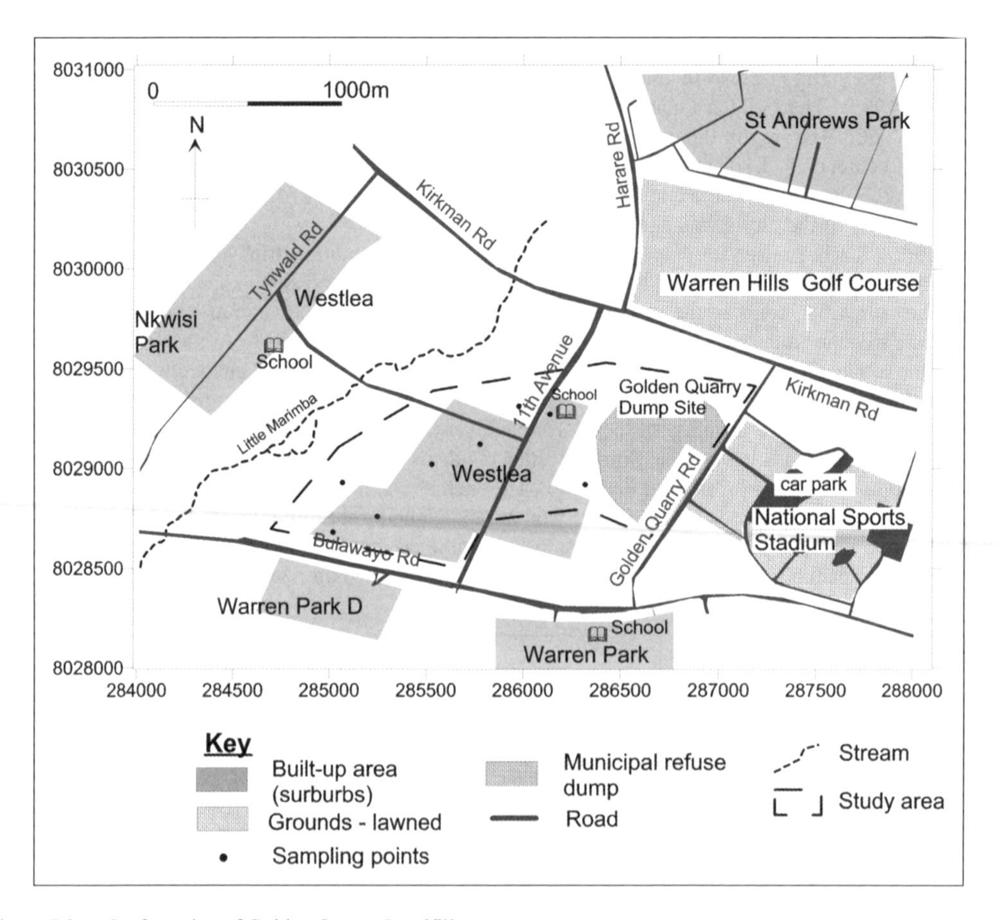

Figure 7.2. Surface plan of Golden Quarry Landfill.

largely ceased. Golden Quarry landfill has received solid waste (see Fig. 7.3) and liquid waste (see Fig 7.4). There has been dumping both by the City authorities (mainly domestic and commercial

Figure 7.3. Solid waste disposal, Golden Quarry.

Figure 7.4. Liquid waste disposal, Golden Quarry.

refuse) and by individual companies. The latter have included food and beverage producers, heavy industries, pharmaceuticals, clothing companies and paint producers (Madimutsa 2000, Nyama 2003). During the late 1990s, average daily waste entering the landfill is more than 700 tones of solid waste and more than 100 m^3 of liquid waste. Domestic and commercial waste disposal was concentrated on the northern side of the landfill, and industrial waste on the southern side.

1.5 *Cemetery study area*

Granville Cemetery is located about 15 km from the Harare city center along the Beatrice road south of the city (Fig. 7.1.). Granville Cemetery was selected for study because it is the largest cemetery in Harare and since it receives the vast majority of the city's burials.

Granville lies in the Harare granites, which are coarse-grained, massive and weathers to give rise to sandy soils (Baldock et al. 1991). The cemetery is in a topographic low, and the water table is high, reaching the surface and causing waterlogging during the rainy season. Shallow groundwater in the area flows southwestwards, into farmland, and feeds a tributary of the Mukuvisi River, which in turn flows into Lake Chivero, Harare's main water source.

Burial at Granville Cemetery started in 1995 and is divided into three sections, 'A', 'B' and a pauper section (Fig. 7.5). 'A' section has the expensive and concreted graves and 'B' section has cheaper non-concreted graves (Fig. 7.6). Graves in Granville 'A" and 'B' are 2.2 × 0.8 m for adults and 1 × 0.5 m for infants in size and have varying depth according to the type of burial (Table 7.1). The depth of the graves is dependent on the anticipated number of coffins and the graveyard section as shown on Table 7.1.

This study focuses on Granville "B". Most of the burials in Granville B are with cheap coffins made of soft wood either covered by black cloth or varnished. In the pauper section (Fig. 7.7) burial is for 3 corpses per grave in separate coffins. Since burials started in 1999, the cemetery section has expanded eastwards (Fig. 7.8).

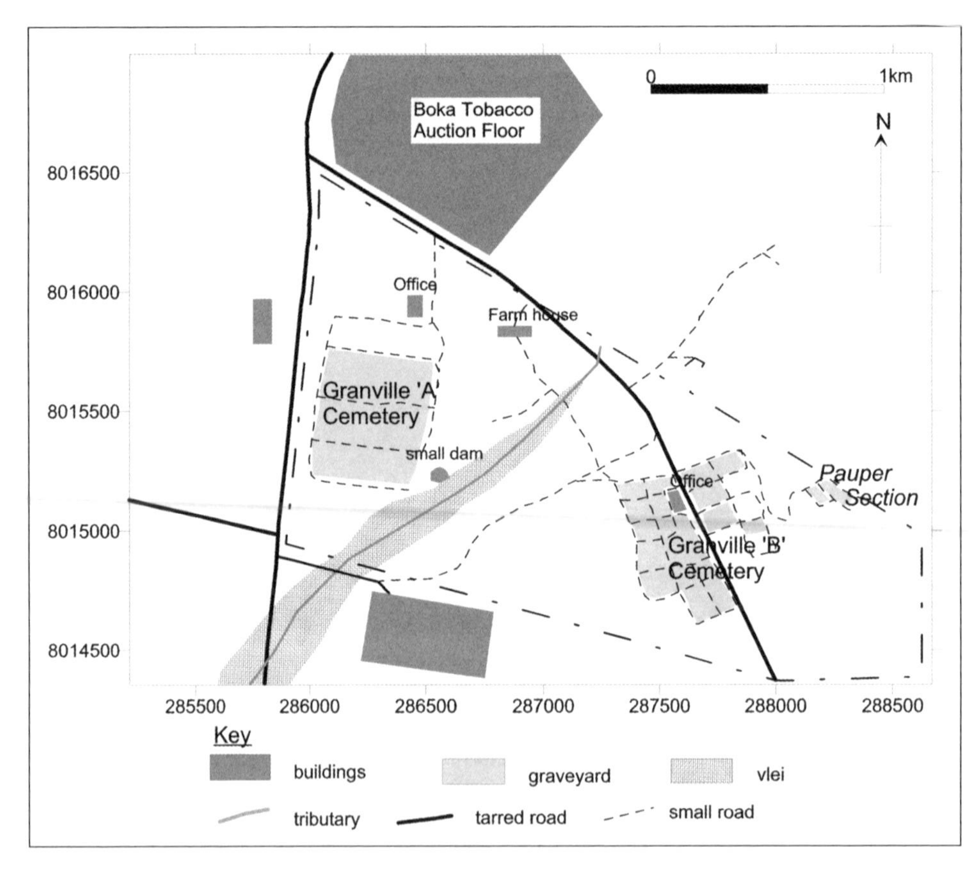

Figure 7.5. Surface plan of Granville Cemetery.

2 METHODOLOGY

2.1 *Location and drilling of boreholes*

At Golden Quarry, nine boreholes were used for groundwater sampling, including a control, located far from the landfill but in the same geological unit (see Fig. 7.2). The boreholes, drilled under a SIDA-SAREC project in the late 1990s, were all drilled using a Vonder Rig, up to a maximum depth of 7.6 m, and cased as described in Chapter 6.

At Granville, seven boreholes were sited and drilled using a Vonder Rig, up to a maximum depth of 10 m. These boreholes were sited in and around the graves (Fig. 7.6). Three boreholes (GBH8, GBH9 and

Table 7.1. Depth of burials in Granville Cemetery.

Section	Number of coffins	Depth of grave from ground level
Granville 'A' & 'B' - adults	1	1.83 m
Granville 'A' & 'B' - children	1	1.40 m
Pauper - adults	3	2.14 m
Pauper - children	3	2.14 m

Figure 7.6. Recent graves in "B" Section, Granville Cemetery.

GBH10) were pre-existing, now property of the city Council but originally installed by the owner of Granville Estate.

A control borehole was chosen from the pre-existing boreholes upslope of and far from the influence of the cemetery in the same geological unit. The new boreholes, with depths ranging between 2.5 m and 7.6 m, were all drilled using a Vonder Rig, and cased as described in Chapter 6.

2.2 *Sampling and chemical analysis*

For both sites, samples were collected for microbiology, physical parameter and chemical parameter analysis. Water samples were collected using glass-sampling bottles sterilized by autoclaving prior to sampling. All the tests carried out were according to standard methods (Clesceri et al. 1989). The physical parameters were determined in the field according to the guidelines outlined in the field sampling kit. Microbiology was done by the membrane filtration method as described in Chapter 6. Nitrate, phosphorous and total metals were determined as described in Chapter 4. Chloride, carbonate and bicarbonate were determined

Figure 7.7. Pauper Section, Granville Cemetery.

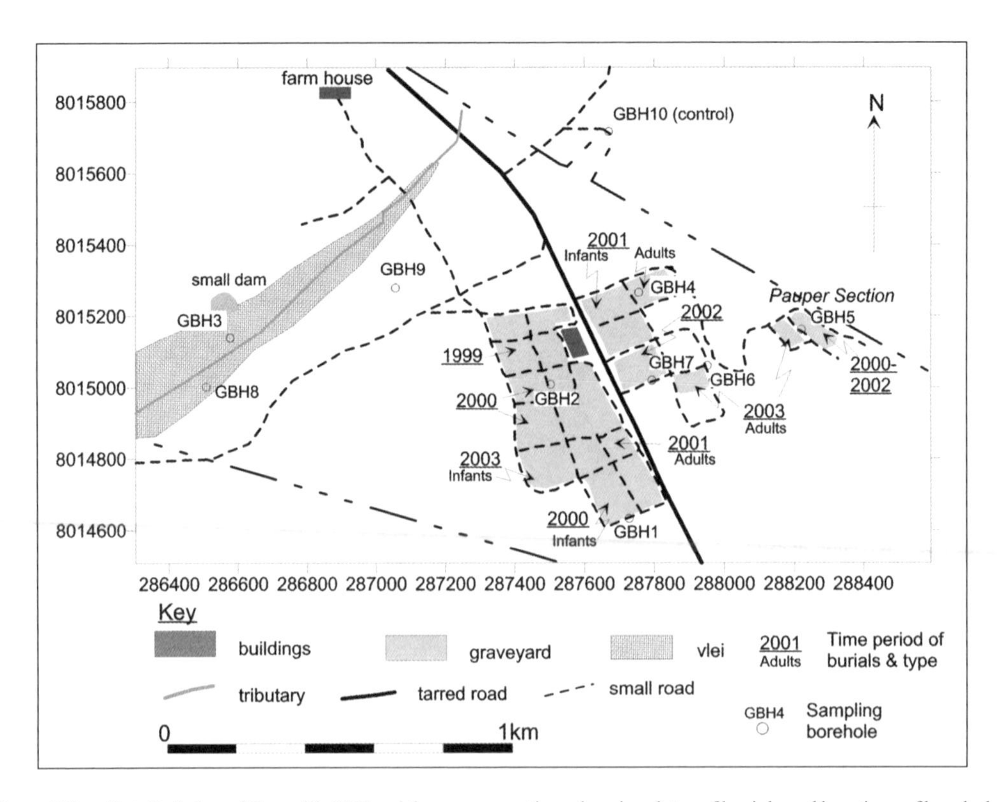

Figure 7.8. Detailed plan of Granville "B" and the pauper section, showing dates of burials and locations of boreholes.

using the titration method and sulphate using the gravimetric method. All analyses were done at the University of Zimbabwe.

3 RESULTS AND DISCUSSION

3.1 *Golden Quarry Landfill*

Levels of coliforms, cadmium, iron, lead, nitrate and pH were above water quality guidelines, hence the water is generally unsafe for domestic use (Table 7.2). Levels of metals decreased westwards-with groundwater flow (Fig. 7.9-7.12). Metal levels drop going westwards from 0.53 mg Cd/l, 11.05 mg Fe/l, 0.21 mg Pb/l and 0.06 mg Zn/l near the south side of the landfill to 0.22 mg Cd/l, 1.38 mg Fe/l, 0.12 mg Pb/l and 0.02 mg Zn/l to the southwest end of Westlea. Levels of Pb, Fe and Cd showed high concentrations in all boreholes, above water quality guidelines, making the groundwater unsuitable for drinking throughout Westlea. The high metal levels do, however, attenuate fairly rapidly downflow (within 1 km), as is expected from numerous studies elsewhere (Christensen et al. 2001).

Coliform levels also decrease westwards with groundwater flow, from 10,000 coliform units (cfu) near the north side of the landfill to 0 cfu at the southwest end of Westlea, making the water unsafe for drinking in northern Westlea. Nitrate levels were highest at 16.55 mg/l near the north side of the suburb and decrease with groundwater flow to 0.40 mg/l with a spot at BH6 (10.20 mg/l) probably being due to urban agriculture, sewage leakages and/or pit latrines (Fig. 7.13).

A comparison with the control borehole suggests that the landfill is a source of groundwater pollution, with metals, nitrates and coliform bacteria all showing elevated levels in the study boreholes. The metal levels could be explained by the disposal of industrial waste on the landfill – except for iron, which is most

Table 7.2. Groundwater quality data, 2004, Golden Quarry Landfill, Harare.

	pH (pH units)	Cd (mg/l)	Fe (mg/l)	Pb (mg/l)	Zn (mg/l)	NO_3 (mg/l)	SO_4 (mg/l)	Cl (mg/l)	CO_3 (mg/l)	Total coli-forms/100 ml	Faecal coli-forms/100 ml
BH1	5.79	0.53	11.05	0.21	0.06	16.55	47.73	7.90	273.00	23000	0
BH2	6.35	0.39	1.88	0.14	0.04	10.13	45.30	3.50	129.00	14000	6000
BH3	6.82	0.32	3.79	0.16	0.05	10.89	28.43	2.11	34.05	0	0
BH4	5.97	0.46	1.48	0.12	0.03	3.92	26.53	2.64	32.00	4000	1000
BH5	6.24	0.37	2.26	0.19	0.02	1.49	19.00	2.35	71.00	0	0
BH6	7.03	0.22	2.52	0.19	0.04	10.20	21.83	4.40	46.00	3000	0
BH7	6.15	0.46	4.72	0.15	0.03	0.40	13.63	3.70	31.00	0	0
BH8	6.37	0.38	1.38	0.19	0.03	4.10	32.54	2.60	19.00	0	0
Control	7.01	0.01	1.21	0.01	0.01	0.20	31.51	13.20	62.00	0	0
Maximum recommended levels for different water uses											
Domestic (WHO, 1993)	---	0.003	0.30	0.01	3.00	50.00	250.00	250.00	---	0	0
Domestic (DWAF, 1996)	6.0-9.0	5.00	0.10	0.01	3.00	6.00	200.00	100.00	---	5	0
Livestock (DWAF, 1996)	---		10.00	0.10	20.00	100.00	1000.00	1500.00	---		200
Crops (DWAF, 1996)	7.5-8.4		5.00	0.20	1.00	5.00	---	100.00	---		1

likely related to the iron-rich bedrock (banded ironstone) under the landfill. Levels of coliforms and nitrate can be related to the disposal of mainly domestic refuse on the north side of the landfill. There may also have been some limited contribution from the residential area, due to the presence of a small number of pit latrines and the practice of urban agriculture in residential backyards.

It can be recommended that the city of Harare should plant more gum trees, especially around the western (downflow) edge of the landfill. Gum trees take up a lot of water, creating a cone of depression, and thus

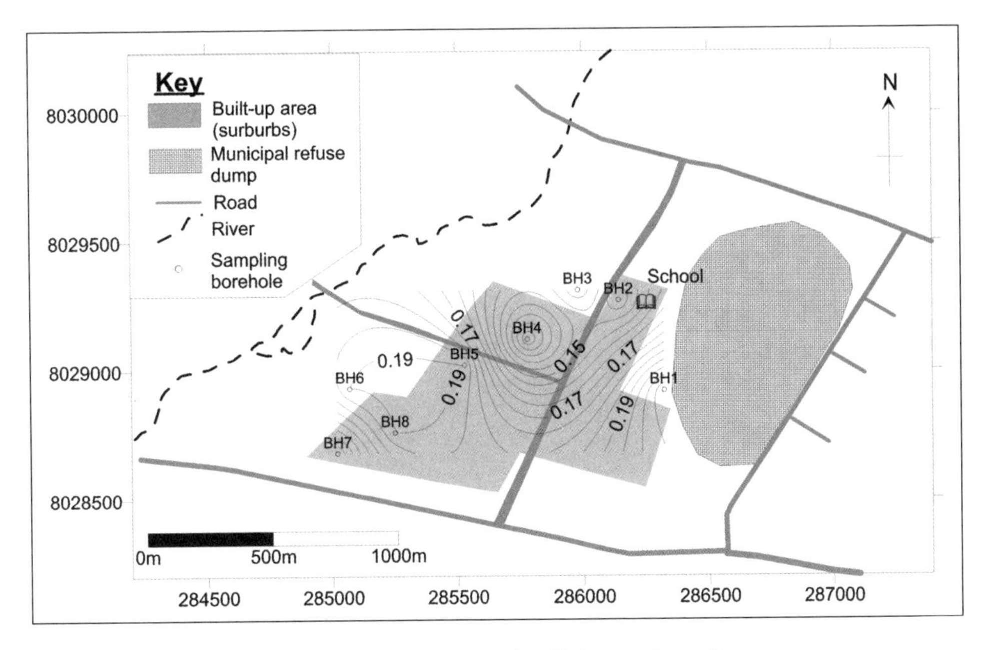

Figure 7.9. Spatial variation of lead in groundwater, Westlea (Values are in mg/l).

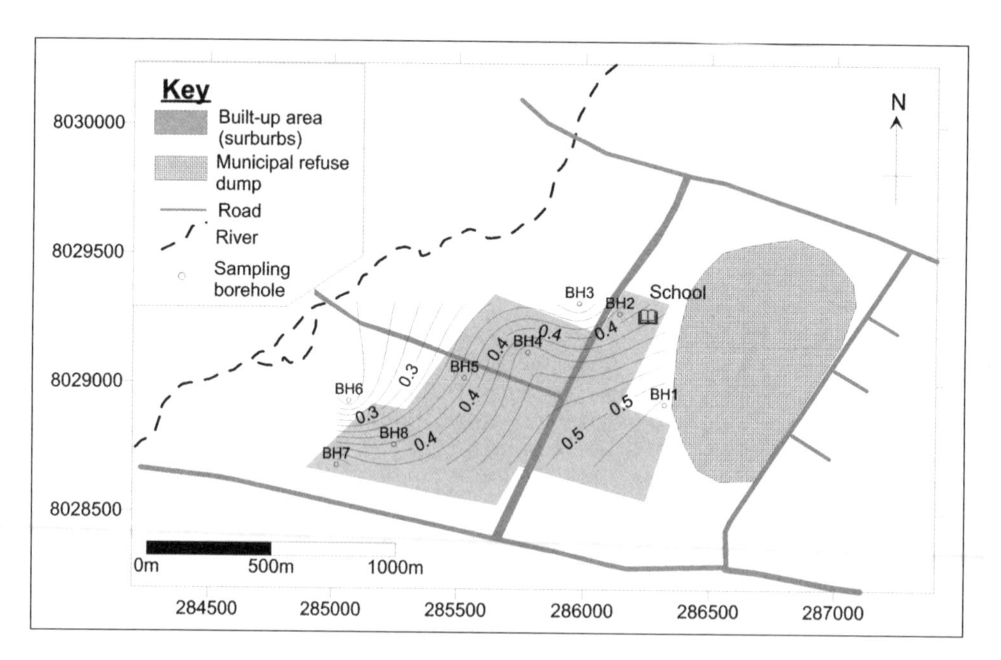

Figure 7.10. Spatial variation of cadmium in groundwater, Westlea (Values are in mg/l).

a pressure barrier to contaminant transport in the direction of regional groundwater flow. Gum trees can also absorb some of the pollutants as they take up water from the aquifer.

Vegetating the dump will help reduce run-off and percolation, therefore decreasing the movement of leachate to groundwater, which is currently encouraged by a perched water table below the landfill.

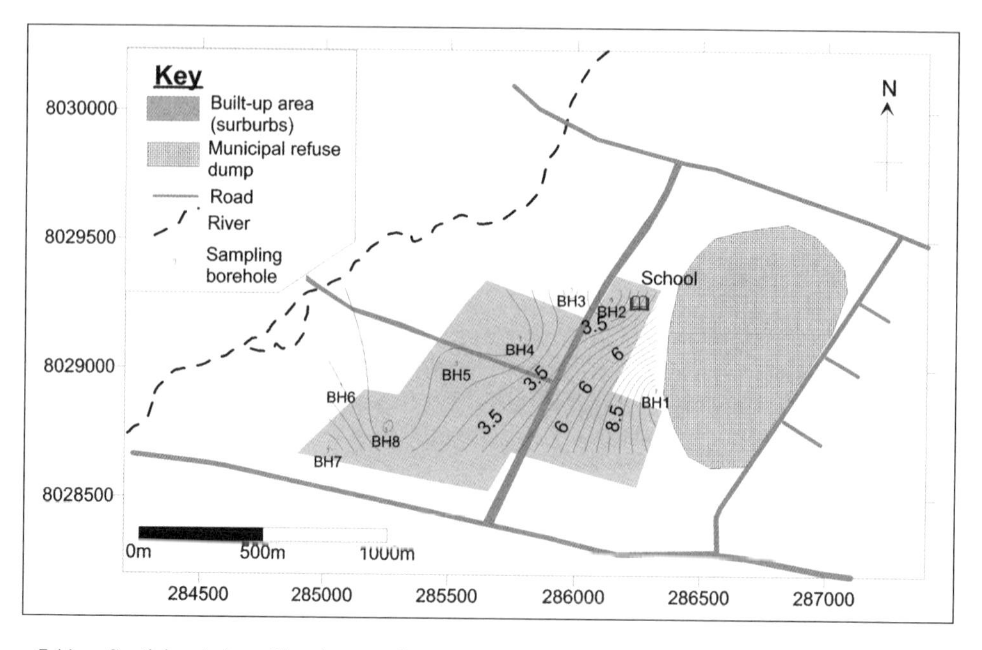

Figure 7.11. Spatial variation of iron in groundwater, Westlea (Values are in mg/l).

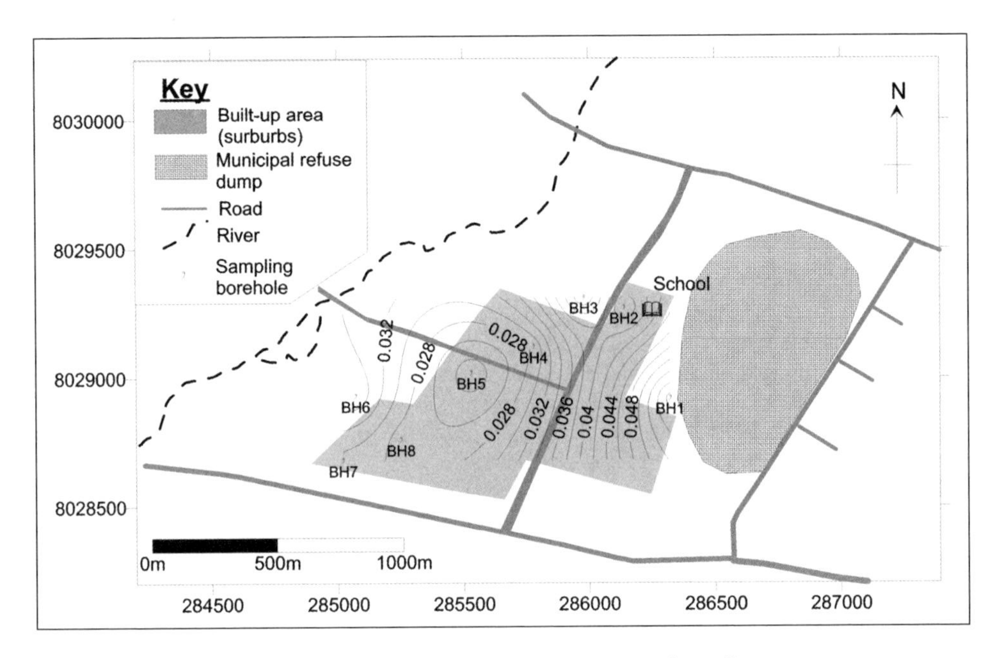

Figure 7.12. Spatial variation of zinc in groundwater, Westlea (Values are in mg/l).

An innovative response to mitigating organic contamination from landfills has been suggested by Kennedy and Everett (2001). It involves the adding of gypsum and iron (III) minerals, such as pyrrhotite, to a landfill. This approach was also shown to decrease releases of toxic gases, such as methane. Such minerals are readily available from industrial waste dumps within the city of Harare, such as Zimphos (Ravengai & Love 2004) and from a number of mines within a 50 km radius of Harare, such as Arcturus (Ravengai

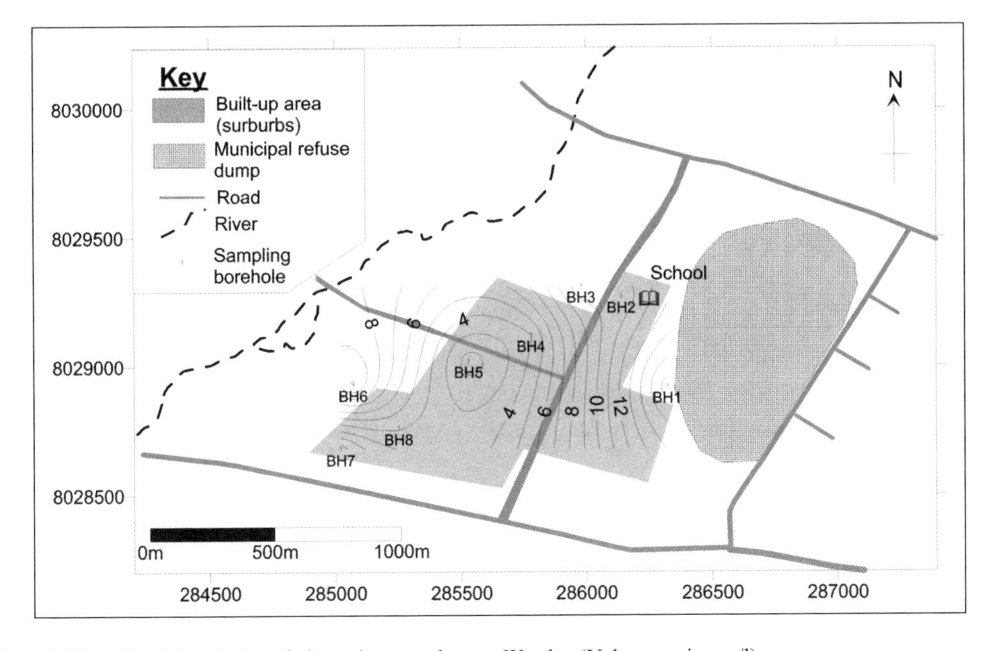

Figure 7.13. Spatial variation of nitrate in groundwater, Westlea (Values are in mg/l).

et al. 2004a), Beatrice (Ravengai et al., 2004b), Madziwa and Trojan (Lupankwa et al. 2004a, 2004b) and Iron Duke (Ravengai et al. 2004c).

Grouting at the margins of the landfill with lime would raise the pH material. This would change the chemistry of water seeping out of the landfill by raising the pH and adding dissolved carbonate, thus precipitating dissolved metals. Future metal contamination would thus be confined to the immediate environs of the landfill.

3.2 *Granville Cemetery*

The results of the study showed the presence of microbiological indicators (total coliform and fecal coliform) in the sampled groundwater. Total coliform (TC) and fecal coliform (FC) bacteria had averages of 1×10^7 coliform forming units (cfu) and 5.2×10^4 cfu respectively. TC were concentrated around the graves, with highest numbers in boreholes in more recent graves (2002) (Fig. 7.14). FC were concentrated around the older graves (1999-2000) and the pauper graves (Fig. 7.15). The results (GH3 and GBH8) indicate that FC have not been carried far in groundwater;

All the boreholes had pH below the control, giving an average of pH 5.9 (Table 7.3). This decrease in pH can be an influence of microbial activity within the graves. Calcium had high levels, greater than 100 mg/l in boreholes GBH7 and GBH9 and the rest were less than 50 mg/l. Iron is relatively high in the control since the borehole was cased using metal, therefore the control borehole cannot be used for comparison of iron values. Boreholes GBH4 and GBH5 had relatively high iron, most probably coming from the graves. Fe, K and Ca were concentrated around the graves. Variation in the concentration of Ca, K and Fe across the study area increases with the groundwater flow. There are two possible explanations for this rise in metal levels with groundwater flow. One possibility is that Ca, K and Fe could be released from decomposing corpses. Alternatively, the increases could be due to water-rock interactions-especially the hydrolysis of feldspars, which are abundant in the local granites (Baldock et al. 1991) and break down to release Ca and K, especially where there is appreciable groundwater flow (Faure 1991).

Table 7.3. Groundwater quality data, 2003-2004, Granville Cemetery, Harare.

	pH (pH units)	K (mg/l)	Na (mg/l)	Fe (mg/l)	Mg (mg/l)	Ca (mg/l)	NO_3 (mg/l)	Cl (mg/l)	P (mg/l)	SO_4^{2-} (mg/l)	Total coliforms /100 ml	Faecal coliforms /100 ml)
GBH1	5.67	3.4	21.26	13.61	0.54	4.94	0.04	0.13	0.42	22.24	0	0
GBH2	5.47	9.83	24.29	13.75	0.69	5.2	0.18	0.09	0.6	38.72	105000	350000
GBH3	5.98	8.1	29.4	7.5	2.92	179.94	0.07	0.22	0.4	<1.65	43000	0
GBH5	5.36	4.17	16.48	12.08	0.57	5.1	0.11	0.09	0.8	1.65	20000	9000
GBH7	5.39	10.67	24.77	0.81	1.94	9.35	0.09	0.09	1.17	118.63	700000	0
GBH8	6.67	6.57	18.87	5.66	2.87	0.57	0.07	0.18	0.02	<1.65	32000	0
GBH9	5.71	1.43	17.66	13.55	13.19	152.24	0.08	0.22	0.08	<1.65	0	0
Control	6.98	3.12	21.93	15.72	16.85	7.24	0.02	0.18	0.01	<1.65	0	0
Domestic (WHO, 1993)	---	---	200	0.30	---	---	50.00	250.00	---	250.00	0	0
Domestic (DWAF, 1996)	6.0-9.0	50	100	0.10	30	32	6.00	100.00	---	200.00	5	0
Livestock (DWAF, 1996)	---	---	2000	10.00	500	1000	100.00	1500.00	---	1000.00	---	200
Crops (DWAF, 1996)	7.5-8.4	---	70	5.00	---	---	5.00	100.00	---	---	---	1

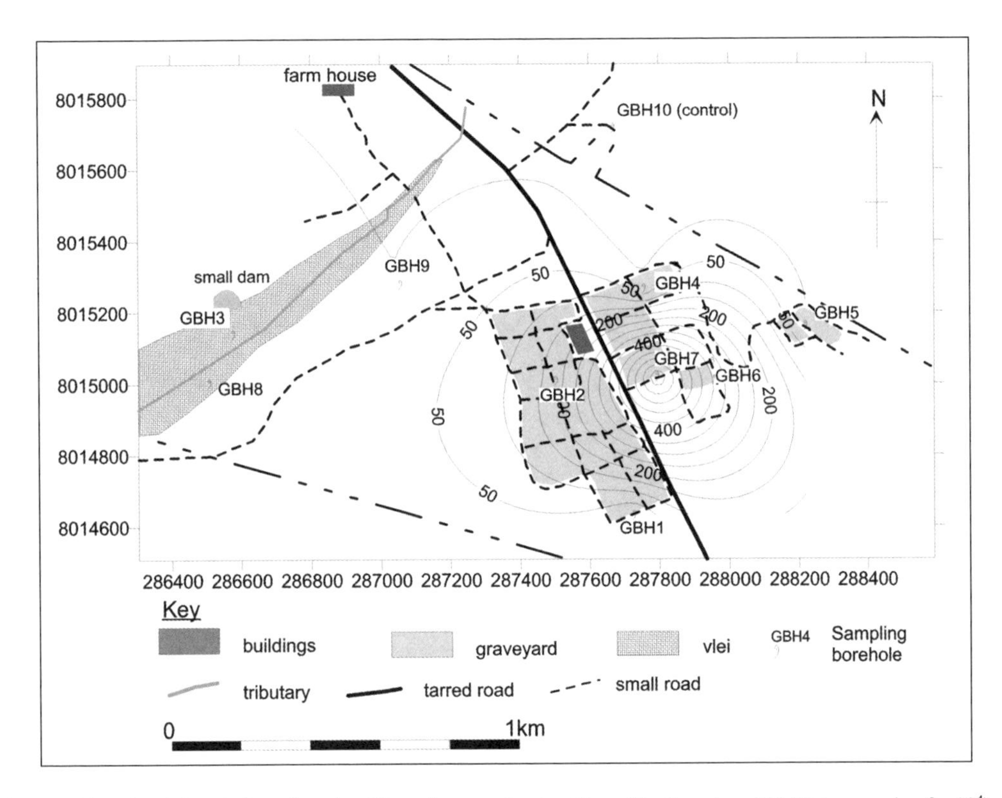

Figure 7.14. Spatial variation of total coliform in groundwater, Granville Cemetery 'B' (Values are in cfu. 10^4).

Magnesium levels at Granville are below those at the control. Sodium had relatively stable concentrations with exceptions in boreholes GBH1, GBH2, GBH3 and GBH7 with levels of above 20 mg/l. The Na and Mg levels were relatively low around the graves and the variation can be a result of ion exchange during water-rock interaction.

Chloride concentrations were more or less constant with higher levels to the north. Nitrates were detected at higher levels in samples from boreholes around recent graves (2000-2003) as shown by Fig. 7.16. Nitrate is formed from ammonia by the nitrification process, catalyzed by nitrifying bacteria. Phosphate concentrations were above the control borehole levels and were concentrated in boreholes around the graves, with highest levels in younger graves (Fig. 7.17). It has been suggested that P is leached to orthophosphates which are locked into insoluble components at $pH < 5$ and $pH > 7$ and are then transformed into soluble components by microorganisms in the soil (Dent et al. 2004). This could explain the high PO_4 levels in fairly recent graves, indicating the source of the phosphate being from the decay of corpses. Sulphate was detected in high levels boreholes GBH1, GBH4 and GBH7, in the fairly recent graves. The high levels of nitrates and sulphate around the graves suggest that they may be formed after dissolution of the gases (e.g. NH_3, CH_4, H_2S), which are released during decomposition in anaerobic conditions (Dent et al. 2004).

Some parameters are higher than the South Africa domestic water guideline (DWAF 1996) and the World Health Organization drinking water guidelines (WHO 1993) as shown in Table 7.3; therefore the water around the cemetery is unsafe for domestic use. Comparison to the control borehole suggests that the cemetery is a source of groundwater pollution, with potassium, sodium, nitrates, sulphate, phosphate and coliform bacteria all showing elevated levels in the study boreholes.

The development of a tree barrier (Fig. 7.18), and the grouting of the downflow side of the site-as discussed for Golden Quarry above-are also recommended for Granville. In addition, the advantages mentioned in the previous section are valid for this section too.

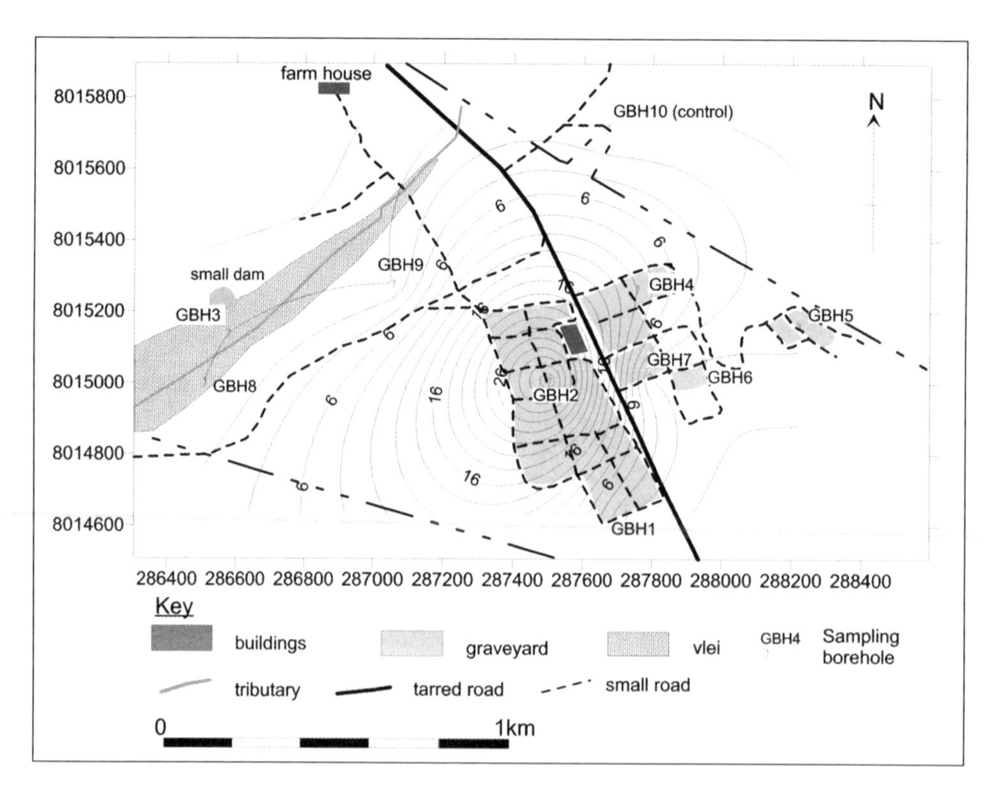

Figure 7.15. Spatial variation of fecal coliform in groundwater, Granville Cemetery 'B' (Values are in cfu. × 10^4).

A clay layer should routinely be put at the base of graves before burial to filter the leachate before it enters the groundwater. This is because clay materials retard bacterial movement.

The local groundwater can only be used to water lawns at the cemetery since the aquifer is already polluted.

4 CONCLUSIONS AND RECOMMENDATIONS

4.1 *Golden Quarry solid waste disposal site*

- The site is a source of groundwater pollution, with levels of metals, nitrates and coliform bacteria all elevated. The source of the metals (except iron, which is naturally high) is likely to be from the disposal of industrial waste on the south side of the landfill. The high metal levels attenuate rapidly downflow.
- Coliform bacteria and nitrate are probably leachate from the disposal of domestic refuse on the north side of the landfill.
- The responsible authority should undertake mitigatory measures to decrease the environmental impact of the landfill. These could include treating the landfill with gypsum and iron (III) minerals, revegetating the landfill surface, planting more gun trees downflow of the landfill and grouting the landfill's western edge.

4.2 *Granville Cemetery*

- The site is a source of groundwater pollution, with levels of some metals, nitrates, phosphate and coliform bacteria all elevated.

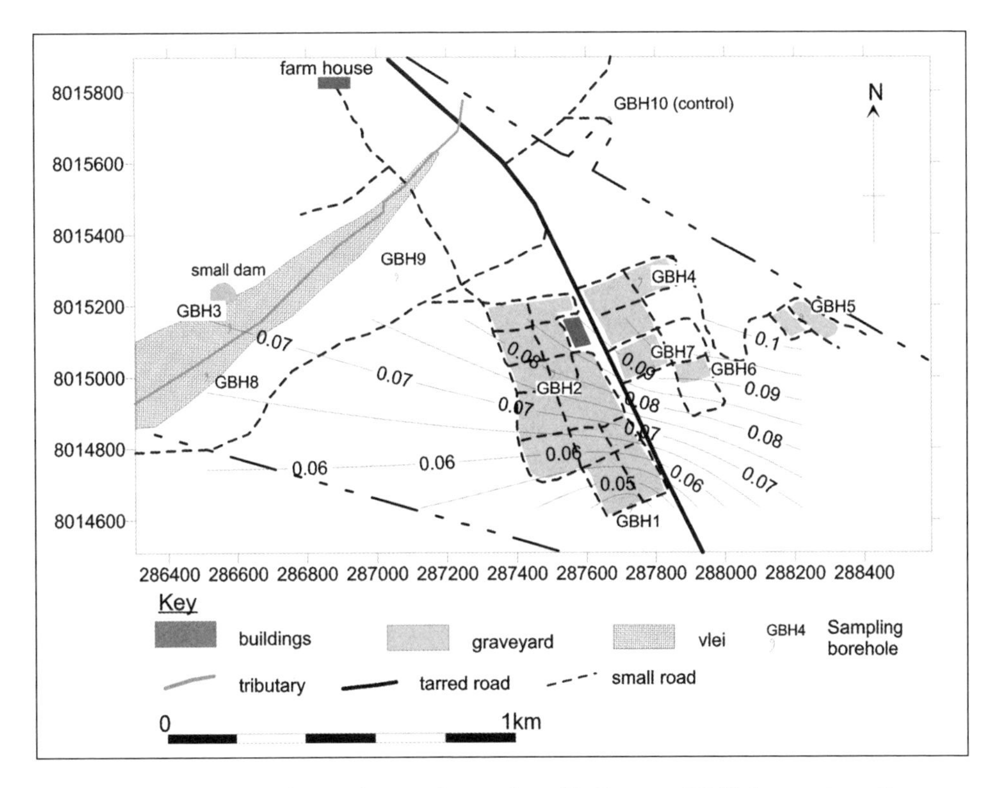

Figure 7.16. Spatial variation of nitrate in groundwater, Granville Cemetery 'B' (Values are in mg/l).

- The major source of these contaminants is likely to be the decomposition of human bodies, with the distribution of parameters relating to the stages of decomposition and associated releases and the ages of graves.
- There is also a natural input from water-rock interactions, especially feldspar hydrolysis.
- The responsible authority should undertake mitigatory measures to decrease the environmental impact of the cemetery, including development of a tree barrier and the grouting of the downflow side of the site.
- Cemetery authorities should ensure that a clay layer is always placed at the base of new graves.

4.3 *Concluding remarks-urban planning*

The main problems that have emerged at the two sites studied are typical of cases of this kind. The problems are, however, exacerbated by poor planning and site selection. The choice of location for both Golden Quarry landfill and Granville cemetery was based upon the availability of land to the city authorities at the time. The choice of Granville farm for a cemetery was also influenced by environmental concerns-the city authorities rejected the first farm offered as it was within a source protection zone for the Mukuvisi River. Nevertheless, the location selected (Granville) remains problematic, as the high water table accelerates groundwater pollution at the cemetery. The location of the Golden Quarry landfill on a topographic high is also problematic as it forms a groundwater recharge zone.

This study has identified some serious environmental contamination problems at the two sites, and although mitigatory measures are recommended, these shall not restore the local groundwater to pristine quality. What can be done is to avoid such problems recurring elsewhere. The only way to avoid the seriousness of the problems reported in this study is for future landfill and cemetery sites to be identified by

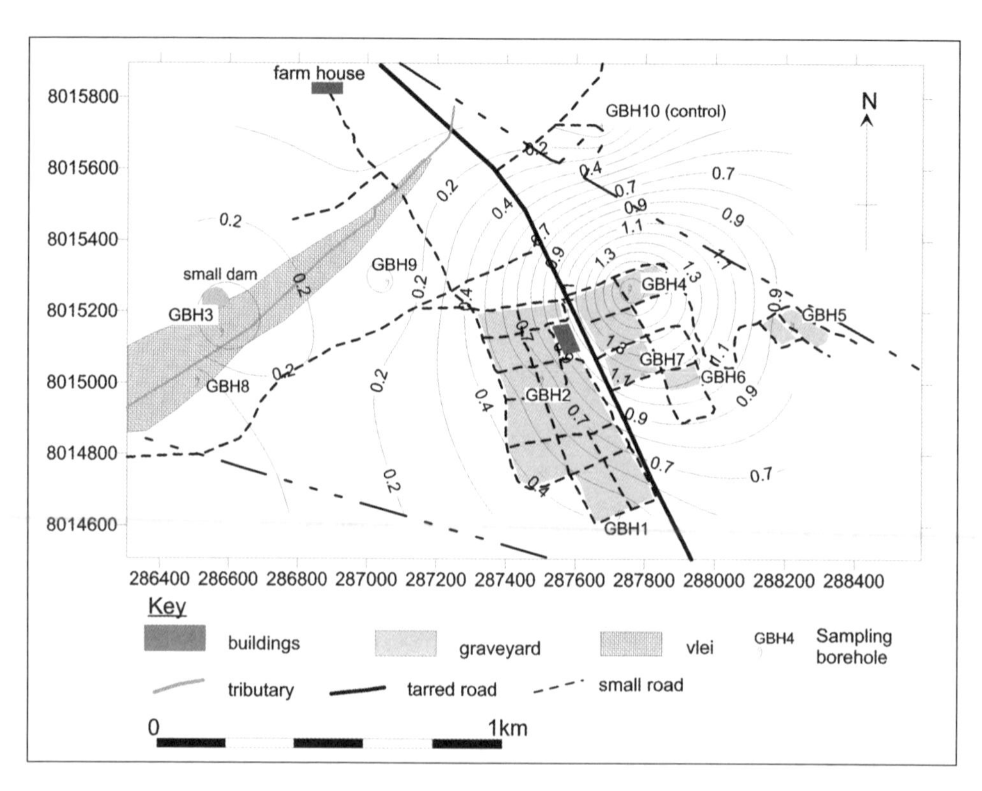

Figure 7.17. Spatial variation of phosphate in groundwater, Granville Cemetery 'B' (Values are in mg/l).

geotechnical mapping. The necessary data on soil types, groundwater levels and so on is already available (Mupaya 2002). Sites that have already been identified for future use (HCMPPA 1989) should be reviewed in the light of geotechnical mapping and the results of this study. For example, the selection of future solid waste disposal sites in the Mount Hampden area, north of Harare, is motivated more by the projected availability of abandoned brick-field quarries than any geotechnical or environmental considerations (Rakodi 1995). Once more environmentally and technically appropriate sites have been identified, then the land could be purchased or acquired by the city of Harare.

Acknowledgements – The data presented and discussed in this paper were collected as part of the "Harare Urban Groundwater Project" funded by WARFSA and incorporate work carried out by Mr. Moyce and Mrs. Nyama as part of their B.Sc. Honours studies at the University of Zimbabwe (Moyce 2003, Nyama 2003). The authors would like to thank the Department of Health of the Harare City Council for information and support. Acknowledgement is also made to Mr. C. Maneya and Mr. F. Zihanzu, technicians in the Department of Geology, University of Zimbabwe for help during data collection.

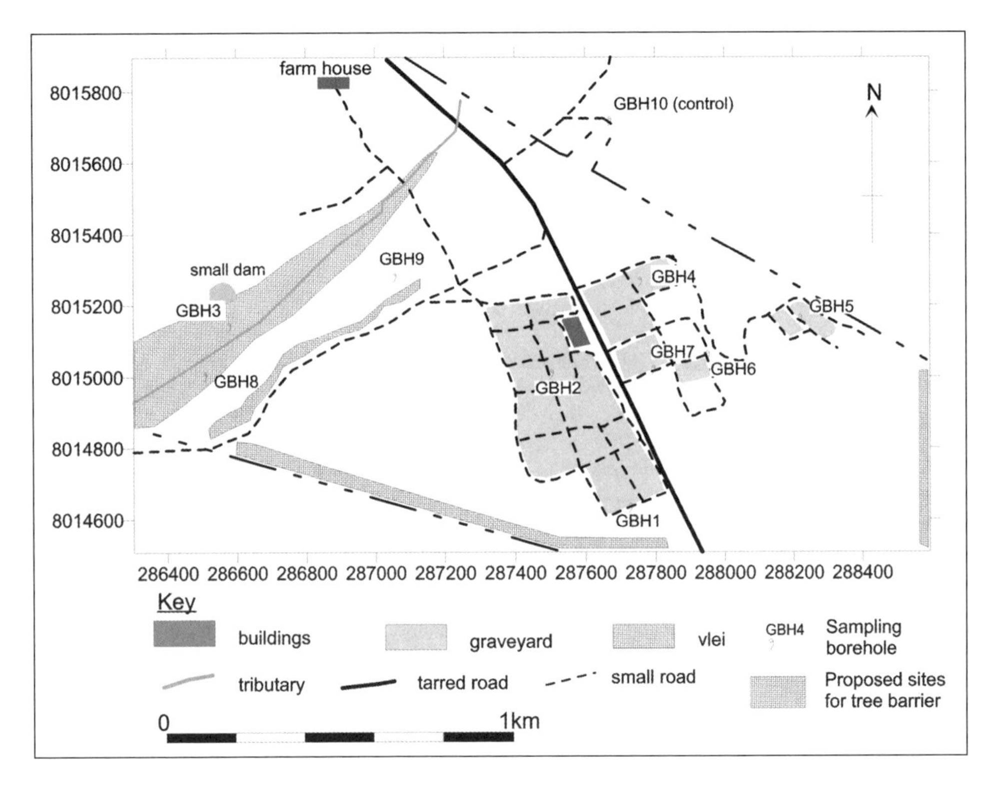

Figure 7.18. Possible position for a gum tree barrier at Granville.

REFERENCES

Aldrick, R.J., Rivett, M.O. and Hepburn, S.L., 1999. Urban groundwater and environmental management: Hull, East Yorkshire. In Chilton, W. (Ed.) *Groundwater in the Urban Environment: Selected City Profiles*. Rotterdam: A.A. Balkema Publishers.

Baldock, J.W., Styles, M.T., Kalbskopf, S. and Muchemwa, E. 1991. The geology of the Harare Greenstone Belt and surrounding granitic terrain, *Zimbabwe Geological Survey Bulletin* 94.

Centre for Disease Control and Prevention, 2000. Bacterial Waterborne Diseases-Technical Information at http://www.cdc.gov/ncidod/dbmd/diseaseinfo/waterbornediseases_t.htm

Christensen, T.H., Kjeldsen, P., Bjerg, P.L., Jensen, D.L., Christensen, J.B., Baun, A., Albrechtsen, H.-J. and Herom, G. 2001. Bio-geochemistry of landfill leachate plumes. *Applied Geochemistry*, 16, 659-718.

Clesceri, L.S., Greenberg, A.E. and Eaton, A.D. (eds.) 1989. *Standard methods for the Examination of Water and Waste Water* (20th Edn). American Public Health Association, Washington DC.

Dent, B.B., and Knight, M.J. 1998. Cemeteries: a special kind of landfill. The context of their sustainable management. In: Weaver TR, Lawrence CR (eds.) *Proceeding of the International Groundwater Conference*, 8 – 13th February, Melbourne. International Association of Hydrogeologist (Australian National Chapter), Indooroopilly, Australia, p451-456.

Dent, B.B., Forbes, S.L. and Stuart, B.H. 2004. Review of Human decomposition processes in soil. *Environmental Geology*, 45, 576-585.

Dutova, E.M., Nalivaiko, N.G., Kuzevanov, K.I. and Koylova, J.G., 1999. The chemical and microbiological composition of urban groundwater, Tomsk, Russia. In Chilton, W. (Ed.) *Groundwater in the Urban Environment: Selected City Profiles*. Rotterdam: A.A. Balkema Publishers.

DWAF (Department of Water Affairs and Forestry) 1996. *South African Water Quality Guidelines, 8: Field Guide.* Government Printer, Pretoria.
Engelbrecht, J.F.P. 1998. Groundwater pollution from cemeteries, *WISA 98 Biennial Conference*, Paper 1C-3. Water Institute of Southern Africa, Halfway House.
Environmental Agency, 1999. *Microbiological Contaminants in Groundwater.* National Groundwater and Contaminated Land Centre, Solihull.
Environmental Agency, 2002. *Assessing the Groundwater Pollution Potential of Cemetery Developments.* National Groundwater and Contaminated Land Centre, Solihull.
Faure, G. 1991. *Principles and Applications of Inorganic Geochemistry.* New Jersey: Prentice Hall.
Fetter, C.W. 1994. *Applied Hydrogeology.* Prentice-Hall, Englewood Cliffs.
HCMPPA (Harare Combination Master Plan Preparation Authority). 1989. *Harare Combination Master Plan: Report of Study.* HCMPPA, Harare.
Hranova, R.K., Amos, A., and Love, D. 2003. Impacts of sewage sludge land disposal practice on groundwater in Harare, Zimbabwe. *Proceedings of the International Water Association Conference*, Cape Town, September 2003.
Kennedy, L.G. and Everett, J.W. 2001. Microbial degradation of simulated landfill leachate: solid iron/sulfur interactions. *Advances in Environmental Research*, 5, 103-116.
Lupankwa, K., Love, D., Mapani, B.S. and Mseka, S. 2004*a*. Impact of a base metal slimes dam on water systems, Madziwa Mine, Zimbabwe. *Physics and Chemistry of the Earth*, 29, 1145-1151.
Lupankwa, K., Love, D., Mapani, B.S., Mseka S., and Smith, V. Influence of the Trojan Nickel Mine rock dump on run-off quality, Mazowe Valley, Zimbabwe. In: *Proceedings of the 5th WaterNet-WARFSA Symposium*, Windhoek, Namibia, November 2004.
Madimutsa, R. 2000. *Assessment of cadmium storage and flows at Golden Quarry Sanitary Landfill.* Unpublished MSc thesis, Water Resources Engineering and Management Programme, University of Zimbabwe.
Moyce, W. 2003. *Groundwater quality around Granville Cemetery, Harare.* Unpublished BSc Honours thesis, Honours Geology Programme, University of Zimbabwe.
Mupaya, F.B. 2003. Urban and environmental geology: new frontiers for the Zimbabwe Geological Survey. In: *Proceedings of the Geological Society of Zimbabwe Summer School*, Harare, Zimbabwe.
Nyama, Z. 2003. *Effects of landfills on groundwater quality: case of Golden Quarry.* Unpublished BSc Honours thesis, Honours Environmental Science Programme, University of Zimbabwe.
Parsons, R. and Jolly, J. 1994. The development of a systematic method for evaluating site suitability for waste disposal, based on geohydrological criteria. *Water Research Commission Report*, 485/1/94.
Rakodi, C. 1995. *Harare: Inheriting a Settler-Colonial City: Change or Continuity?*, Chichester: John Wiley & Sons.
Ravengai, S. and Love, D. 2004. Acidic groundwater pollution from heavy industry: the case of Zimphos, Zimbabwe. In: *Proceedings of the Geological Society of South Africa Conference Geoscience Africa*, Johannesburg, South Africa, July 2004.
Ravengai, S., Love, D., Love, I. and Kambewa, C. 2004*a*. The impact of gold mine dumps on water quality around the Arcturus group of mines, Mazowe Valley, Zimbabwe. In: *Proceedings of the IWA Specialist Group Conference on Water and Wastewater Management for Developing Countries*, Victoria Falls, Zimbabwe, July 2004.
Ravengai, S., Love, D., Mabvira-Meck, M.L., Musiwa, K. and Moyce, W. 2004*b*. Water quality in an abandoned mining belt, Beatrice, Zimbabwe. In: *Proceedings of the 5th WaterNet-WARFSA Symposium*, Windhoek, Namibia, November 2004.
Ravengai, S., Owen, R.J.S. and Love, D. 2004*c*. Evaluation of seepage and acid generation potential from evaporation ponds, Iron Duke Pyrite Mine, Mazowe Valley, Zimbabwe. *Physics and Chemistry of the Earth*, 29, 1129-1134.
Stone, A.W. 1996. Landfill leachate generation and groundwater contamination in arid and semi-arid areas. *Imiesa*, 5, 9-17.
Tatsi, A.A. and Zouboulis, A.I. 2002. A field investigation of the quantity and quality of leachate from a municipal solid waste landfill in a Mediterranean climate: Thessaloniki, Greece. *Advances in Environmental Research*, 6, 207-219.
Tevera, D.S. 1991. Solid waste disposal in Harare and its effect on the environment: some preliminary observations. *Zimbabwe Science News*, 25 9-13.
Trick, J. and Klinck, B.A. 2001. Grave concerns – health risks from human burials. Geology and Health, *Earthwise*, 17.
Trick, J.K., Klinck, B.A., Coombs, P., Chambers, J., Noy, D.J., West, J. and Williams, G.M. 2002. Groundwater impact of Danescourt Cemetery, Wolverhampton. *National Groundwater and Contaminated Land Centre Report* NC/99/72.
USEPA (United States Environmental Protection Agency) 1993. Criteria for solid waste disposal facilities: a guide for owners/operators. *EPA Report* EPA/530-SW-91-089.

Usisik, A.S. and Rushbrook, P. 1998. The impact of cemeteries on the environment and public health: An introductory briefing. European Centre for Environment and Health, Nancy Project Office, *EUR/HFA Target* 23.
WHO (World Health Organisation) 1993. *Guidelines for Drinking Water Quality, Volume 1: Recommendations* (2nd Edn). World Health Organisation, Geneva.
Zaranyika, M. 1997. Sources and levels of pollution along Mukuvisi River: a review. In Moyo (ed), *Lake Chivero: A Polluted Lake*, 35-42, University of Zimbabwe Publications, Harare.
Zhu, X.-Y., Xu, S.-H., Zhu, J.-J., Zhou, N.-Q. and Wu, C.-Y. 1997. Study on the contamination of fracture-karst water in Boschan District, China. *Ground Water*, 35, 538-545.
Zingoni, E., Love, D., Magadza, C., Moyce, W. and Musiwa, K. 2004. Groundwater use zones and water supply options for Epworth semi-formal settlement, Zimbabwe. In: *Proceedings of the 5th WaterNet-WARFSA Symposium*, Windhoek, Namibia.

CHAPTER 8

Sewage sludge disposal on land – impacts on surface water quality

R. Hranova & M. Manjonjo

ABSTRACT: Basic principles of sewage treatment and sludge disposal options, common for the region have been presented, together with regulatory aspects for a safe practical application. A case study of digested sludge and effluent mixture, disposed beneficially for pasture irrigation on Crowborough farm in Harare, has been investigated during 1998 – 1999, in terms of loading rates and the impacts of such type of disposal method on surface water after 30 years of operation. Results show considerable adverse impacts on natural streams, draining the pasture, with respect to TP, TKN, nitrate and Ni, with pronounced seasonal variation. TKN and TP impacts were pronounced during the dry season, while nitrates impacts were pronounced throughout the year with higher impacts during the wet season. Recommendations with respect to the future operation of the pasture have been made.

1 INTRODUCTION

1.1 *Reusing wastewater and sludge for irrigation*

Urban population centers are a major source of wastewater generated by domestic, commercial, institutional and industrial activities. In the vast majority of the cases throughout the world, wastewater is collected transported and treated in centralized wastewater systems, which discharge the effluents to natural water bodies and are considered as point sources of pollution. In countries with limited water resources, these effluents could be reused beneficially for different purposes. One of the most common types of beneficial reuse is the application of wastewater in agriculture for irrigation purposes. It could be applied after different levels of treatment of the raw wastewater. During the treatment process, the main flow of wastewater is clarified and pollutant constituents reduced at acceptable levels. As a result of this process, a considerable amount of sewage sludge is generated, which contains in concentrated form the pollutant constituents removed from the wastewater during the treatment process. Despite the fact that the sludge could contain harmful contaminants, such as pathogens or toxic elements, it could be considered as valuable product to be reused in agriculture due to the high level of organic constituents and nutrients. If treated and applied properly, the sludge could be reused as a fertilizer in different types of agricultural activities.

In arid climatic conditions, which are typical for many countries in the Southern African region, where water resources are scarce, the wastewater reuse practice in general, and the wastewater reuse for irrigation purposes in particular, should be encouraged and should be viewed as an alternative source of water. Such practice has the following benefits:

- Reduces the volumes of fresh water resources needed for different irrigational activities;
- Reduces the need of fertilizers;
- Improves the soils' quality due to the addition of organic materials;
- Reduces the cost of wastewater treatment, in cases where high effluent standards are required for discharge into natural water bodies.

The practice of wastewater and sludge reuse has limitations, which need to be considered in order to prevent undesirable impacts on human health and the environment, which could be summarized as follows:

- The volumes of wastewater and sludge produced are relatively constant with no time interruption throughout the year and with a relatively low diurnal and seasonal variation, while the demand for any beneficial use of the treated effluent and sludge usually varies, and this is very much pronounced in the case of irrigation. Therefore, this practice requires a proper consideration and management of the irrigation process and the provision of storage volumes or other alternatives.
- The wastewater reuse practice should comply with specified requirements with respect to the quality of the treated effluent and sludge, corresponding to each specific type of use. The compliance with these requirements must be very strict and should be backed up by an appropriate monitoring process. Thus, the practice requires a high standard of operation and maintenance during the wastewater treatment process.

1.2 *Land application of wastewater and sludge*

Land application of municipal sludge is one of the beneficial and successful options for sludge disposal, which has been practiced since ancient times, and recently in many cities in Europe and North America. The banning of ocean disposal and strict regulations on landfills has seen land application, particularly for agricultural use, increasing (Pescod 1992). USEPA (1995) gives agricultural land, strip-mined land, forests, parks and gardens as some of the land application options, whereas Pescod (1992), provides four major categories: agricultural utilization, forest utilization, land reclamation and land disposal.

The advantages of land application systems are:

- Water conservation;
- Easy and inexpensive maintenance;
- Provision of plant nutrients.

The limitations of land application methods are connected with risks to public health and environmental pollution. However, the disposal of sludge on land is also a type of sludge treatment method, close to natural treatment processes. Studies have shown that wastewater parameters such as BOD_5, suspended solids and fecal indicator organisms are nearly completely removed within the aerobic, unsaturated zone of medium-to-fine textured soil, and within a depth of 0.6 to 1.5 m, and when a neutral pH of the soil is maintained during the land treatment system operation (USEPA 1992). The limitations of such practice are connected to:

- Potential groundwater pollution and transmission of pathogens;
- Dependence on soil conditions and climate;
- Large land area requirements.

In cases of inadequate design procedure or operational practice, the method of wastewater/sludge land application might become a source of diffuse pollution to ground and surface water bodies, and a potential public health hazard. This chapter and the following two chapters present specific cases of sludge and wastewater land disposal methods and their impacts on soils, surface and ground water.

2 SEWAGE TREATMENT AND SLUDGE DSPOSAL

2.1 *Sewage treatment*

The efficiency of any land application system is dependant on a reliable wastewater treatment technology and the understanding of the different processes involved. This section presents briefly some wastewater treatment technologies commonly applied in the countries of the region.

2.1.1 Conventional treatment methods

They usually consist of two separate and consecutive treatment stages – primary and secondary treatment. Primary treatment consists of preliminary treatment, which has the purpose to remove course materials, sand and grit by screening and grit sedimentation. The materials separated during this stage are disposed as solid waste and do not form part of the sludge generated during the treatment process. After being preliminary treated, the wastewater is directed to primary sedimentation tanks, which have the purpose of separating suspended particles from the main flow by means of gravity sedimentation. The separated material is retained at the bottom of the tank and constitutes the primary sludge. It is taken out of the tank at prescribed time intervals. Primary sedimentation reduces the suspended solids' content of the wastewater to about 50%-60% and the organic matter to about 30-40%. Pathogen removal is limited to the extent of these microorganisms, which are attached to suspended particles. About 90% of the heavier pathogenic microorganisms, such as helminth eggs and protozoa cysts, are retained in the primary sludge. Helminth eggs have a long survival period of several months and could be a potential public health hazard, when the sludge is disposed on land.

The secondary treatment stage consists in biological treatment of the wastewater, where engineered biological treatment processes are applied to reduce the organic material remaining after the primary stage. Biological treatment processes applied during the secondary stage are aerobic, which means that the complex culture of different microorganisms used to consume the organic matter present in the wastewater, require oxygen for their metabolism. The system consists of a biological reactor and a secondary sedimentation tank to separate the biomass from the final effluent. Biological reactors could be subdivided into two major categories – attached growth reactors and suspended growth reactors.

In attached growth systems (Fig. 8.1) the biological culture, known as "biofilm", is attached to a fixed media, which could consist of natural material – stone or gravel, or plastic blocks with a high specific surface area. The reactor, where the process takes place is known as trickling filter (biofilter). The wastewater is spread over the filter media and on its way through it, gets in contact with the biofilm. The air current, formed due to the temperature difference between the reactor and the surrounding environment, supplies oxygen for the process. Figure 8.1 shows a treatment scheme commonly used in the region.

In suspended growth systems, also called activated sludge systems, the biological culture (biomass), is mixed with the wastewater and held in suspension within open tanks (aeration tanks). The oxygen needed for the process is supplied by means of mechanical aerators or through the provision of compressed air, dispersed by diffusers in the form of bubbles. A large variety of different types of aeration systems exist, which have the purpose not only to supply oxygen for the process but also to keep the mixture of biomass

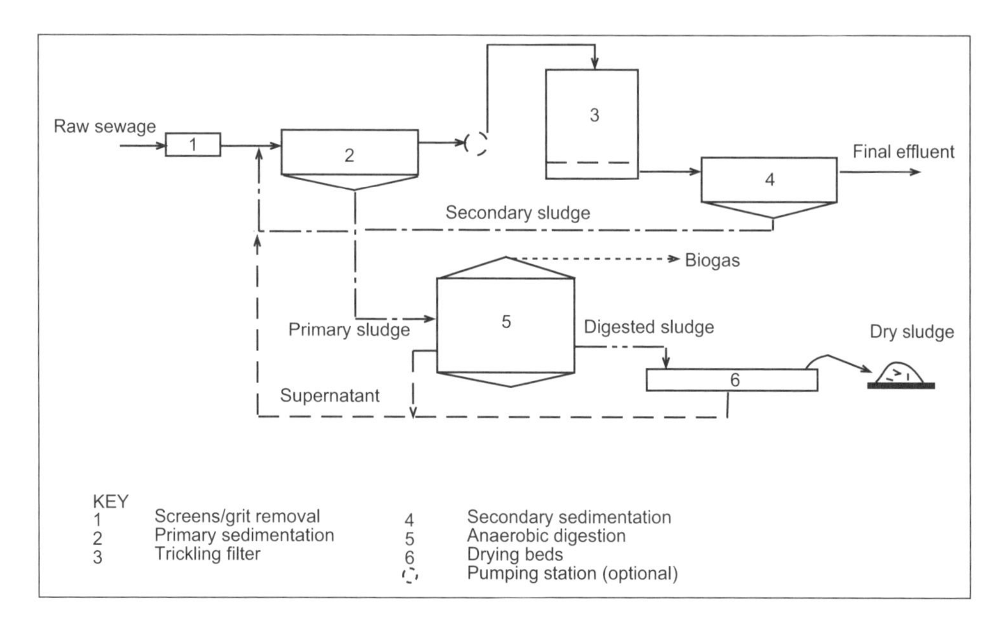

Figure 8.1. Sewage treatment scheme.

and wastewater in suspension. The required concentration of the biomass in the reactor is maintained by the recirculation of a portion of the sludge, retained in the secondary sedimentation tank, back to the reactor. The treatment effect with respect to the removal of the organic material, expressed as BOD_5, would depend on the design of the system and the configuration of the reactor in terms of hydraulic conditions, retention times of the wastewater and the biomass, and organic loading rates.

Attached growth systems require less energy consumption and are simpler to operate and maintain. For this reason, they have been a preferred secondary treatment alternative in many conventional wastewater treatment plants in the region. The first plants of such type were introduced during the 1950's and one typical treatment scheme is shown in Figure 8.1. The influent (raw sewage) is treated in preliminary treatment facilities, followed by primary sedimentation and biological treatment in trickling filters. After the secondary sedimentation (humus tanks), the effluent is discharged to natural water bodies. The secondary sludge is returned to the inlet of the primary sedimentation tank and separated from the flow together with the primary sludge, which is stabilized in anaerobic digesters, and after that is dried in sludge drying beds. The supernatant released during the sludge stabilization process and from the drying beds is returned to the primary sedimentation tank inlet and treated together with the sewage.

Conventional treatment systems are designed for wastewater of a specific quality and hydraulic regime. Large variations in the hydraulic loading and the quality of the wastewater disrupt the efficiency of conventional systems. When hydraulically overloaded, conventional treatment systems fail to absorb the additional load. This results in poor effluent quality. A balance between the biomass concentration and organic loading is required for optimum treatment. High pollutant loads could disrupt the functioning of the system and may lead to a breakdown of the biological process.

Conventional treatment systems cannot handle concentrated industrial wastewater from specific industries, because such type of wastewater often contains toxic elements and other constituents, which could have a negative impact on the biological process or cause a total loss of the biological activity of the microorganisms. In addition, conventional treatment systems require sludge treatment and handling facilities. Since the treatment processes tend to concentrate the pollutants in the sludge, sludge treatment and disposal is an important aspect of the plant design and operation.

In general, the main constituents removed by conventional treatment processes are suspended solids and BOD_5. Pathogen removal has rarely been considered as a special objective in conventional wastewater treatment processes. In contrast, waste stabilization pond (WSP) systems are efficient in pathogen removal, making the effluent suitable for irrigation. A more detailed description of this alternative of the conventional sewage treatment processes is given in Chapter 10.

2.1.2 Biological nutrient removal methods

Conventional sewage treatment systems do not remove nutrients at a high level and consequently effluents from these plants, discharged into natural water bodies, could be considered as a source of pollution with respect to nutrients, leading to enhanced eutrophication. This is a serious problem for the countries in the region, given the high average annual temperatures, the abundance of sunlight and the relatively limited volumes of surface water resources. For this reason, during the 1970s, effluent discharge regulations were introduced, which required a high standard of treatment of sewage with respect to nutrients, before their discharge to natural water bodies. Alternatively, the effluents from conventional treatment plants and WSP systems could be disposed on land or reused for irrigation. As a result of these regulations, the new treatment plants were designed to incorporate nutrients removal techniques and many conventional treatment plants were upgraded to include such treatment methods.

Nutrients could be removed by chemical or biological treatment methods. In the Southern African region, the biological nutrient removal (BNR) treatment was adopted. It consist of a modification of the conventional activated sludge systems and provision of special conditions within the reactor for:

- Phosphorous release and consequent biological consumption by the biomass – the release of phosphorous is enhanced in anoxic conditions, which precede the oxygen reach stage, where the phosphorous is consumed by the biomass;

- Nitrification of ammonia to nitrates – it is an aerobic process, held parallel to the removal process of the organic material in the aeration tank;
- Denitrification of nitrates to nitrogen gas and its release to the atmosphere – this is an anaerobic process, which requires a carbon source.

In the activated sludge system, microbes are held in suspension and perform a wide range of biological processes on the wastewater as it passes through the aeration tank. The varying aerobic, anoxic and anaerobic conditions in the system encourage the growth and activity of different microbes with specific action on the sewage. This results in a higher rate of nutrients' removal during the biological process. The major advantages of BNR plants are nutrient removal and compactness, while the major disadvantages are high energy input for aeration, large number of mechanical equipment that requires skilled labor and a high cost for operation and maintenance.

2.2 *Sludge treatment and disposal methods*

2.2.1 Sludge stabilization

Raw sludge contains a high concentration of organic matter and pathogenic microorganisms. It is prone to degradation and putrefaction, which leads to malodors and insect breeding. In addition, it provides a perfect environment for the growth of pathogen microorganisms, thus posing a substantial public health problem in terms of diseases' spreading. Sludge stabilization is the process of mineralization of the organic matter contained in sludge. Usually, about 50-60% of the organic material is mineralized during this process. Stabilized sludge does not degrade further when disposed on land. During this process, partial removal of pathogen organisms is achieved as well.

The most widely applied method for sludge stabilization is the anaerobic digestion, although in some cases aerobic stabilization methods, similar to the process of activated sludge treatment, are applied too. The anaerobic digestion of sewage sludge is a process, involving a complex mixture of different types of facultative and strictly anaerobic bacteria. A schematic representation is shown in Figure 8.2. During the acid forming stage, known also as acetogenesis, the end products are CO_2 and volatile organic acids, which are extremely malodorous, producing the specific putrefaction smell. The organic acids formed during the process of acidogenesis are further degraded to CH_4 by methanogens. The limiting stage is the methanogenic one, as the strictly anaerobic methanogens are very sensitive to the environmental conditions and require a narrow range of pH variation (between 6-8) for their survival.

In cases of high organic loading or other unfavorable conditions, the acid-formers produce large amounts of organic acids, leading to the accumulation of acids in the reactor and lowering the pH, which additionally inhibits the process and if not controlled on time could lead to its complete failure. Under such conditions, the sludge produced has the specific putrefaction smell, indicating that the process of stabilization has been compromised. Thus the process reaction rate and correspondingly the reactor organic loading rates are determined by the rate of biological degradation of the methanogens.

In most cases, anaerobic digestion is performed at a recommended temperature of 33°C to36°C (mesophilic regime), but in some cases, the design might provide a termophilic regime of 53°C to56°C, which has the advantage of a higher organic loading rate, lower reactor volumes required, and achieves an almost complete pathogen removal. The mesophilic regime provides a partial pathogen removal effect.

One of the most attractive aspects of the anaerobic digestion process is the production of biogas – a mixture of methane and carbon dioxide (Fig. 8.2), as a result of the organic mineralization. Under normal process conditions, the methane content in the biogas is about 70%. Another advantage is the low volume of excess sludge formed during the process. The biogas is a valuable energy source, which could be reused in the plant for maintaining the necessary temperature in the reactor. The excess amount could be transformed to electricity and used to supplement the energy supply in the vicinity.

The design of the reactor and the corresponding reactor volumes are determined on the basis of loading rates, which account for both – the acid formation stage and the methanogenesis to be performed in the same reactor. Most of the organic matter is usually hydrolyzed before the sludge reaches the reactor, along

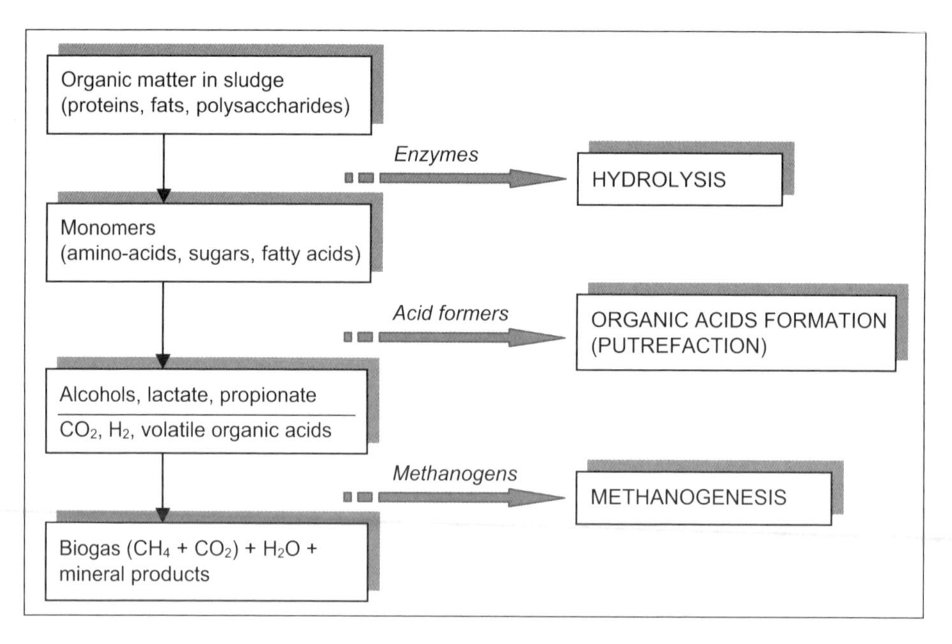

Figure 8.2. Fermentative pathways of anaerobic digestion of organic matter in sludge.

the sewer lines and during the sedimentation process. For a proper operation, a delicate balance needs to be maintained between the acid formers and the methanogens, which usually is controlled by the organic and hydraulic loading rates, the pH, the temperature and the volatile acids concentrations in the reactor.

Different types of reactors have been constructed and are operational, but the design approaches could be broadly classified, based on the different stages involved in the process:

- One-stage reactors provide the conditions for the anaerobic digestion to take place in one reactor volume, together with a partial thickening of the stabilized sludge and the supernatant removal.
- Two-stage reactors consist of two separate reactors, working in sequence, where the first stage is a completely mixed reactor, with intensified conditions for the fermentation process to take place, followed by a second reactor, which has the main function to thicken the sludge and provides conditions for the methanogenesis to be finalized. This configuration is the most widely applied in the practice, specifically for medium to large treatment plants.
- The latest developments with respect to the research and investigation of the process suggest multystage reactors, which provide separate compartments for the different stages of the fermentation process. This type of configuration has not found a wide practical application yet.

2.2.2 Sludge disposal options

Disposal options for sewage sludge are land application and disposal, incineration and ocean disposal (USEPA 1995, Attewell 1993, Pescod 1992). In 1982, sludge production in the USA was estimated at 7 million tones (dry weight), whereas in the European Community (EC) it is estimated at 6 million tones (dry weight) of raw sludge and is expected to increase in the future (Pescod 1992). In 1987, 40% of the 1.2 million tones dry sludge produced in the United Kingdom were utilized on agricultural land while it was 30% of 6 million tones in the EC countries (Lotter & Pitman 1997). The same authors mention that in South Africa, 47% of the sewage sludge is disposed on sacrificial land. The major sludge disposal options include:

- Beneficial land application – this refers to the use of sewage sludge in agriculture as a fertilizer and also, as a source of water for irrigation;

- Landfills – refers to the cases when sludge is treated together with solid waste, usually in wet form. In some cases, sludge is dried during the treatment process and after that used as inert material to fill natural or artificial depressions;
- Incineration – this practice applies to the case, where land is scarce and climatic conditions do not allow sludge disposal on land options, and as a result, incineration plants are needed to burn the sludge;
- Surface disposal – applies to the disposal of sludge in dry or wet form on specifically designated areas, without the sludge being beneficially used;
- Sea (ocean) disposal – this practice has been applied in the past, but has been recently restricted elsewhere, and in the EU it has been banned since 1999, due to environmental considerations.

Beneficial land application and disposal on landfills are the most popular practices for sewage sludge disposal. In the EC and USA, these disposal practices contribute 70% and 74% of the total sludge generated respectively (Pescod 1992)

The objectives of effluent and sludge disposal are an extension of wastewater treatment, which are to protect public health and the environment. Sludge disposal techniques should be considered as the final stage of the sewage treatment process. It is widely accepted that the volume and characteristics of the effluent and sludge affect the choice of disposal options. USEPA (1995) specifies the concentrations of trace metals, the level of pathogens and the attractiveness of the sludge to disease vectors, as the three main sludge characteristics considered in disposal options. Lotter & Pitman (1997) recommended several factors to be considered with respect to an optimal and safe choice of sludge disposal techniques in South Africa, and recommended several disposal options, appropriate for the conditions in the country:

- Factors, to be considered for an optimal choice of disposal method:
  - Environmental acceptability;
  - Cost effectiveness;
  - Operational feasibility;
  - Political acceptability.
- Sludge disposal options recommended:
  - Co-disposal with solid waste in landfills;
  - Remote farm disposal;
  - Composting and beneficial use as fertilizer;
  - Incineration.

USEPA (1995) recommends a variety of factors to be considered for an optimal choice of a sludge disposal technique, which could be grouped into three major categories:

- Regulatory considerations,
- Technical aspects and;
- Economic factors.

Thus sludge disposal becomes a task that requires a multi-disciplinary approach, which necessitates the participation of different types of specialists, such as environmentalists, industrialists, water engineers, agricultural specialists, and water resources managers, to work in a team with common objectives, in order to obtain optimal and cost effective results.

USEPA (1995) defines surface disposal as the placing of sewage sludge in a mono-fill, in a surface impoundment, on a waste pile, on a dedicated disposal site or in the form of dedicated beneficial use. A surface disposal site is defined as one where sludge remains on the ground for more than two years. Types of surface disposal sites are mono-fills, surface impoundments and lagoons.

Mono-fills are trenches where sewage sludge with at least 15% solids content is disposed and covered periodically. The application of cover distinguishes mono-fills from piles and dedicated disposal sites, and from surface impoundments, which receive cover only at closure of site. Mono-fills can be categorized into trenches where sludge is applied into excavated areas, and area fills where sludge is applied on the surface.

Surface impoundments are located on the ground surface and use dikes to contain the sludge. In cases where such impoundments are located below ground surface they are named lagoons. Recommended solids content of the sludge is 2% to 5%. When full, they are either covered or dredged. High disposal rates, ranging from 9100 to 28,400 m^3/h, are normal.

The land application of sludge in the case of beneficial use has the advantage of providing plant nutrients and improving soil fertility, in addition to the main objective to dispose of it. Macro-pollutants such as nitrogen, phosphates and potassium, and micro-pollutants such as Cu, Fe and Zn, are undesirable characteristics in the case of effluent discharges, but at the same time they are essential for plant growth. The organic matter improves the soil structure, enhancing the water holding capacity, cation exchange capacity and aeration. However, the land application of wastewater could be a potential source of surface and groundwater pollution, phytotoxicity, salinity, and change in soil structure and could be a potential health hazard if polluted surface or ground water is used for drinking, washing or recreational purposes. In order to prevent such scenarios, proper design, operation and maintenance of sludge disposal sites are essential. Metcalf & Eddy (1991) provide the following steps in the development of sludge application systems:

- Characterization of sludge quantity and quality;
- Review of pertinent federal, state and local regulations;
- Evaluation and selection of site and disposal option; and
- Determination of process design parameters – loading rates, land area, application methods and scheduling.

3 REGULATING THE PRACTICE OF SLUDGE LAND DISPOSAL

3.1 *Hydraulic loading rates*

In cases when the sludge is applied to land in a wet form, one of the main design parameters, which would determine the necessary size of the field required for sludge disposal, would be the hydraulic rate of sludge application. The following factors would affect the recommended hydraulic rates:

- The nature of soil – sandy, loamy overlaying gravel soils could have higher hydraulic rates than fine clay soils.
- The nature of sludge effluent – the concentration and level of treatment of sludge would affect hydraulic loads, with higher loads recommended for a higher level of treatment of the sludge.
- Climatic conditions – in a dry and hot climate, hydraulic loads could be much higher than in a wet and cold climate.
- Vegetation cover – hydraulic loads are higher for land with no crops, compared to cases of a beneficial use of sludge for crop irrigation when restrictions are imposed with respect to the water requirement of the specific crop grown. However, in cases of pastures or indigenous vegetation, evapotranspiration rates are higher, compared to bare lands, thus allowing for higher hydraulic loads.

The sludge effluent has to be applied intermittently since lands require some rest at intervals. Also, for the conditions of the South African region, hydraulic loads would differ during the wet and dry season. Then the size of the land to be dedicated for this purpose would be determined by the recommended loads for the wet season, considering the additional load from rainfall and the corresponding evapotranspiration rates. It should be noted that the determination of the size of the area dedicated, should also consider the need for intermittent irrigation by applying the rotational principle. It consists of dividing the area in plots and irrigating a prescribed number of plots only, allowing the rest of the plots to absorb the pollutants, infiltrate the moisture and recover until the next portion of sludge is disposed on the same plot.

Hydraulic loads are an important design parameter to be considered during the design stage and controlled during the operation of sludge land disposal sites, but they should always be applied in conjunction with the pollutant loading rates. The final loading rate, which would be used for the design of the field, should be based on the limiting factor for each specific case. If the pollutant absorption rate is lower than the rate of application, which has been calculated based on the hydraulic rate, then it would be the limiting factor and the hydraulic application rate should be reduced.

In South Africa, the level of treatment of sewage sludge, which would influence the hydraulic loads, determine four types of sludge: A, B, C and D in a decreasing order of potential to cause odor nuisance and fly-breeding as well as the potential to transmit pathogenic organisms to man and the environment (Lotter & Pitman 1997).

- Type "A" Sludge: Unstable with a high odor and fly nuisance potential; high content of pathogenic organisms.
- Type "B" Sludge: Stable with low odor and fly nuisance potential, reduced content of pathogenic organisms.
- Type "C" Sludge: Stable with insignificant odor and fly nuisance potential, containing insignificant numbers of pathogenic organisms.
- Type "D" Sludge: It is similar in hygienic quality as type C, but is applied for an unrestricted use on land at a maximum application rate of 8 dry tones per hectare per year. The metal and inorganic content are limited to acceptable levels.

Type "A" sludge is not allowed to be applied on pastures with grazing animals, types "B" and "C" sludge could be applied but should be mixed or covered with soil, whenever possible.

3.2 *Pollutants loading rates*

Pollutants loading rates reflect the mass of pollutant constituents, which will be directed to the designated site during a given time interval. They are recommended, based on the absorption capacity of the soils and the plant uptake of the vegetation cover or crops, in the case of beneficial applications. The pollution loading for each specific case is determined by the mass (volume) of the sludge and the specific constituent concentration. Usually, the limiting constituents are toxic metals, which could be accumulated in the soil during the exploitation period of the site. The determination of the necessary land requirement for any disposal site is based on the selection of the pollution load of a critical parameter, which would lead to the maximum land requirement. In cases of sewage sludge from purely domestic origin, toxic metals concentrations in the sludge are usually low, then the limiting parameter might become nitrogen, which is easily transported to ground water and might lead to the pollution of ground water resources.

The recommended criteria vary considerably in the different countries, with USA recommendations being more relaxed, while EC recommendations are more stringent. In South Africa, the criteria given in Table 8.1 have been recommended with respect to toxic elements (WRC 1997).

It should be emphasized, that guidelines and regulatory instruments usually prescribe the maximum permissible concentrations of selected toxic elements in the soils. Therefore, decisions regarding the closing of a given land disposal site should be based on such criteria after a thorough monitoring program has shown that the prescribed values are exceeded.

The Zimbabwean regulations (WWEDR 2000) treat sludge land disposal on land as a specific case of effluent discharge and requires a permit for the activity. Permit limitations and corresponding charges are based on two major criteria – the annual application rates with respect to cadmium and nitrogen. The maximum allowable annual application rates are 4 kg Cd/ha and 600 kg nitrogen/ha, which present high environmental hazard risk. The maximum allowable soil concentration of Cd is 10 mg/kg dry soil.

In the case of wet sludge used for irrigation the pollutant loading rates are dependent on the pollutant concentrations in the sludge. Guidelines regarding these criteria are presented in more detail in Chapter 10.

3.3 *Safety requirements*

The risks, which are associated with the reuse of wastewater and sludge for irrigation, could be classified into three categories:

Table 8.1. Recommended criteria for sewage sludge land application in SA.

Metal	Maximum Permissible Content in Soil (mg/kg)	Total Load in kg/ha for 25 years of operation	Main reason why elements are limited	
			Phytotoxicity	Zoo-toxicity
Cadmium (Cd)	2.0	3 140		x
Cobalt (Co)	20.0	20 000		x
Chromium (Cr)	80.0	350 000	x	
Copper (Cu)	6.6	10 100	x	
Mercury (Hg)	0.5	2 000		x
Molybdenum (Mo)	2.3	5 000		x
Nickel (N)I	50.0	40 000	x	
Lead (Pb)	6.6	10 100		x
Zinc (Zn)	46.5	70 700	x	
Arsenic (As)	2.0	3 000	x	
Selenium (Se)	2.0	3 000		x
Boron (B)	10.0	16 000	x	
Fluoride (F)	100.0	80 000		x

- Risks, associated with the consumption of the products of the irrigated field;
- Risks, associated with the personnel, operating the field;
- Risks, associated with the soil conditions of the irrigated field.

The primary concern when sludge is used for irrigation is related to the health risks for close-lying neighborhoods and for the consumers of the products, irrigated with reused sewage. Sludge land disposal sites should be well protected to avoid the spreading of sludge or effluent to the neighborhood or access of unauthorized persons to the site. For these reasons, surface irrigation methods are preferred to spring irrigation, as the wind could spread wastewater or sludge outside the protected areas. The risks vary greatly depending on the local state of sanitation in the area, farming methods, customs and climate. In general, sewage should not be used on or near crops that are eaten raw. The use of sewage on hay meadows does not seem to pose any great problems, while for grazed pastures, a proper rotating scheme of irrigation could considerably reduce the risk. Cereals, beets and oleaginous crops are the types of cultivation more suited.

The main health risk of the sludge disposal on land technique is associated with the personnel, which would operate the site and is exposed to direct contact with the sludge or wastewater. A proper training program in respect to safety requirements, together with proper protective clothing and good hygienic facilities at the site would help to prevent the spreading of diseases among the personnel.

The main risk to soils, irrigated with treated sewage are clogging, increased salinity, and the introduction of toxins. Land disposal can alter the physical properties of the soil. In particular, the introduction of an excessive amount of sodium and the absence of leaching can destroy the soil structure. An effluent that has a salinity level exceeding 2 g/l also causes trouble and leads to a stricter control on the quantity of the water that is spread and the level of salinity in the soil. The carbon/nitrate ratio remains one essential crop requirement. One of the drawbacks of using sewage for irrigation is that it may raise the nitrate level of ground water. Therefore, a specific monitoring program to control soil characteristics during the operational life of the site is important and could prolong its design life substantially. More information with respect to the control parameters and maximum permissible criteria are presented in Chapter 9.

4 SLUDGE DISPOSAL ON PASTURES – IMPACTS ON SURFACE WATER QUALITY

4.1 *The study area*

Crowborough Wastewater Treatment Works (CWTW) is the second largest sewage treatment works of Harare, serving more than 400,000 people. The land use pattern of its catchment area includes medium and high-density residential areas, commercial and industrial enterprises. The location of CWTW and the dedicated farmland for sludge and effluent disposal are shown in Fig.8.3. The Crowborough farm receives part of the pond's effluent and the whole amount of sludge mixture (primary digested sludge and waste activated sludge) daily throughout the year. The remaining part of the pond effluent is pumped to another dedicated area – Ingwe Farm.

The Crowborough farm has a total area of 594 ha, but the active farmland comprises of 17 paddocks with a total area of 427 ha covered mainly by star grass (*Cynodon plectostachus*) and Kikuyu grass (*Pennisetum clandestinum).* The farm is an example of the beneficial reuse of sludge and effluent, where this mixture is applied to land, which is used as a grazing pasture. It has been operated since the early 1970's. The terrain has a gentle slope of about 0.008 from north to south. The topography of the farmland defines a saddle like shape with a distinct watershed along the access road to CWTW. About 40% of the farm area drains into the Little Marimba River and the rest into Marimba River.

The soils are variable in derivation, texture and depth, with clays in the north, sandy clay and sandy clay loams in the center, sands and sandy loams in the south. Due to long term irrigation with sludge and wastewater, a thick humus blanket covers the surface of the farm. The lower southern part of the farm area and the areas along the two rivers are wetlands with a high groundwater table, which do not dry out during the dry season and some parts are inaccessible during the wet season. The farm is provided with night storage ponds and an under-drain system for collection, transportation and discharge of the seepage,

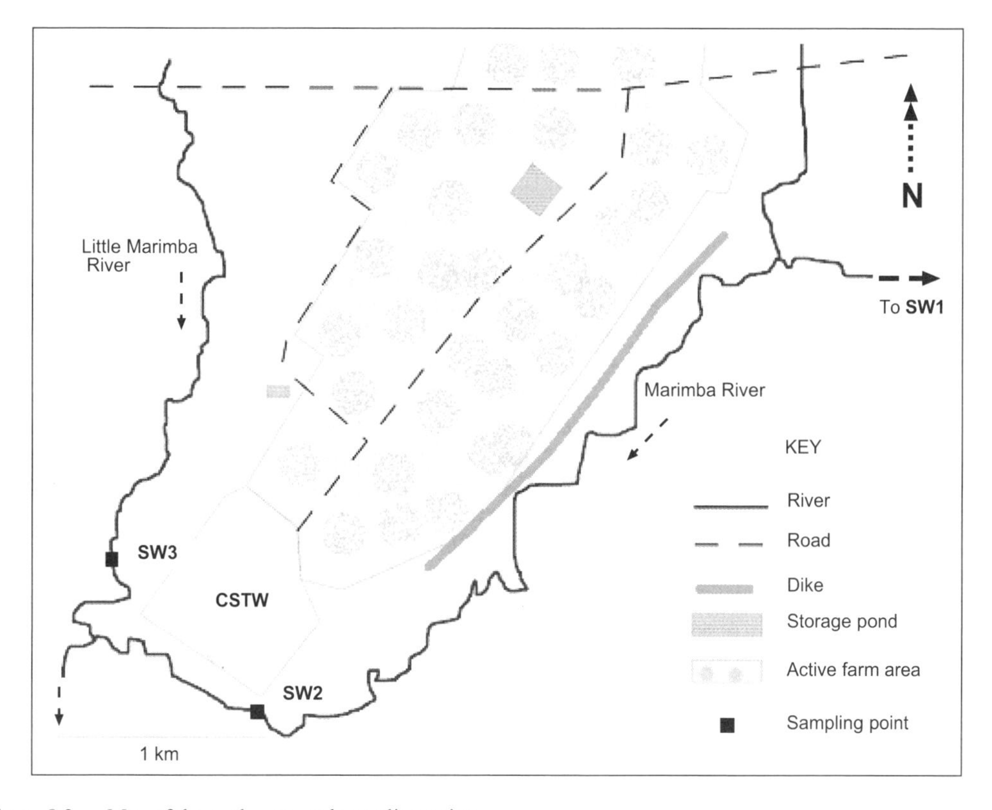

Figure 8.3. Map of the study area and sampling points.

Figure 8.4. Distribution chamber and open ditch at Crowborough farm.

infiltrated through the soil. The spreading of the sludge and effluent mixture on the land is done manually by a system of open main and secondary ditches, connected to pressure release chambers (Fig. 8.4) and to the night storage ponds. The sludge and effluents mixture is released directly to selected paddocks for 7 hours per day. During the rest of the day, only ponds effluent is pumped out of the treatment plant, and could be released directly to the land or discharged into the storage ponds.

The farm area is used for pasture grazing of about 3000 heads of cattle with about 1000 heads allocated per paddock. This allows keeping the grass at a reasonable height of 100 mm.

The CWTW was originally designed as a trickling filter plant, with a treatment scheme similar to the one shown on Figure 8.1. During the 1970's the plant was upgraded with a separate BNR treatment line. The new line treats the increased flow volumes and discharges them in Marimba River, downstream the location of SW_2. The effluent from the old trickling filters' line is directed to a maturation pond and after that pumped to the farm. The stabilized primary sludge from the anaerobic digesters is mixed with waste sludge from the BNR plant in a secondary digester and after partial thickening is pumped to the same pumping line which delivers the pond effluent. A schematic presentation of the treatment scheme is shown in Figure 8.5.

The rainfall pattern over the study area is similar to the one described in the previous chapters. The rainfall pattern during this specific study (October 1999 to April 2000) is presented in Figure 8.6. The monthly rainfall figures (Fig.8.6b) were measured at a gauge station located near the study area. During the period May to September 1999 no rainfall was recorded. The annual rainfall in Harare is about 800 mm/year, but there are only 130 rainy days with about 70-80 of them generating runoff.

The rainfall events are mostly in the form of thunderstorms with an average of 73 numbers per year recorded for a twenty-year period. Figure 8.6a presents data about the Marimba River's average flow rates at the gauge station CR24, which collects the runoff of a total catchment area of 189 km^2, and is located near the discharge point of the river into Lake Chivero. It presents the runoff from the whole catchment area of the River, which includes the farm runoff, the streams collecting runoff from high-density low income areas in Kuwadzana (Chapter 5), the runoff from the Coventry road storm water channel

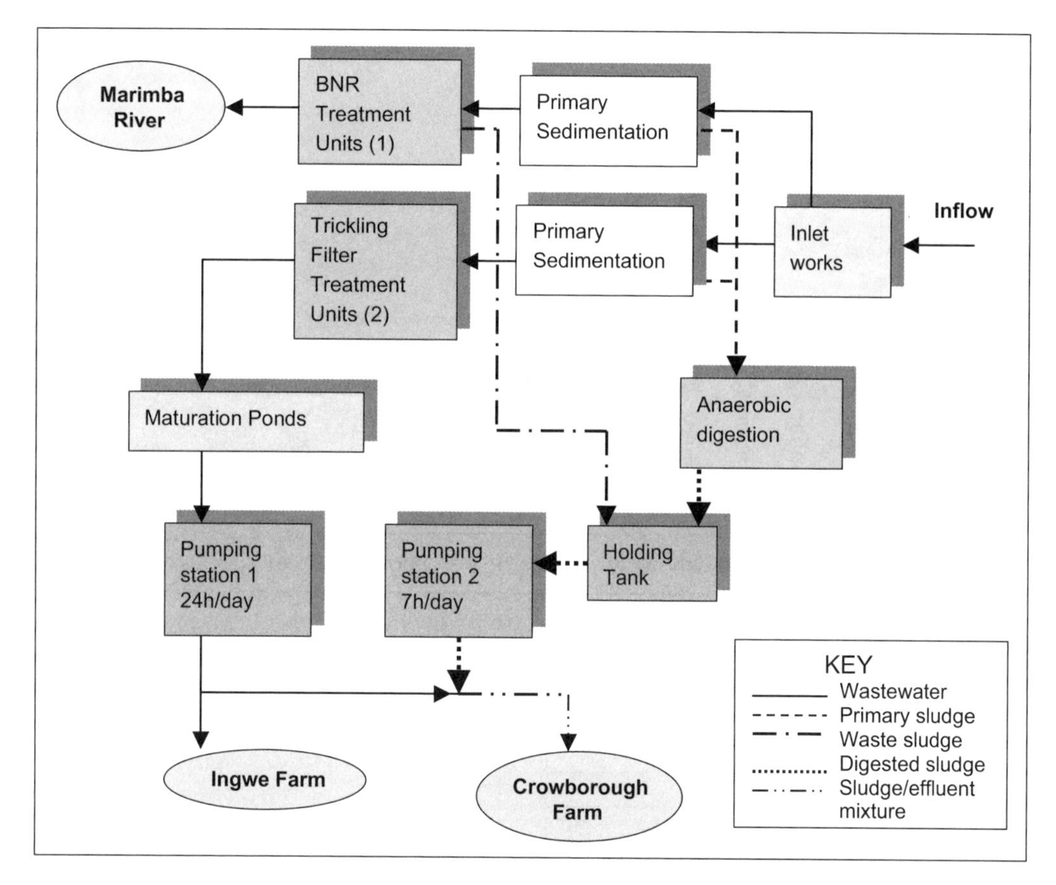

Figure 8.5. Scheme of the CWTW flow chart.

(Chapters 3 & 4), as well as the runoff from medium density areas upstream this discharge point. In addition, it collects the BNR effluent from CWTW. The design capacity of this treatment unit is 18 000 m^3/d (Hranova 2002), which constitutes between 60-80% of the base flow during the dry season.

The surface runoff and the seepage from the study area are drained into the Marimba River and into Little Marimba River. The farm area, which drains into the Marimba River, is protected by a dike (Fig. 8.3) and is also provided by an under-drain system for the collection of the excess seepage. The portion of the farm area, draining into the Little Marimba River, is not protected by a dike, but is provided with an under-drain system too. During the study, it was not possible to obtain data regarding the operational condition of the under-drain system. It might be blocked at some locations due to the long period of operation. All this makes it difficult to accurately evaluate the expected surface runoff and seepage contributed from the farm to the River.

Considering the different ways of pollution transport from the farm area into the natural streams, we could differentiate four possibilities:

- Farmland seepage discharged by the underground drainage system;
- Direct runoff;
- Polluted groundwater discharge to the river;
- Direct leak from the wet areas.

The pattern of pollution transport to the river depends on hydraulic loading and is influenced by seasonal conditions. During the dry season, the precipitation is very low and the run-off during this period could

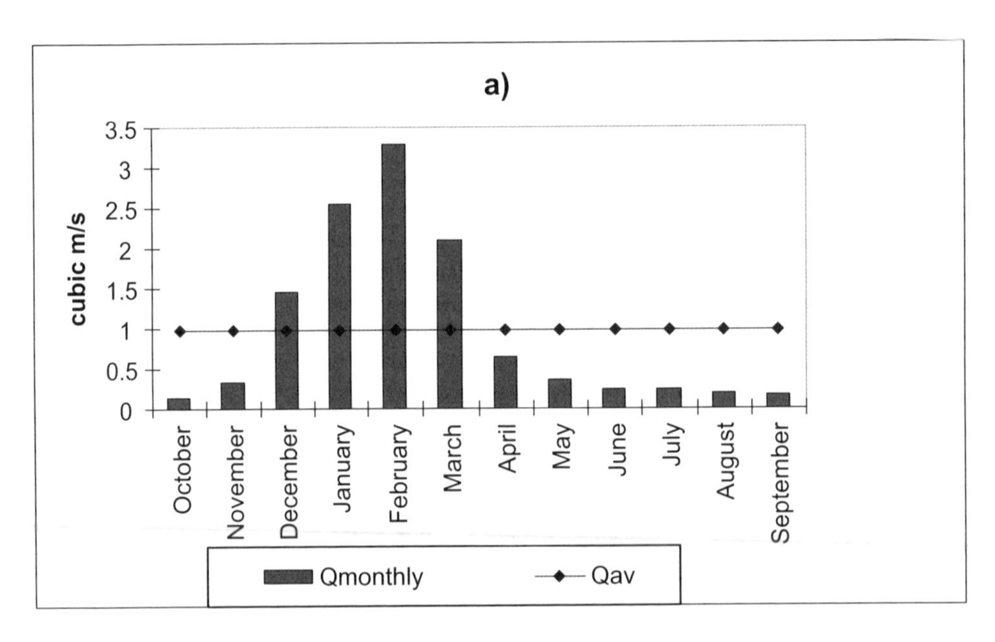

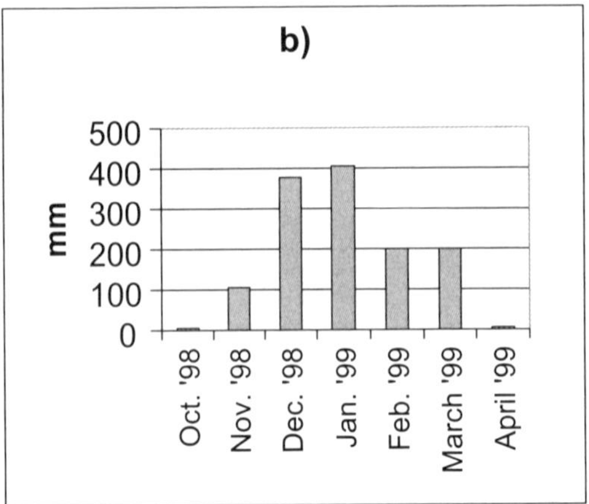

Figure 8.6. Rainfall and runoff characteristics of the study area.
a) Average flow rates of Marimba River at C_{24} for the period 1953-2000.
b) Total monthly rainfall at Crowborough gauge station during the study period.

be neglected. During the 1998 dry season the evapotranspiration capacity was 847 mm (AQUASTAT 2003) while the hydraulic load due to sludge and effluent mixture was 1019 mm as an average annual figure.

4.2 *Methodology*

The location of the sampling points is shown in Figure 8.3. SW1 is located on Marimba River upstream the farmland and serves as a control point; SW2 is located on Marimba River at a distance of about 6 km downstream of point SW1 but before the effluent discharge of the treatment works; SW3 is located on Little Marimba River, downstream the study area. It is a small stream, but it was not possible to estimate its flow rate due to time and personnel limitations.

Considering the possible pathways of pollution transport, the worst impact on river water quality could be expected during the beginning of the rainy season, because of the relatively low flows, and because of expected higher pollution concentrations in the runoff and seepage, due to the flushing effect of the first storms. During a typical wet season, the Marimba River flows increase tenfold (see Fig. 8.6a), leading to a considerable potential for dilution effect. In addition, pollution concentrations of water flows from the pasture are expected to decrease, which contributes to the effect of dilution of pollution concentrations in the river. For this reason, it was decided to study seasonal variations at three distinct periods – dry season (October), beginning of the wet season (November) and wet season (February). Sampling was done twice for each season. Rainfall data during the period of study to support the choice of seasonal variation is shown in Fig 8.6b. For each sampling occasion duplicate samples were analyzed.

On each visit, the following parameters were examined in the field: pH, temperature, EC, turbidity and nitrate (determined as nitrate plus nitrite). Throughout the text, the term "nitrate" is used for this parameter and results are expressed in mg N/l. TP, TKN, BOD_5, TS with mineral (TSm) and volatile (TSv) fractions, Pb, Zn, Cd, Cr (determined as Cr^{+6}) and Ni were analyzed in laboratory conditions. The field measurements were executed on site by an ELE-Paqualab modular field system, incorporating a photometer, a portable conductivity/temperature meter and a pH meter. Turbidity was measured on site by a portable HACH turbidimeter. Samples for laboratory analyses were collected as the composite of four grab samples, 500 ml each at 30 minutes intervals, then the field tests were done on the composite sample. During collection and transport samples were kept in cooling containers filled with ice. Immediately after the composite sample was taken it was distributed in different polyethylene containers and preservations for TKN (addition of H_2SO_4), TP (addition of $HgCl_2$) and metals (addition of HNO_3) were executed according to Standard methods 1989. Samples were transported for analysis within 4 hours after sampling. The laboratory analysis were executed as described in chapter 4, except for metals, which were determined as total concentrations and analyzed, using a SHIMADZU – AA6401F Atomic Absorption Spectrophotometer calibrated by standard solutions after a nitric acid-sulfuric acid digestion.

Data with respect to the effluent and sludge mixture-quality and quantity characteristics were provided by the municipality.

Statistical calculations (mean, standard deviation and one-tiled Student's t-test at 95% of confidence) were performed using the Microsoft EXCEL statistical package. Throughout the text "significant difference" means that $p < 0.05$, when comparing the mean values of the data sets.

4.3 *Hydraulic and pollution loads*

The flow values were obtained by averaging the daily flows from the plant operation records during the study period. The dry weather flows were obtained by averaging the flow rates for October and November 1998, while wet weather flows were averaged for the period December 1998 up to March 1999. During the dry months an average incoming raw sewage of 50,230 m^3/day was measured, which is still within the design limits of the treatment plant (54,000 m^3/day ADWF). An average of 19,406 m^3/day effluents and 921 m^3/day wasted and digested sludge, were pumped to the pastures. There was no overflow from the ponds into the Marimba River during the dry season, which is in line with the actual observations on site. An average flow of 3790 m^3/day was being released to the river from the BNR treatment units (Fig.8.5).

The wet weather flow was found to be 123,120 m^3/day on average. The volumes pumped to the pastures were 21,030 m^3/day and did not vary considerably from the dry weather values. The ponds showed a very high average overflow value into Marimba River of 50,183 m^3/day, while the BNR plant was releasing 15,905 m^3/day. The flow to the trickling filters treatment units was 53,246 m^3/d on average, which was quite near the design wet weather flow value of 54,000 m^3/d. On the ground, however, during heavy rains, the actual peak flows exceeded the design flows resulting in overflows occurring at the distribution box. The average annual hydraulic load to the farm was 17,670 m^3/ha.year, with a dry component of the mixture between 3.5-4.5%.

In Harare, the sewer system receives wastewater from residential, commercial and industrial sites, and in the vast majority of the cases, the industrial discharges do not undergo preliminary treatment. This leads to a considerable amount of heavy metals in the row water, which are accumulated in the sludge during the treatment process. The data presented in Table 8.2 supports this statement. The values have been calculated on the basis of the average pumped rates and the average concentrations of ponds effluent and sludge mixture during the study period. Consideration has been given to the operational practice of different pumping rates and times of the effluent and sludge respectively. The limit values (maximum permissible values) with respect to South Africa and USA are based on literature data (WRC 1997), while the Zimbabwe limits are based on WWEDR (2000).

According to Reed et al. (1988) the annual nutrients uptake rates for different types of grass vary between 150 and 670 kg/ha for nitrogen and between 24 and 84 kg/ha for phosphorous. Results in Table 8.2 show that the phosphorus load exceeds the maximum value of plant uptake capacity by about 50%.

Nitrogen rates, expressed as TKN load, exceed the maximum permissible value (high hazard) for sludge land application of 600 kg/ha.year (WWEDR 2000). Comparison with Canadian and USA sludge quality criteria for land application (Roy & Couillard 1998) show that Ni concentrations exceed twice the maximum USA limits of 400 mg/kg and Cd concentrations exceed the Canadian maximum limit of 15 mg/kg. The application rates regarding Cd are within the WWEDR (2000) requirement, but Ni application rates exceed all stated guidelines and limits. It has been reported that sludge containing more than 7 mg Cd/kg may present a risk of groundwater contamination even at an annual sludge application rate of 200 kg available N/ha (Roy & Couillard 1998). The stated application rate of nitrogen is expressed as TKN but not as available N and it could be expected that the loadings with respect to available N should be lower. A study performed on the same area in 2002 examined the impact of this method of sludge application on soils and ground water and is discussed in Chapter 9.

4.4 *Impacts evaluation*

This study focused on surface water quality assessment in terms of the impacts from the farm area, which reflect the combined effect of surface runoff, seepage from the under-drain system and possible recharge from ground water. The impacts have been discussed in terms of seasonal variations of water quality at each sampling point, as well as in terms of spatial variation between the control point and the other points, reflecting the impact of the farm.

Results of field measurements are presented in Table 8.3. Seasonal variations of turbidity show increased values during the wet season at all points. The significant spatial increase from 18 NTU to 33 NTU at SW2 compared to SW1 is an indication of the washout of suspended colloidal substances from the farm area. Spatial and seasonal variations of pH were not significant. Results on conductivity do not show a significant spatial variation due to pasture impact, but seasonal variations show an expected reduction of values due to the dilution in February.

Results of the rest of the tested parameters are presented on Figures 8.7, 8.8 and 8.9. Seasonal variations were studied by testing for a statistically significant difference of the mean values with respect to each one of the studied parameters. The mean values for a given season were compared with the ones, representing the other seasons at the same sampling location, e.g. with respect to SW_1 the TP mean values for October were compared with those for November and for February, and those for November were

Table 8.2. Average annual sludge application rates to Crowborough Farm.

	TKN	TP	Pb	Zn	Cd	Cr^{+6}	Ni
Sludge & Effluent (mg/kg.d.w)	12899	2133	73	133	20	72	873
Application rate (kg/ha.year)	684	113	4	7	1	4	46
Limit SA (kg/ha.year)	-	-	0.41	2.83	0.13	14*	21
Limit ZW (kg/ha.year)	600			-	4	-	-
Limit USA (kg/ha.year)	-	-	15	140	1.9	150*	1.6

* Total Chromium

Table 8.3. Field measurements results.

	October	November	February
SW1			
pH	7.8	7.5	7.6
T (°C)	18	21	21
Conductivity (μS)	547	542	433
Turbidity (NTU)	11	18	25
SW2			
pH	7.9	7.0	7.3
T (°C)	21	23	23
Conductivity (μS)	580	529	470
Turbidity (NTU)	8	33	21
SW3			
pH	*	7.4	*
T (°C)	20	21	22
Conductivity (μS)	*	602	*
Turbidity (NTU)	6	11	14

*-Data not collected due to technical fault.

compared with the values for February. All statistically significant values within each sampling location are denominated by different letters on the corresponding figures.

Spatial variations, reflecting the impact of the farm area during each distinct season, were tested for a tatistically significant difference by comparing the results for the control point SW_1 with SW_2 and SW_3 separately. Statistically significant values within each season are denominated by different numbers. The averaged results for the whole period of study are presented in Table 8.4 and are compared with the present recommended limits for safe environmental water quality characteristics.

4.4.1 Nutrients

Variations of TKN (Fig. 8.7a) and TP (Fig. 8.7b) have a similar pattern. TKN seasonal variation at SW_1 shows significant differences (aa, bb, cc) for all three seasons. In February, the pollutant concentrations are much lower compared to the ones in October and November. This indicates to the presence of river pollution upstream the control point, which is pronounced only during the dry season due to the relatively low base flow in the River. At SW_2 there was no significant difference between October and November observations, but in February the concentrations were significantly lower. At SW_3 the seasonal variation was very pronounced (ff, gg, hh), showing a peak in October, which is reduced in November, reaching very low values in February. Spatial variations during October are significant for all three points (1, 2 and 3) and are of a larger magnitude with respect to SW_3, indicating to a considerable impact of this part of the farm area, contributed to the Little Marimba River.

In November, significant difference was found only between SW_1 and SW_2 (4 and 5), indicating to a relatively low impact from the portion of the farm draining into the Marimba River. During the wet season, no impact from the farm area was found.

The seasonal variation of TP (Fig.8.7b) shows a significant difference only at SW3, with maximum concentrations during October, which decrease with the start of the rains. In February, the TP concentrations at SW_3 have been reduced up to ten times, compared to the ones in October. The spatial variation indicates to an impact from the farm area at SW_3 only. A sharp increase in the TP concentrations at SW3 was found during the dry and the beginning of the wet season, while during the wet season, they are lower than the control point.

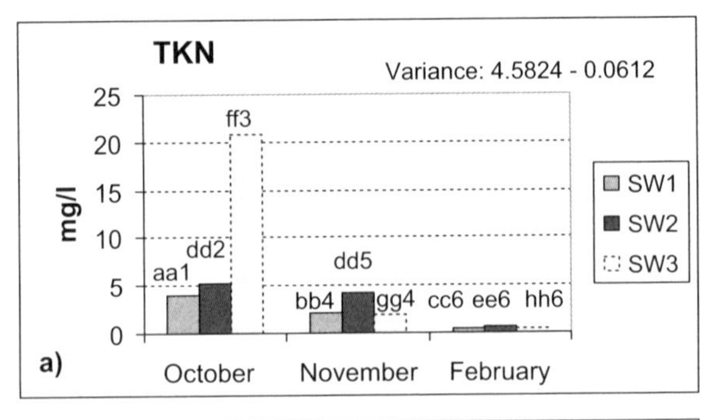

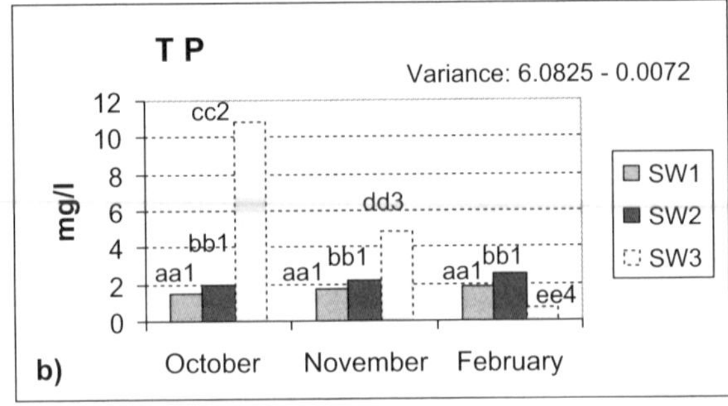

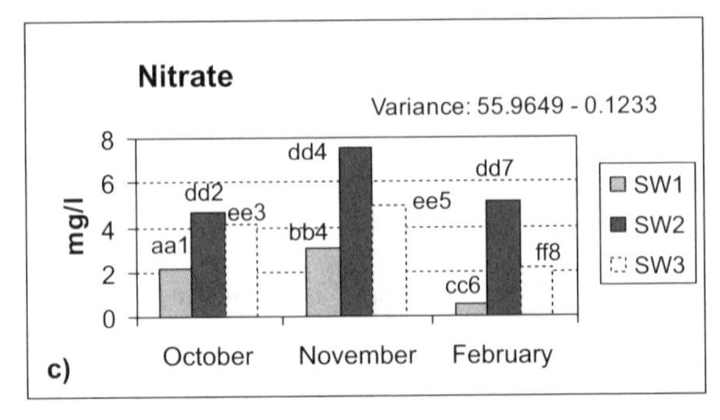

Figure 8.7. Spatial and seasonal variations of TP, TKN and Nitrate.
Mean values between the time series in Oct., Nov., and Feb. at each sampling point denominated by the same letter, and values within each time group followed by the same number are not significantly different at the 0.05 level.

Nitrate seasonal variation (Fig. 8.7c) at SW_1 shows an increase during November and a sharp decrease during February. At SW_2 a pronounced increase was found in November, but it was not statistically significant, most probably due to the high variance of the data set. At SW_3 a significant decrease was found during the wet season. Spatial variations indicate to a significant contribution from the farmland at both points.

Analyzing the possible pathways of nutrients transport, it should be considered that mobility of phosphorous (P) is chemically controlled and is dependent on the oxygen content of the upper soil layer. Anaerobic conditions on the soil surface are contributing to a significant release and leaching of dissolved P. Rydin (1996) reported that under anaerobic conditions, 30% of the TP could be released from sludge. Irrigation with wastewater and sludge of high organic content tends to create anaerobic conditions and contribute to a P leaching (David & David 2000). Particulate P can be transported by the drainage water or with eroded suspended particles. A well-defined correlation between suspended solids and P concentrations in runoff has been found (Djodjic et.al. 2000, Umemoto et al. 2001). Djodjic et al. (2000) reported that more than 50% of the TP in drainage water was in particulate form, but a two times higher leaching was observed from no-tillage plots in the form of dissolved P, and also that ponding conditions promote P leaching. Grass plots tend to retain P better than other cultivated plots (David et al. 2000).

Due to chemical and biological transformations, organic nitrogen is transformed to ammonia, and ammonia is nitrified to nitrite and nitrate consequently. Those processes occur simultaneously at different depths of soil texture. Nitrate moves easily through soil, because of its negative charge, and can be leached, while ammonia is retained in soil texture due to its positive charge (Sukreeyapongse et al. 2001).

In a laboratory study of wastewater infiltration through a sand column (Reemtsma et al. 2000), it was found that about 85% of ammonia was transformed to nitrate at a depth of 1.2 m and the process of nitrification leads to the acidification of the upper layers of the soil.

Considering the site conditions, it should be noted that approximately half of the farm area drains into the Marimba River (SW_2) and the rest into The Little Marimba River (SW_3). However, the former has a smaller flow rate and a dyke does not protect its banks, leading to formation of wetlands. By visual inspection it was found that septic conditions have been developed throughout the formation of wet areas, which are persisting during the dry season. They provide an environment, conducive for the P and ammonia release into the wetland water, which finds its way into the Little Marimba River, thus contributing to significant increase of the nutrient at this point. Also, the banks of the stream near and upstream SW_3 are not covered by grass, as it is the case near SW_2, thus enhancing erosion, which is reflected by the significant increase of TS at SW_3 during the dry and beginning of wet seasons, as shown in Fig. 8.8.

The contribution of the farm area to SW_2 is much lower in terms of the increase of the nutrients concentrations compared to the control point. It could be associated with the protective function of the dike. In addition, considering the higher flow rate in the River, the farm contribution did not cause a highly pronounced impact during the study period. Another factor, which could explain the low nutrients' concentrations at this point, is the better grass cover near SW_2, leading to a higher uptake of nutrients. In addition, the lower, unprotected part of the farm area, on the side of Marimba River, is converted to a marsh due to the hydraulic overloading. It could be expected that this area has a considerable nutrient uptake rate, thus preventing nutrients to reach the river water.

High rainfall figures (Fig. 8.6b) during December, January and February lead to the dilution of the pollutants' concentrations in the wet areas and to a partial oxygenation, thus alleviating the septic conditions and decreasing the release of P and ammonia. This fact, together with the increased flow rates in the stream could explain the lack of impact at SW_3 from the farm area, regarding TP and TKN during this season. Studies on phosphorous seasonal variations in rivers mention that maximum loads are associated with heavy rainfalls (Boar et.al. 1995, House & Denison 1998). During the period of study, intensive

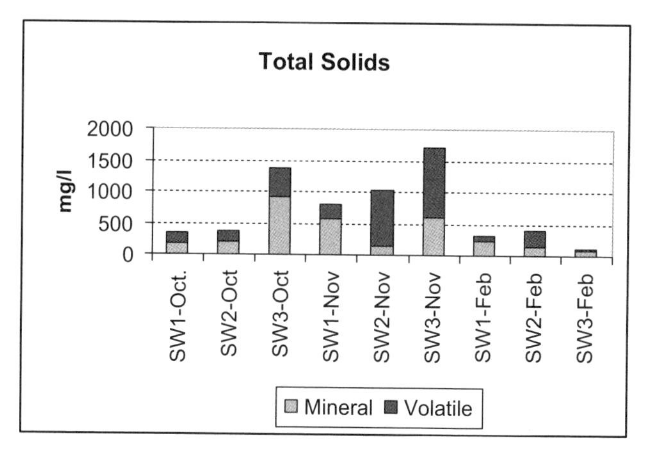

Figure 8.8. Spatial and seasonal variations of Total Solids – mineral and volatile fractions.

rainfalls characterize the wet season, leading to an increased turbidity, but no significant increase in P concentrations was found at the 3 points. Actually, the TP concentrations were reduced in November and February, while the turbidity increases, compared to the control point. This indicates that, the most probable pathway of phosphorous transport to the river is through direct leak from the wetland in diluted form.

The significant contribution of nitrates from the farmland during all seasons and at both points suggests a different pathway of transport, associated with a regular load from seepage and ground water discharge. It is more pronounced at SW_2, which could be explained with the fact that the existing drainage structure has four points of discharge to Marimba River and only one point of discharge contributing to SW_3. The fact that this impact was present during all seasons indicates that the soil is saturated with nitrates, resulting in a constant leaching to both rivers through groundwater or tillage water. However, a study performed later (Chapter 9), which investigated the ground water quality at a level below the under drain system, shows a considerably lower nitrate concentrations of the aquifer over the whole farm area. This suggests that the leaching of nitrates from the topsoil layers is pronounced with respect to the drainage water, which discharges into the surface water bodies, thus protecting the aquifer. The impact is more pronounced during the wet season, due to increased tillage, caused by increased hydraulic loads.

The degradable organic pollution, measured as BOD_5, did not show a significant increase in spatial variations, which indicates that biodegradable organic matter is retained on the farmland and is not influencing the river water quality. Variations of TS and fractions (Fig. 8.8) show a significant change in the fractional composition, where the TS volatile fraction during November was drastically increased, compared to the dry and wet season data. This suggests a contribution from the farmland, where the organic materials are higher in percentage compared to the mineral fraction. However, this data was not supported by a significant increase of BOD_5 during the same period. A significant decrease in concentrations was detected in February from 7.7 mg/l to 3.1 mg/l as the average of the mean values of all the points, which could be attributed to dilution.

4.4.2 *Metals*

The results obtained during this study point out to a significant seasonal variation of the concentrations of heavy metals. Cd concentrations did not show a significant spatial variation (Fig. 8.9a) at SW_2 and SW_3 indicating that despite the high loading rates the pasture does not affect river water quality.

A well pronounced seasonal variation at all three points was measured, with a significant difference from 0.03 mg/l in October to 0.07 mg/l and 0.02 mg/l in November and February respectively, given as average values for the study period.

Ni spatial variation during October shows a significant increase at SW_2 and SW_3 compared to SW1 (Fig. 8.9b). No significant spatial difference was found during November at the two examined points, and in February, a statistically significant but not pronounced increase was found only at SW3.

The seasonal variation shows a significant difference during the three occasions. A sharp increase in concentrations from 0.56 mg/l to 2.39 mg/l was measured in November, with a significant decrease in February to 0.35 mg/l, given as the average values for the three sampling points.

Significant increase in spatial variations of Zn, Cr^{+6} and Pb was not observed during the study period, indicating that the farmland is not affecting river water quality regarding these metals. However, a significant seasonal variation of these parameters was detected at all points. Variations between November and February, given as average of the means of the three points, show a significant difference for Pb (decrease from 0.33 mg/l to 0.13 mg/l) and for Cr^{+6}, which increased from 0.05 mg/l to 0.11 mg/l. Cr^{+6} is the only measured metal, which shows significant increase of concentrations during February.

Leaching of heavy metals from soils, irrigated with effluents and sludge, depends on the characteristics of the soil, application rates and the period of application, and the availability of water movement to transport them to ground or surface water bodies. Under a proper managerial practice, metals are accumulated in the soil surface and immobilized in the humus layer by adsorption on the surface of particulate organic matter. The leaching of metals from soil horizons is enhanced by a low pH, reduced organic matter, low cation exchange capacity of the soil, water movement (which is likely to occur when hydraulic application rates exceed evaporation rates), and is more likely in sandy soils (Roy & Couillard 1998).

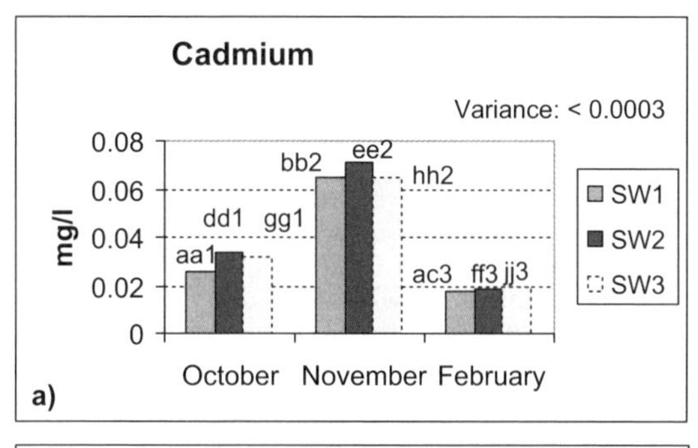

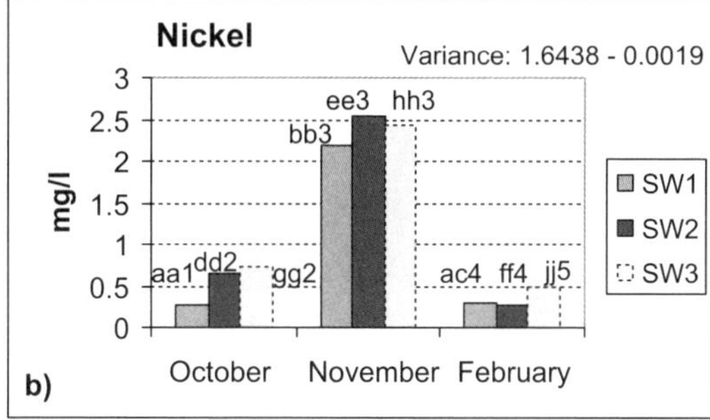

Figure 8.9. Spatial and seasonal variations of Cd and Ni.
Mean values between the time series in Oct., Nov., and Feb. at each sampling point denominated by the same letter, and values within each time group followed by the same number are not significantly different at the 0.05 level.

The possible pathways of metals from the soil to surface water are through surface runoff, containing particulate matter with adsorbed metals, or through seepage and ground water, contaminated with leached metals. Impact from the farmland was found at both SW_2 and SW_3, regarding Ni only and it was pronounced during October. It could be associated with the acidic conditions described above and suggests that the most likely pathway is through a direct leak from the wetland. The high application rates, shown in Table 8.1, support this finding, indicating the high pollution loading rates with respect to this metal. It could be expected that the retention capacity of the soil, with respect to Ni, should have been exhausted. In November and February, the impact of the farmland is reduced and this could be associated with the increased concentrations at the control point, indicating to a higher pollution load, which has been contributed upstream of the control point. The farmland contribution is relatively low compared to the upstream load.

Heavy metals concentrations show distinct seasonal variations at all points, with peak values for Cd, Ni and Pb during November and a decrease of concentrations in February, which could be associated with a first flush effect over the drainage area. The river is relatively small and shallow, thus turbulent flow conditions in November allow for the lifting and transport of sediments and mixing them with the water flow, which could contribute to increased metal concentrations as well. Similar seasonal patterns of diffuse pollution, associated with heavy metals, have been reported earlier (Baily et al. 1999, Gray 1998) and specific attention is given to the fact that seasonal variations in concentrations, together with other factors as hardness and the ratio of different metals concentrations, are leading to an increase of the toxic effect of these cations. Therefore, it is advisable that regulations for heavy metals concentrations in effluents or in natural water bodies should consider event-focused toxicity, which more accurately represents the acute toxicity of wet weather events (Brent & Herricks 1999).

Table 8.4. Summary of water quality characteristics of Marimba River during the period of study.

Parameter in mg/l	SW1	SW2	SW3	Safe regulated limit
1	2	3	4	5
Cadmium	0.036 (0.065)	0.042 (0.072)	0.039 (0.065)	0.01[0.3]
Chromium (VI)	0.07 (0.1)	0.07 (0.11)	0.072 (0.1)	0.05[0.5]
Lead	0.26 (0.33)	0.19 (0.3)	0.24 (0.37)	0.05[0.5]
Nickel	0.92 (2.18)	1.16 (2.54)	1.21 (2.43)	0.3[1.5]
TP	1.67 (1.76)	2.19 (2.5)	5.42 (10.8)	0.5[5]
TKN	2.2 (4.05)	3.83 (5.2)	7.77 (20.9)	0.5*[2]
Zinc	0.37 (0.51)	0.29 (0.36)	0.3 (0.55)	0.5[15]
BOD_5	4.61 (7.35)	5.05 (10.65)	3.31 (5.3)	30[120]
Nitrates	1.93 (3.05)	5.8 (7.55)	3.75 (5.15)	3[10]
Total Nitrogen**	4.13 (7.1)	9.63 (12.75)	11.5 (37.6)	10[50]

(..)-Maximum measured values; [..]-Maximum permissible limit (high hazard)
* The value is given as free and saline ammonia
** Total Nitrogen = TKN+ Nitrates

4.4.3 Assessing surface water quality

The discussion regarding the river water quality status is based on the data presented in Table 8.4. The critical parameters, exceeding the stipulated values in the acting regulations, are TP and TKN. The average value of TP at the control point exceeds significantly the safe value but is lower than the high hazard value. Contribution from the farmland is significant at SW_3, where the high hazard values have been exceeded. The average TKN values are high at all three points, with a significant contribution from the pasture, indicating that the high hazard prescribed concentrations for ammonia could be significantly exceeded at SW_2 and SW_3. The average and maximum Total Nitrogen values at SW_2 and SW_3 exceed the safe limit, but are lower than the high hazard limit.

The average values of Ni, Cr^{+6} and Pb at all points exceeded the safe limits, but only Ni average concentrations exceeded the high hazard limits as well Cd average values exceed up to four times the safe limit, but are lower than the high hazard value. The acute toxicity threshold criterion for the protection of aquatic life of 0.002 mg Cd/l and 0.08 mg Zn/l (Roy et al. 1998) has been exceeded more than ten times for both elements.

Earlier studies on the water quality of Marimba River (Mathuthu et al. 1997), conveyed at the location of SW_1, show concentrations of TKN ranging from 40 to50 mg/l and ortho-phosphates ranging from 2 mg/l to 15 mg/l. Concentrations of Cd, Cr, Pb and Ni varied from 0.002 mg/l to 0.07 mg/l, and Zn varied from 0.21 mg/l to 0.35 mg/l. The comparison with the results of the present study indicates the same trend of high TKN and P concentrations, while the concentration of heavy metals is considerably increased regarding Pb, Ni and Zn.

5 CONCLUSIONS

The practice of sludge land application on pastures has a considerable treatment effect with respect to the assimilation of organic pollution and nutrients. Also, it has the advantage of providing a buffer capacity during periods of low irrigation water requirements, thus the need of large storage requirements is avoided. However, the present case study shows that due to the high hydraulic and pollution loads, the pasture affects natural surface water quality adversely, and is a source of diffuse pollution to surface water. TP and TKN contributions from the pasture were significant during the dry season mainly, and were very high with respect to SW_3 only, while the contribution of nitrates was significant during all seasons, and was more pronounced at SW_2. significant contribution from the pasture was found only with respect to Ni at both points and is attributed to the very high application rates. Seasonal variations for the three points show a significant increase during the beginning of the wet season for Ni, Pb and Cd, which could be associated with diffuse pollution from runoff upstream the study area. The average measured concentrations of the Marimba River water quality at the three points exceed the safe regulation limits regarding all parameters, but the TP vales exceed the high hazard regulated limit at SW_3, indicating a serious level of pollution.

With respect to the general practice of sewage sludge application on land, it could be recommended that:

1. Hydraulic, nutrients and metal loading rates should not exceed the recommended criteria for long-term application and should be based on projections with respect to a future increase in the number of the population served. They should be based on a thorough investigation of the soil and ground water characteristics at the designated site.
2. Engineering measures to protect surface runoff from the area or direct leak (in the case of wet applications) should be provided, in order to protect surface water pollution.
3. The design and the operation of the wastewater treatment plant and the sludge disposal site should be conveyed in a systematic way, considering that the quantity and quality of the sludge produced, as the output of the treatment process, would be the input to the disposal site and would determine its effective operation.
4. A continuous and regular monitoring program to control surface and ground water quality and soil characteristics should be applied, in order to provide public health and environmental protection.

Acknowledgements – SIDA/SAREC Water Project (grant SR15/98) provided the funding for this study. The author would like to thank the sponsors for the financial support offered, the City of Harare authorities for providing information and access to the study area, and the technical staff of the Department of Civil Engineering and the Department of Chemistry, University of Zimbabwe, for their constant support during the sample collection and analysis process.

REFERENCES

AQUASTAT 2003 FAO's Information System on Water and Agriculture at http://www.fao.org/ag/AGL/aglw/aquastat/countries/zimbabwe/index.stm

Attewell, P. 1993. Ground Water Pollution. *Environment, Geology, Engineering and Law.* London: Spon Press.

Bailey H.C., Elphick J.R., Potter A., & Zak B. 1999. Zinc toxicity in storm water runoff from sawmills in British Colombia. *Water Research,* 33, 2721 – 2725.

Boar R.P., Lister D.H., & Clough W.T. 1995. Phosphorus loads in a small groundwater-fed river during the 1989-1992 East Anglian drought. *Water Research,* 29, 2167 – 2173.

Brent R.N. & Herricks E.E. 1999. A method for the toxicity assessment of wet weather events. *Water Research,* 33, 2255 – 2264.

David M.N. & David J.H. 2000 Tracing phosphorous transferred from grazing land to water. *Water Research,* 34, 1975-1985.

Djodjic F., Ulen B. & Bergstrom L. 2000. Temporal and spatial variations of phosphorus and drainage in a structured clay soil. *Water Research,* 34, 1687-1695.

Gray N.F. 1998. Acid mine drainage composition and the implications for its impact on lotic systems *Water Research,* 32, 2122 – 2134.

House W. & Denison F. 1998. Phosphorous dynamics in a lowland river *Water Research,* 32, 1819 – 1830.

Hranova R. 2002 Water Reuse in Zimbabwe – an Overview of Present Practice and Future Trends. In: *Proceedings of 3rd International Conference on Integrated Environmental Management in Southern Africa", August 27-30, 2002. Johannesburg, RSA. (*Proceedings on CD-ROM format).

Lotter, L.H. & Pitman, A.R. 1997. Aspects of Sewage Sludge Handling and Disposal. WRC Report 316/1/97. Pretoria: WRC.

Mathuthu A.S., Mwanga K. & Simoro A. (1997) Impact assessment of industrial and sewage effluents on water quality of the receiving Marimba River in Harare. In: Moyo, N.A.G. (Ed). *Lake Chivero: a polluted lake*. Harare: University of Zimbabwe Publisher.

Metcalf & Eddy Inc. 1991. Wastewater Engineering-Treatment, Disposal and Reuse. New York: McGraw-Hill Inc.

Pescod, M. B. 1992. *Wastewater Treatment and Use in Agriculture*. FAO Irrigation and Drainage Paper No. 47. Rome: FAO, UN.

Reed S.C, Middlebrooks E.J. & Crites R.W. 1988. Natural systems for waste management & treatment. USA: *McGrow-Hill Book Company.*

Reemtsma T., Gnirs R. & Jekel M. 2000 Infiltration of combined sewer overflow and tertiary municipal wastewater: an integrated laboratory and field study of nutrients and dissolved organics. *Water Research,* 34, 1179-1186.

Roy.M. & Couillard D. 1998 Metal leaching following sludge application to a deciduous forest soil. *Water research*, 32, 1642-1652.

Rydin E. 1996. Experimental studies simulating potential phosphorous release from municipal sewage sludge deposits. *Water Research,* 30, 1695-1701.

Standard Methods for the Examination of Water and Wastewater 1989. 17th edn, American Public Health association/ American Water Works Association/Water Environment federation, Washington DC, USA.

Sukreeyapongse O., Panichsakpatana S. & Thongmarg J. 2001. Nitrogen leaching from soil treated with sludge. *Water Science and Technology,* 44, 146-150.

Umemoto S., Komai Y. & Inoue T. 2001. Runoff characteristics of nutrients in the forest streams in Hyogo Prefecture, Japan. *Water Science and Technology,* 44, 151-156.

USEPA (United States Environmental Protection Agency). 1995. Process design manual: surface disposal of sewage sludge and domestic septage. *USEPA Report* EPA/625/R-95/002.

WRC (Water Research Commission) 1997 Permissible utilization and disposal of sewage sludge *Report TT85/97,* Pretoria: WRC.

WWEDR 2000. *Water (Waste and Effluent Disposal) Regulations*, Statutory Instrument 274 of 2000, Republic of Zimbabwe.

CHAPTER 9

Sewage sludge disposal on land – impacts on soils and groundwater quality

R. Hranova, D. Love & A. Amos

ABSTRACT: Pollutant transport mechanisms in soils and ground water, together with regulatory aspects and design criteria for a safe application of sewage sludge on land have been presented. The impacts of long-term sludge land disposal on soils and groundwater quality have been assessed by means of a specific case study of digested sludge and effluent mixture, disposed beneficially for pasture irrigation on Crowborough farm in Harare, performed during the period 2002-2003. Results show that soils were acidified at selected portions of the pasture, subject to excess irrigation loads, but in general, the soil's organic carbon and nutrients content has been improved. The metals content of soils were far below recommended maximum values. Adverse impacts on ground water were found with respect to metals – Cr, Cd, Pb and Ni, which present a high environmental risk, and restrict the use of this aquifer for any future beneficial purpose. Recommendations to improve the present practice have been made.

1 INTRODUCTION

The beneficial reuse of treated wastewater and sludge for different types of agricultural activities is one of the most widely applied options in the world. Sewage has been applied to fields since ancient times. During the ninetieth century and the beginning of the twentieth century, cities throughout Europe and North America installed water-borne sewage systems and many of them established sewage farms, adopting crop irrigation as their preferred means of sewage disposal. In a survey carried in 1987 in the United Kingdom, it was found that agricultural use was the most economic option accounting for 40% of the 1,210 million tones dry sludge disposed annually on 1-2% of farmland in England and Wales. In 1987, 30% of the 6 million tones dry sludge in the European Community was used in agriculture, whereas in Hong Kong sewage sludge was utilized for landscaping and topsoil regeneration (Lotter & Pitman 1997).

The agricultural use of sewage is an attractive method of diverting nutrients from rivers (where they result in eutrophication) to agriculture where they are used as nutrients by plants for biomass production. This option can be cost-effective if land is readily available, near to the treatment plant, and also, if it is properly applied and managed. Apart from the nutrients, the moisture content of treated wastewater and sludge is used to replace the use of fresh natural water for the same purpose. It is well known that irrigation water demand forms by far the most substantial water use activity in a catchment. Therefore, the reuse of treated wastewater and sludge for irrigation of different types of agricultural products could contribute to an improved water demand management practice and could save a considerable amount of fresh water resources for other beneficial activities. This is of specific importance for arid countries, which have limited water resources. Irrigation with treated sewage and sludge, as an effluent and sludge disposal option, has the advantages of:

- Polishing the effluent through the application of natural soil filtration and purification methods;
- Being an additional source of nutrients to the plants,
- Providing a soil building capacity and improving the organic content of soils;
- Being a cheap, easy to operate and maintain, and a cost effective technology;
- Requiring less energy, compared to other alternative sludge treatment and disposal options;

- Conserving natural water resources; and
- Attaining an economic return from crops.

Use of sewage in agriculture is limited by several public health and environmental risks, such as possible transmission of diseases through pollution of water resources or through the food chain of the crops irrigated with sludge. Also, the pollution of ground water and soils could limit their future use and pose an environmental threat for future generations. However, an adequate design and operation practice, together with regular monitoring could reduce to a minimum level such risks. Other limitations for the application of such practice could be that:

- Large areas are required, due to the low reaction rate of the purification process and the plant water requirements;
- Soil conditions are not suitable;
- Long transportation and energy cost might be involved, if the irrigation field is located faraway from the treatment plant, correspondingly more intensive sludge treatment disposal options might be economically viable, compared to disposal on land or irrigation;
- Large storage facilities are required, especially in countries with temperate and cold climates, in order to store the volumes, which will be released, when water is not required for irrigation.

Sludge and effluent disposal on land, together or as separate entities, is a preferred practice of sludge treatment and disposal in the region. This is due to the specific climatic conditions, which allow disposal throughout the year. Sewage effluent, treated in conventional treatment plants, could be a source of enhanced eutrophication of surface waters, due to a high nutrient content of the effluents and sludge. Reusing the effluent and sludge for irrigation purposes has the double effect of utilizing in an optimal way, the already used water, and protecting the natural water resources. Land availability is also a conducing factor, together with the relatively low-skills and low-cost operational requirements.

The problems associated with the risks of application of this practice and possible adverse effects on the environment have been discussed in the previous chapter. In this study, specific attention is given to the possible risks associated with soil contamination and ground water pollution. In order to understand the regulatory instruments and to apply them correctly during the design and operational stages of this practice, basic principles regarding the mechanisms of the transport of pollutants in soils and groundwater are described. A specific case study of long-term irrigation of a pasture with effluents and sludge is presented, and the impacts of such practice assessed in terms of soil and groundwater quality. The study area is the same as the one discussed in Chapter 8, but here the assessment focuses on ground water quality, discussed in the context of soils characteristics. In addition, a second assessment of pollution loads directed to the area has been done during the year 2002, which allows for the comparison and trend evaluation with respect to the material (sewage sludge and effluent mixture) used for irrigation. Thus, in terms of the diffuse pollution of water resources, this chapter focuses on ground water pollution as a result of long-term irrigation with sewage sludge. Usually, such a type of diffuse pollution is classified as originated from agricultural activities and is associated with rural areas. However, in the case of sludge land application, the most common case is to develop the farm area close to the location of the wastewater treatment plant, usually in the sub-urban areas. Thus, in cases of improper design or operation, groundwater pollution from such activities becomes one separate aspect of the management of aquifers in urban areas, and the related impacts on surface water through base flow and recharge.

2 POLLUTION TRANSPORT IN SOILS AND GROUNDWATER

2.1 *Soil treatment mechanisms*

The soil has a large capacity to retain, transform and recycle some of the pollutants found in municipal wastewater and hence is frequently used to provide both treatment and disposal of wastewater (Ayers & Westcot 1985, Yadav et al. 2002). Wastewater treatment in the soil occurs due to physical, chemical and biological processes that include retention, transformation and destruction of pollutants. Some of the

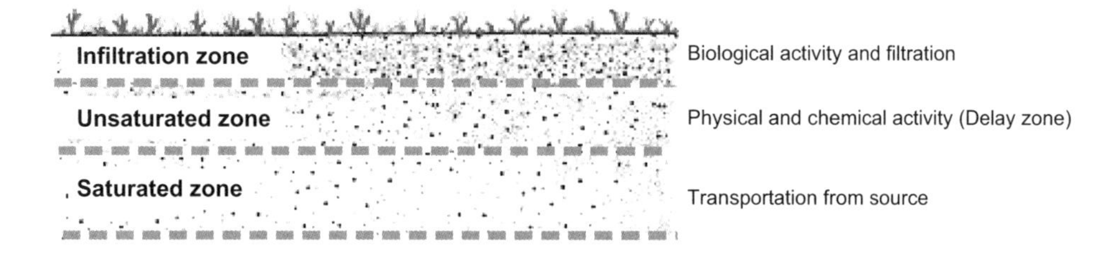

Figure 9.1. Soil matrix zoning.

physical processes are infiltration, dispersion, and dilution, while vitalization, adsorption, complexation, precipitation and photodecomposition are chemical processes that occur in the soil matrix. Biological processes include biological oxidation (mineralization), nitrification, denitrification, immobilization, and plant uptake (USEPA 1992).

The fate of pollutants in the soil matrix can be traced in the three zones of the subsurface, which are: the infiltration zone, the unsaturated zone and the saturated zone (or aquifer). The last two form the vadose zone. A schematic representation is shown in Figure 9.1.

Most of the sewage treatment is performed in the first two zones with the saturated zone offering dilution and transportation of pollutants away from the source. The infiltration zone is biologically active and acts as a physical, chemical and biological filter to remove suspended solids and organics from the wastewater. The filtered solids are a source of nutrients for the active biomass in the zone. Significant amounts of chemicals are biologically degraded by microorganisms, or transformed by chemical and physical processes, and plants take up some of the end products of these processes. The unsaturated zone acts as a delay barrier for the dissolved pollutants before they reach the saturated zone. The main processes in the unsaturated zone are solution, dilution and hydrodynamic dispersion, which result in the attenuation of pollutants. Pathogens die off in this zone due to anoxic conditions, temperature, starvation and predation. Wastewater movement in the unsaturated zone is generally vertical with limited dispersion. Pollutants leave the site in the saturated zone through horizontal flow of groundwater (Freeze & Cherry 1979).

These natural processes do not attenuate all contaminants, as some remain unchanged and others are changed to more harmful products. An example for such a transformation in aquatic ecosystems is the highly toxic methylmercury formed from elemental mercury by aquatic microorganisms (Fetter 1994).

2.2 *Pollutants transformations in soils*

2.2.1 Nutrients

2.2.1.1 Nitrogen

The major plant nutrients in sewage sludge are nitrate, phosphate and potassium compounds. Typical amounts of nitrate and phosphorous in digested sludge are 3% and 2.5% of TS (Pescod 1992). In most land application systems, sludge provides sufficient nutrients for good plant growth. Nutrient uptake rate by plants is a key parameter for sludge loading rates in land application systems. A typical rate of nutrients plant uptake for selected crops is presented in Chapter 10 (Table 10.2).

Nitrogen is present in soils in different forms. The most stable and predominant form of nitrogen in soil and groundwater is nitrate (NO_3^-), although it also occurs in other forms as dissolved ammonium (NH_4^+), nitrite (NO_2^-), ammonia (NH_3), nitrous oxide (N_2O) and organic nitrogen. The conversion of organic nitrogen to ammonia is known as ammonification. Nitrification is the process of converting ammonia to nitrate by biological means. Both processes occur in the soil above the water table in the presence of oxygen and organic matter (Freeze & Cherry 1979). Organic nitrogen is readily and rapidly nitrified biochemically in aerobic soils. Nitrification is a two-stage process, executed by two different types of microorganisms in the presence of oxygen, as shown by equations 9.1 and 9.2. Nitrification rates are reported to be in the range of 0.010-0.016 gN/m^2 per day (Polprasert 1996).

$$NH_4^+ + 1.5O_2 \xrightarrow{Nitrosamonas} NO_2^- + H_2O + 2H^+ \quad (9.1)$$

$$NO_2^- + 0.5O_2 \xrightarrow{Nitrobacter} NO_3^- \quad (9.2)$$

Denitrification is the reduction of nitrate to nitrogen gas, and occurs in anaerobic conditions. Conditions that are favorable to denitrification are:

- High organic content;
- Limited free oxygen;
- Fine textured soils;
- Moisture;
- Neutral to alkaline pH;
- Vegetative cover; and
- Warm temperature.

For the overland flow treatment system, denitrification and volatilization losses can range from 20% to 30% of the applied nitrogen (Polprasert 1996). Nitrite is significantly more toxic to humans than nitrate (DWAF 1996a).

Due to its anionic form, the soil exchange complex (which is mainly a cation exchanger) very weakly sorbs nitrate and its movement in soil is affected to a limited extent by exchange reactions (Foth 1984). Therefore nitrates are easily transported in groundwater and leach freely, with their relative concentration and distribution in the soil matrix being largely determined by the leaching fraction (DWAF 1996c). However, the organic content and pH of soils also influence the nitrate concentration in the soil solution (Barry et al. 1995). The major control on its transfer from soil to aquifer is the permeability of the vadose zone (McLay et al. 2001). Nitrate is very mobile in groundwater, because of its solubility. In a strongly oxidizing environment, nitrate is stable in dissolved form, and moves in groundwater with no retardation or transformation. Thus, it can migrate large distances in highly permeable soils or fractured rock.

Nitrogen is an essential plant nutrient in the soil. However, at elevated concentrations, in its various forms, it becomes a pollutant. Elevated concentrations of nitrate in water consumed for potable purposes may result in methaemoglobinaemia (infantile cyanosis) and carcinogenesis (gastric cancer). Ruminants are sensitive to nitrogen and trace application of nitrogen to pastures can cause its accumulation to levels that are hazardous to animals (DWAF 1996b).

2.2.1.2 Phosphorous

Phosphorous is one of the essential plant nutrients and is frequently a limiting factor in vegetative productivity (Foth 1984). Phosphorous in natural treatment systems is extracted by both biotic and abiotic processes. Biotic processes include plant uptake and mineralization by biological means, while the abiotic processes include sedimentation, sorption, precipitation and exchange processes between soil and water interface. Applied phosphorous is either taken up by plants and incorporated into organic phosphorous or becomes weakly or strongly adsorbed onto aluminum, iron and calcium salt surfaces depending on pH. Under acidic conditions ($pH < 6$), phosphorous is precipitated by aluminum and iron salts. Under alkaline conditions ($PH > 8$), the presence of calcium and magnesium cations in soils lead to the formation of phosphorous precipitates in the form of salts of these ions. Thus the control of phosphorous concentrations in groundwater depends predominantly on the control of solubility of low soluble phosphate minerals present in the soil layer. Long-term application results in the top 30 cm of the soil profile becoming saturated with phosphorous due to adsorption, greater bioactivity and accumulation of organic matter (Metcalf & Eddy 1991). In a land treatment system, the major removal processes of phosphorous are precipitation, sorption and plant uptake.

2.2.2 Pathogens

The potential for pathogen transmission exists During irrigation with sewage, if the system is not properly designed or managed. Domestic sewage may contain four types of pathogens, namely, eggs of helminthes (worms), protozoa, bacteria and viruses. The modes of transmission of pathogens include groundwater, surface runoff, direct contact and consumption of raw crops. However, very few diseases outbreaks have been associated with sewage application on land, except where inadequately treated sludge was applied to gardens or other crops that were eaten raw (Pescod 1992). Bacteria and viruses may be transported with percolating water into groundwater and subsequently ingested, causing infection. This is in agreement with WRC (1997) observation that this pathway posed the greatest risk factor to human health.

The fate and transport of microorganisms in groundwater is not well documented, with the little documentation available indicating that some microorganisms are able to leach into groundwater, albeit with reduced infection capability (USEPA 1992). Pathogens are removed by predation, entrapment, adsorption and die-off in the soil matrix as the wastewater flows through the soil matrix. Many researchers have used advection and dispersion mathematical models to predict the movement of pathogens (Fetter 1994). The models have a limitation in simulating the behavior of microorganisms in groundwater, as they assume that microorganisms behave as a dissolved pollutant, while their behavior actually is closer to a colloidal model (Domenico & Schwartz 1997).

Microorganisms are retained in the soil matrix by filtration, sedimentation and adsorption, whereas their movement is directly related to the hydraulic infiltration rate and inversely to the size of soil particles and concentration of cations in the solution (Pescod 1992). The survival of pathogens depends on different factors such as:

- Type of organism;
- Temperature;
- Soil moisture and type of soil;
- Organic matter, and
- Presence of other organisms.

Pathogens survive longer in moist soils with a neutral pH, low temperature and in the presence of appropriate organic matter as a source of food. In the first 0.5 m of the soil, pathogens are retained by soil filtration, inter-grain contacts, sedimentation and adsorption by soil particles (USEPA 1995, Pescod 1992). Clay soils remove pathogens through adsorption whereas for sandy soils it is through filtration. Pathogenic and coliform organisms generally do not penetrate more than several meters in medium-grained sand or finer material, while in fractured rocks where groundwater velocities can be high, pathogens can be transported hundreds of kilometers (Freeze & Cherry 1979). Fine textured soils tend to adsorb viruses more than coarse textured soils and clay. Studies have shown that in heterogeneous aquifers of sand or gravel, sewage borne pathogens can be transported tens or hundreds of meters. Laboratory investigations reveal that viruses are relatively immobile in granular materials with adsorption being the major retardation mechanism (Freeze & Cherry 1979). Wright (1999) reports on gross bacterial contamination in karstic limestone terrain. Unsaturated fine-grained soils reduce bacteria to acceptable levels and contamination of aquifers underlying unconsolidated soils is unlikely.

2.2.3 Metals

High concentration of cadmium, lead, iron, manganese, aluminum, copper and nickel pose a potential health hazard to humans and animals, and copper, zinc and nickel are phytotoxic (Pescod 1992). Some metals such as Cu, Fe and Ni are essential micronutrients for plants but become toxic at higher concentrations. Factors affecting the plant uptake of trace metals are soil characteristics, toxic elements in the wastewater, type of crop grown and background toxic elements in the soil and their distribution. Metals are also removed from water through adsorption, ion exchange, hydrolysis, complexation with organics, and precipitation reactions. Most metals in water tend to form hydrolyzed species by combining with inorganic anions such as (bi) carbonate, sulfate, fluoride or nitrate. Organic complexes may also be formed. Thus, the mobility of metals in groundwater depends on the concentrations of the most important complexes formed by the element in the water. In non-acidic groundwater, with high concentrations of dissolved carbonates, metals like cadmium, lead and iron are maintained at low concentrations due to their precipitation as insoluble hydroxides or carbonates (Freeze & Cherry 1979). Thus lime application on soils to increase pH can be an effective way of reducing the mobility and uptake by plants of such metals (USEPA 1992). Physical adsorption of metals in groundwater occurs via clays, organic matter and colloids. Fixation of cobalt, nickel, copper and zinc in soils and freshwater sediments is controlled by the presence of hydrous oxides of iron and manganese. In an oxidizing environment, the oxides of these metals occur as coatings on grains enhancing the adsorptive capability of the medium (Freeze & Cherry 1979).

2.2.4 Major cations

The major ions, which undergo ion exchange in soil solution, resulting in changes to soils characteristics are Ca, Mg, K and Na. Ion exchange sites are found in all soils, though they are predominant in clays and soil organic materials (Fetter 1994). The cation exchange capacity is defined as the sum of exchangeable cations adsorbed per unit weight of soil. The Na and Ca exchange reaction is of special importance when it occurs in smectite clays because it can cause large changes in permeability (Freeze & Cherry 1979). The relative size of Ca and Na is such that two Na ions require more space than one Ca ion; hence the replacement of Ca with Na causes an increase in the dimension of the crystal lattice, resulting in a decrease in permeability. This can cause degradation in the agricultural productivity of soils. The sodium adsorption ratio (SAR) is a measure of the capacity of a given irrigation wastewater to induce sodic soil conditions (DWAF 1996b). More information with respect to the application of SAR as a control parameter is given in Chapter 10.

An essential plant nutrient, K is abundant in many soils and is readily available for plant uptake in large quantities, second to nitrogen. In soils, K exists as water soluble, exchangeable, non-exchangeable (fixed) and mineral potassium. Due to its positive charge, K is readily bound by negatively charged clays with a high cation exchange capacity and water holding capacity. This retards the K movement down the soil profile. Thus the movement of K can be a few centimeters from the point of application (Foth 1984). K is the most abundant cation in plants and is taken in larger quantities than any other cations. The fixed form of K is found in the interlayer spaces of soil minerals. However, for sandy soils with a low cation exchange capacity, leaching can be as high as 90% of the added K, after 150 kg/ha were added followed by an irrigation of 400 mm (Harter et.al. 2002). A similar study showed that following 40 years of K application, no K accumulated in the top 75 cm of a sandy soil, due to leaching. There are no reported adverse effects of K on plants and the environment.

2.3 *Pollutants transport in groundwater*

In considering transport mechanisms in groundwater, a distinction is made between the transport mechanism of groundwater and the solutes. At very low concentrations, the dissolved phase of the solute may assume the same transport mechanism as the groundwater, but at higher concentrations the pollutants can sometimes move faster than the groundwater (Domenico & Schwartz 1997). Examples are the highly soluble nitrates and conservative chloride solute. In many cases, as the solute moves through the porous media, physical and chemical interactions between the solute and the porous media affect the transportation process of the pollutant. Thus, transportation of pollutants in groundwater is more complex than the transportation of the ordinary groundwater.

Different processes that include advection, dispersion, diffusion, adsorption and decay influence pollutants transport in groundwater. These processes can act in combination or separately to influence the transport process. Physical factors such as moisture content and water balance in the unsaturated zone, together with the hydraulic gradient and the water balance in the saturated zone, influence the pollutants transport. Wright (1999) concurs with these factors, and notes some geological properties, such as the adsorption capacity and hydraulic conductivity of soils, as additional factors. The two major transport mechanisms of pollutants in groundwater are advection and dispersion.

Advection is defined as the flow of water due to a hydraulic gradient between two points. The velocity of flow is proportional to the hydraulic gradient. Advective flow is determined by factors such as the terrain slope and soil permeability, and is typically in the order of cm/d. It is described by the Darcy's equation (9.3):

$$q = k\ I \tag{9.3}$$

where: q = specific discharge (m/d), k = hydraulic conductivity (m/d), I = hydraulic gradient

Pollutants transported by advection travel at the same rate as the average linear velocity of the groundwater flow (Fetter 1994). The hydraulic conductivity, "k", is a function of the properties of both – the fluid and the porous media through which it is flowing. In its general form, Darcy's equation is valid for flows at a very low velocity and through soils with isotropic structure (sands). A schematic representation of the transport mechanisms of pollutants in groundwater is shown in Figure 9.2. It shows that the shape, which the polluted flow would take after leaving the source, is conical and is known as a pollution plume. After different time intervals the plume would change its shape, covering larger parts of the aquifer but with reduced concentrations of pollutants along the basic direction of the advective flow.

Dispersion represents the transport of pollutants in all directions, including the main flow direction. Dispersion is caused by heterogeneities in the aquifer that result in groundwater pollutants flowing through different pores at different rates and various flow paths. This results in diffusion and mixing due to velocity variations in the flow of the groundwater or the pollutant. We could differentiate between mechanical and hydrodynamic dispersion. Mechanical dispersion is the mixing of the pollutants resulting from the physical movement through the porous media. When a pollutant is diluted in the groundwater, it follows the path of the main flow, which is called lateral dispersion (Domenico & Schwartz 1997).

Due to the soil matrix heterogeneity in terms of soil particles size and distribution, dispersion will occur and will distribute pollutants not only in the direction of flow but in other directions as well (Fig.9.2).

Molecular diffusion is the movement of both ionic and molecular species dissolved in water, from a region of higher concentration to a region of lower concentration. Diffusion could be a major transportation process in less permeable media and where the pollutant has a higher concentration than adjacent areas (Fetter 1994). For an one-dimensional analysis under steady state conditions, Fick's first law governs diffusion, where the mass flux of solute is proportional to the concentration gradient. In practice, diffusion and mechanical dispersion cannot be separated and their combined effect is termed hydrodynamic dispersion.

Retardation of pollutants depends on the nature of the pollutant activity in the soil matrix. Conservative pollutants, like the chloride ion, do not react with the soil or do not undergo biological or radioactive decay. The process of absorption is the major cause of retardation of pollutants in groundwater (Domenico & Schwartz 1997). Clays tend to be stronger absorbers owing to their high specific area and high negative surface charge. Although positive adsorption sites are present in clays, they are not as abundant as negative sites. For this reason, usually negatively charged ions such as bicarbonate, sulfate and nitrate, are not affected by retardation. Also, they are too large to be adsorbed, hence the high mobility of these ions in groundwater. Soil adsorption capacity could be described by the Freundlich adsorption isotherm, which presents a straight-line relationship in logarithmic format between the mass of pollutant adsorbed by the

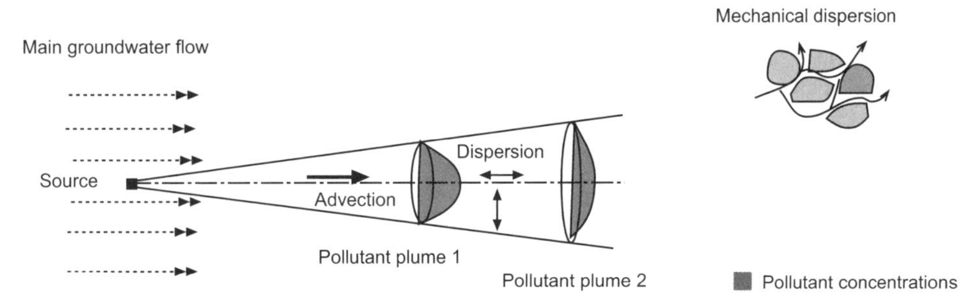

Figure 9.2. Scheme of pollutant transport mechanisms from a continuous source.

soil, and the concentration of pollutant in groundwater. The soil adsorption capacity could also be estimated by using the Langmuir adsorption isotherms, where the ratio of the equilibrium concentration of a given pollutant to the amount absorbed by the soil is given as dependent on specific coefficients characterizing the soil adsorption capacity. This equation assumes heterogeneous soils conditions, constant energy of adsorption and a fixed number of binding sites.

3 IMPACTS ON SOILS

3.1 *Hydraulic and pollution loads*

3.1.1 Methodology

The study was performed during the period December 2002 to March 2003. The study area – Crowborough Farm – is located to the north of Crowborough Sewage Treatment Works (CSTW) as shown on Figure 9.3, and has been in use since 1972. A detailed description is presented in Chapter 8. During the period of study the actual area under irrigation at any given time was 3 paddocks, or 51.3 ha, which were supposed to be rotated. However, a visual inspection and interviews with the personnel in the field show that this principle has not been followed strictly. The most frequently irrigated part of the farm lies along the road to CSTW. A preliminary investigation of the present hydraulic loads directed to the farm was carried out, based on interviews and data obtained from the managing authority – the Municipality of the City of Harare.

Sludge and effluent mixture (SEM) samples were collected as time integrated samples at 30 minutes intervals for 2 hours on 2 sampling occasions at a discharge chamber on the farm (Fig. 9.4). Parameters tested included TDS, TSS, T, TP, ortho-P, Nitrate, pH, Cd, Cr, Pb, Zn, Ni, Cu, and Fe. All tests were performed according to Standard methods (1989). The analytical procedures are the same as described in Chapter 4, except for nitrates. A "Spectronic 21D" spectrophotometer was used for nitrates determination by the ultraviolet spectrophotometric screening method at wavelengths of 220 nm and 275 nm. Duplicate samples were included in all analyses for quality control.

3.1.2 Results and discussion

During the period March 2002 to March 2003 the effluent and sludge flow rates averaged 23,020 m^3/d and 1033 m^3/d respectively. The monthly variation is shown in Figure 9.5. The average value is about 15% higher than the flows pumped to the farm during 1998-1999 (Chapter 8). The variations of the flow to the farm relate primarily to fluctuations of the inflow rate to CSTW.

The trickling filters line and the holding ponds receive the majority of any incoming excess flow (the BNR plant works at a load, close to its design capacity). The highest figures are recorded during the wet season and are attributed to the infiltration of run-off into the sewer system and recharge to portions of the sewer system from groundwater in saturated areas. On some occasions of very intensive storms, part of the pond effluent is discharged from the ponds to the Marimba River.

A water balance of incoming and out-coming flows during the period of study, based on the total active farm area is presented on Figure 9.6. It shows that the SEM hydraulic load to the farm exceeds the evapo-transpiration rate based on average annual loads and the assumption that the total active pasture area is used. During the wet season, the farm area is hydraulically overloaded. However, the actual hydraulic load, calculated based on the annual average daily flow rate and the area of three paddocks, which are used for irrigation most often, is 47 mm/d. This shows a significant hydraulic overloading due to an inadequate operational practice, which is reflected in ponding and the creation of anaerobic conditions on selected portions of the farm area. The under drain system could alleviate this problem in the portions, where soils are more permeable (the southern lower part of the pasture). However, due to the long period of exploitation, the under drain system might be blocked. All these factors reflect in a very irregular hydraulic loading pattern along the pasture.

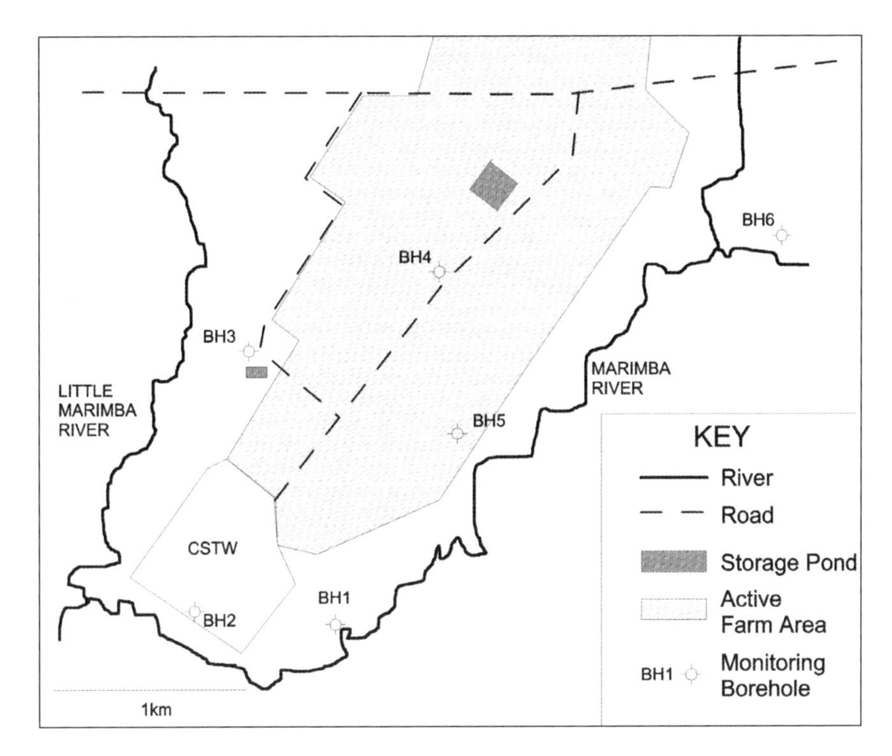

Figure 9.3. Map of the study area and location of sampling points.

Figure 9.4. Crowborough pasture – a distribution chamber for SEM.

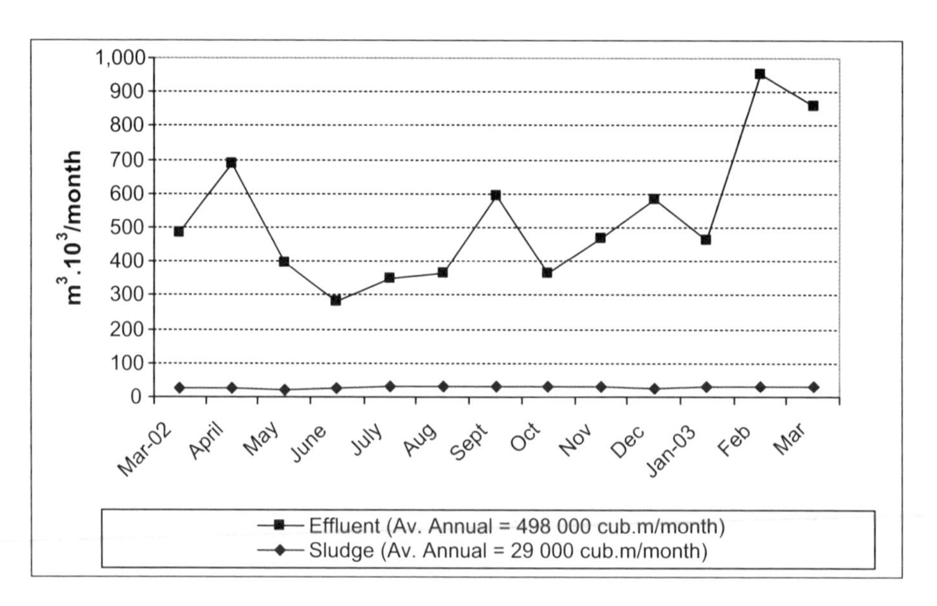

Figure 9.5. Variation of monthly flow rates of ponds effluent and digested sludge.

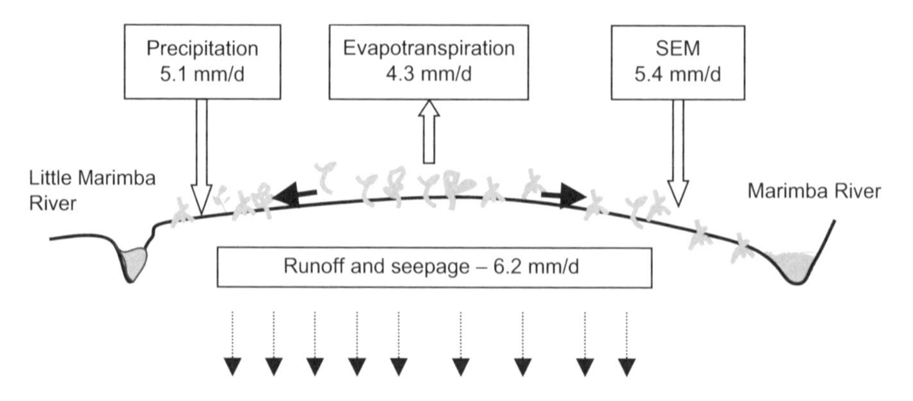

Figure 9.6. Water balance at crowborough pasture (average a mual loads)

In terms of SEM concentrations, reflecting the qualitative aspect of the media disposed on the farm, results are presented in Table 9.1. Pollution loads to the farm are calculated based on the average annual SEM flow rates and the corresponding average concentrations during the period of study.

The effluent and application limiting values (maximum permissible values) with respect to South Africa, France, EU and USA are based on literature data (WRC 1997, Degremont 1991), and the Zimbabwe limiting values are based on WWEDR (2000).

Comparing SEM concentrations with recommended criteria, values near or exceeding the recommended criteria have been found with respect to Cd, Cr and Ni only. During this study, nitrogen concentrations have been determined as nitrates only and the comparison with the criteria shows the concentrations are lower than the recommended values. However, if the TKN concentrations were to be considered, the total nitrogen load would be much higher. A direct comparison of the results of this study with those performed earlier (Chapter 8, Table 8.2) with respect to SEM metal concentrations, is difficult because of the different units, but it shows the same trend of increased Ni concentrations.

Table 9.1. Concentrations of various parameters in SEM and application rates to Crowborough farm during 2002-2003.

	Pb	Fe	Cd	Cr	Zn	Cu	Ni	TP	NO_3
SEM concentrations (mg/l)	0.83	7.80	0.11	5.94	0.54	0.15	1.99	6.48	31.93
Effluent limit ZW	5.00	-	0.01	0.10	-	0.20	0.20	-	-
Effluent limit ZW	20.00	-	0.05	20.00	-	5.00	2.00	-	50.00
Effluent limit SA	2.00	20.00	0.05	1.0	5.0	5.00	2.00	-	30.00
Effluent limit France	5.00	5.00	0.01	0.10	2.00	0.20	0.20	-	-
Application Rates (kg/ha.year)	16	154	2	117	11	3	40	127	629
Application Limit ZW	-	-	4	-	-	-	-	-	-
Application Limit EU	15	-	0.15	-	30	12	3	-	-
Application Limit SA	0.41		0.13		2.83				
Application Limit USA	15	-	1.9	150	140	75	21	-	-

With respect to the application rates, the TP annual loading rate considerably exceeds the phosphorous uptake rate for grasses of 24 to 85 kg/ha.year (Reeds et al. 1988), and the same result was obtained previously (Chapter 8, Table 8.2). During this study, the TP load was 11% higher than the load in 1999, but the increase should be associated with the increased hydraulic load, as SEM concentrations did not vary considerably. The application rate with respect to nitrate is comparable to the one due to TKN loading in 1999 and this indicates that the total nitrogen load to the farm is very high. However, it should be mentioned that the pollution loads during this study were obtained based on two measurements only with a high standard deviation, which might not be representative for the annual pollution load. Unfortunately, the plant operators were not able to provide additional data regarding this media, as the plant-monitoring program during this period was restricted due to financial constraints. Zimbabwean regulations limit the nitrogen application rate to 600 kg TN/ha.year (high environmental risk) and in this case it has been exceeded considerably.

The metals loads, given as application rates (Table 9.1), are of major concern due to the possible hazardous effect on ground and surface water. The application rates regarding Pb, Cr and Ni are alarmingly high. Comparing with the previous study during 1999, a considerable increase of the loading rate has been found with respect to Pb and Zn, while Ni rates are consistently high.

Comparing the different international criteria, the significant difference between EU criteria and USA criteria is of interest, indicating the need for additional research on this topic. As the Zimbabwean regulations recommend a limitation with respect to Cd rates only, the rates obtained during this study are within the safe recommended limit, however, they are comparable to the USA maximum permissible value and exceed more than 10 times the EU limits. No signs of phytotoxicity on the farm were observed, reflecting either high crop tolerance or high capacity of the soil to reduce toxicity. Madyiwa et al., 2002 had similar observations.

3.2 *Assessing soils characteristics*

3.2.1 Methodology

The location of the different boreholes was selected to represent different soil types and different irrigation routine. The six boreholes (BH) were drilled specifically for the purpose of this study, with BH6 serving as a control point. An illustration of a sampling borehole is shown in Figure 9.7. The study area has been surrounded by new residential developments and some of the undeveloped open spaces in close vicinity have been used for informal agriculture. This considerably restricted the possibility to choose a pristine control point and might have affected the natural groundwater quality at such a location. The final choice of the control borehole was based on the following considerations:

Figure 9.7. Sampling borehole at Crowborough farm.

- The site is near the study area and has similar geological conditions, which would be the basis for similar background pollution characteristics;
- It lies on the opposite bank of an adjacent stream from the study area. This stream acts as a pressure barrier to groundwater flow from the area beneath the farm.

During the period of study, the area near the control borehole (BH6) was used for informal agriculture (maize crop) and a new high-density residential suburb was under construction nearby. BH1 is located at the lowest part of the study area, which collects the excess hydraulic load from the pasture. It is wet throughout the year, and has not being used for irrigation; correspondingly it has not being provided with a drainage system for collection of seepage. BH2 is located within the premises of the treatment plant near to the unlined ponds and was selected in order to investigate the possible effect of infiltration of the ponds effluent and the impacts on soils and ground water. BH5 and BH4 are located at the most frequently irrigated areas, with BH5 being located in the lowest part of the area, near the protection dike. BH3 represents the impact of the sludge and effluent disposal practice on the part of the pasture, which does not have a protective dike.

Soil samples were collected during borehole drilling at three depth profiles: 0-0.3 m, 0.30-0.60 m and 0.60-1.2 m. The collected soil samples were placed in polyethylene bags clearly marked with a borehole number, depth profile and date. The location of the sampling boreholes is shown in Figure 9.3. Samples were also collected at depths where there was remarkable change in soil texture and color. Borehole logging was done during drilling at each site by 'finger' assessment as recommended by Foth (1984). The following parameters were tested: pH, Cation Exchange Capacity (CEC), Organic Carbon (OC), TP, Available Phosphorous (Pav) and TN at the Analytical Laboratory Services by using standard procedures (Page et al. 1982). Nitrate was extracted using potassium chloride, followed by steam distillation. The determination was done using the Salicylate method employing an uv/vis spectrophotometer. Ortho-P was determined by the Bray 1 method (for acidic soils) and Olsen method (for basic soils), followed by a UV/VIS spectrophotometer determination. The extraction of TP followed the dry-ashing and dissolution method. The determination of total phosphates was as for Orthophosphates. Exchangeable bases (Ca, Mg, K and Na) were extracted using acidified ammonium acetate as an index. Ca and Mg were then determined by an atomic absorption spectrophotometer (AAS) employing nitrous/acetylene as fuel. Na and K were determined by a flame photometer employing a low-pressure butane gas as fuel. The determination of pH was done in a 1:5 soil: $CaCl_2$ suspension, using a pH meter.

Metals were analyzed at the laboratories of the Geology Department and the Institute of Mining Research, University of Zimbabwe by an atomic absorption spectrophotometer (model VARIAN TECHTRON SPECTRAA 50B-110 SOFTWARE) employing an air-acetylene fuel. Duplicates were prepared and included in the analysis for quality control as per recommendations of WRC (1992). Soil samples for metal analysis were dried at room temperature, grounded, and then passed through a 180 μm sieve until the entire sample had passed. About 0.5 g (actual weights were noted) of each sample was transferred into a 30 ml test tube and digested using nitric acid following recommendations of Page et al. (1982). Then, 10 ml of the supernatant was transferred into 100 ml flasks and mixed with distilled water for the AAS analysis.

3.2.2 General parameters and nutrients

Chemicals and pollutant constituents exist in soils in different phases: solid, liquid or gaseous. Constituents in solid phase are associated with soil particles – adsorbed on their surfaces or precipitated in different forms and chemical complexes. Liquid constituents are dissolved in water or soil moisture. Gaseous fractions may result as the product of chemical or biological processes. In general, only the constituents in the solid phase are immobilized in the soils and are non-available for biological degradation by microorganisms or plant uptake. The parameters, which characterize the soil capacity to immobilize pollutants, also known as "capacity controlling parameters" are OC and the CEC. OC represents the organic material in soils, which has a strong binding capacity, while CEC is dependent on the surface area and the nature of soil particles. The smallest particles, typical for clay soils, have the highest adsorbing capacity. CEC is also associated with soils' permeability, and represents the capacity of soil to retain its structure.

Redox conditions and pH are other factors, reflecting the soil environment and are considered capacity and intensity controlling parameters. These parameters are strongly related to the different biological and chemical process associated with biological transformations of constituents in soils. Increased OC concentrations enhance biological transformations and the oxidation of organic and chemical constituents and lead to anaerobic conditions and the decrease of pH. These changes result in the reduction of selected parameters from their original status of immobile precipitates to soluble and thus mobile forms, which could be leached with the seepage to ground water. Typical examples are the cases of phosphorous and most metals. Thus decreased pH values are signaling a decrease in the soil retention capacity. The other factor, which influences metal mobility in soils is salinity. Increased salinity concentration enhances the release of metals in solution and their mobility.

Nutrients are important constituents of the soil structure and from an agricultural point of view are important for crop growth and yields. However, due to the intense agricultural practices applied worldwide, a substantial amount of nutrients has been accumulated in soils. Phosphorous is easily accumulated in soils, with a relatively low fraction available for plant uptake (Pav). The process of immobilization of phosphorous depends on the following factors:

- Al and Fe oxides contribute to phosphorus fixation in acid soils;
- Calcium compounds control the solubility of phosphorous in calcerous soils;
- Organic matter contributes to phosphorous adsorption.

Most of the applied phosphorous is contained in particulate form, when soil becomes saturated, significant leaching might occur. Novotny (2003) reports that in the Netherlands, the annual load of phosphate on maize-growing areas, defined as the application rate minus the crop intake has been in the order of 150-450 kg/ha, which is exceeding by far the loading rates determined in this study.

In contrast to phosphorous, nitrogen does not accumulate in soils and is easily transported to ground water in the form of nitrate. In the case of sludge application rates, a considerable portion of the nitrogen might be in the form of organic nitrogen or ammonia, which are prone to biological degradation and usually are retained in the topsoil layer. Thus, in these cases, immobilization of nitrogen is due to these two fractions and the factors, which enhance this process, are connected with ammonia fixation to clay particles or soil organic matter and biological immobilization.

Results reflecting the Crowborough farm soil characteristics after 30 years of operation are presented in Tables 9.2 and 9.3. General soil characteristics and nutrients at the different sampling locations and along the depth profiles are presented in Table 9.2.

The pH values show acidification of the soils along the whole depth profile with respect to the most often used areas (BH3, BH4 and BH5), but BH1 and BH2, which are not under irrigation, have neutral pH values similar to the control point (BH6). The effect of long-term irrigation is well pronounced at BH4 and BH5. It results in very high OC, TP, TN concentrations at the top layer, compared to the control point and the rest of the boreholes.

Increased OC concentrations at this location show increased CEC as well. The significant difference between TP and Pav show that only a small portion of the accumulated phosphorous will be reduced by plant uptake, the rest would accumulate in the future until the soil sorption capacity is exhausted. Thus phosphorous accumulation in the soil could be considered as a time bomb, because after the retention capacity of the soil is fully exhausted, it would be leached to the ground water.

Table 9.2. Soils characteristics-general parameters and nutrients.

Well	Depth (m)	Parameter					
		pH	CEC me%	TP mg/kg	Pav. mg/kg	Org.C mg/kg	TN mg/kg
BH1	0-0.3	6.3	30.9	99	7	197	19
	0.3-0.6	7.2	28.9	338	14	161	14
	0.6-1.2	6.9	34.2	347	5	26	3
BH2	0-0.3	6.2	22.8	529	21	160	11
	0.3-0.6	6.4	14.7	198	4	75	7
	0.6-1.2	5.6	23.3	115	3	61	4
BH3	0-0.3	4.6	7.0	793	64	90	9
	0.3-0.6	5.5	17.3	330	31	66	5
	0.6-1.2	6.1	11.8	149	8	43	4
BH4	0-0.3	4.4	25.4	793	57	1400	42
	0.3-0.6	5.0	20.2	463	53	51	7
	0.6-1.2	4.7	22.8	628	55	726	24
BH5	0-0.3	4.6	40.1	1718	156	2290	91
	0.3-0.6	5.6	7.7	793	58	70	7
	0.6-1.2	5.5	14.0	99	3	51	5
BH6 Control	0-0.3	5.5	10.9	446	30	60	6
	0.3-0.6	7.2	37.2	66	4	46	4
	0.6-1.2	6.6	35.3	99	5	17	2
Limit USA[a]		No limit	-	-	-	-	-
Limit EC[b]		>6	-	-	-	-	-

BH1 has up to three times higher concentrations of OC at a depth of 0.6 m below the surface. This could be due to the fact that being at the lowest point, it receives the seepage from the farm, which affects the soil characteristics and forms marshy conditions, thus leading to the transportation of accumulated pollutants to the lower parts of the depth profile (0.6 m – 1.2 m). The soil characteristics (TP, TN and OC) along the first two depth profiles are higher compared to the control point and only at the lowest soil profile (0.6 m-1.2 m) the characteristics are comparable with the control point with respect to the measured parameters, except for TP.

BH2 presents the soil characteristics at CWTW, which could be affected by the infiltration of wastewater from the ponds. In general, no impact was found, except for pH, which is lower at a depth of about 1 m, indicating acidification, influenced most probably by the sludge accumulated in the ponds.

Increased organic content of soils is directly related to increased CEC, leading to improved soil quality and improved capacity for absorption and immobilization of metals and micro pollutants. This link is reflected in the results and illustrates one of the positive consequences of long-term irrigation. Nyamangara et al. (2001) have obtained similar results with respect to Firle farm. However, results with respect to metals concentrations in the pastures soils (Table 9.3) show that actually, the highest metals concentrations are found in the second depth layer (0.3 m-0.6 m), while the highest OC concentrations, are found in the top 30 cm, as could be expected.

3.2.3 Metals

Metals, and more specifically, toxic metals are immobilized in soils by precipitation and formation of metal complexes. This process might be enhanced by the presence of sulfides, which results in the formation of precipitates of metal sulfides. Also, immobilization of Fe, Zn, Ni and other metals could be achieved by precipitation under increased pH conditions in alkaline media. Decreasing the pH of the soil creates conditions of dissolution of such precipitates and leads to a reduction in the concentrations of sulfide ions, thus leading to increased solubility of metal ions and their leaching to ground water.

The results of this study with respect to the metal concentrations in soil (Table 9.3) show that at the different locations, the metals' concentrations are comparable to the control point. In general, the concentrations of all tested parameters are far below the recommended limits. This reflects the fact that long – term irrigation has not influenced adversely the soil characteristics. A possible explanation could be that the metals have been flushed from the soil by the continuous percolation of water from the surface to the groundwater due to the high hydraulic load and the acidic conditions at the most frequently irrigated sites.

Fe concentrations are much higher compared to the rest of the tested metals and could be associated with the natural soil conditions. Comparing BH2, which has never been irrigated to the control point, we could notice relatively similar depth profile distributions of metals.

With respect to Pb there is a distinctive pattern of increased concentrations with depth at BH1, BH3, BH4 and BH5. The highest Pb concentration was observed at a depth of 0.3-0.6 m at BH1, BH3 and BH5. At BH4, which is the most frequently irrigated site, the highest Pb concentration has been measured at the lowest depth profile (0.6 m-1.2 m).

Cd has not been detected at BH1, which is wet throughout the year. In respect to Zn the highest concentrations have been found in the second depth profile for BH1 and BH5 only. Ni soil concentrations did not show variation along the depth profile. Comparing the spatial variation of metal concentrations in the soil, BH5 shows slightly higher values with the maximum concentrations at a depth of 0.3 m-0.6 m, with respect to Pb, Zn and Cu.

Based on literature data and the discussion made in the previous sections, it could be expected that metal concentrations in the farm soils should be much higher than the actual observations during this study. Such results could be explained with the high hydraulic load and the soils acidification and supported by studies at other wastewater irrigated sites (Chapter 10). However, considering the importance of the decision in respect to the future use of the farm area as a disposal site, and the costs involved in eventual alternative solutions, it could be recommended to repeat the procedure, with a more rigorous quality check of the laboratory testing procedures.

Table 9.3. Soils characteristics – metals.

Well	Depth (m)	Pb mg/kg	Fe mg/kg	Cd mg/kg	Cr mg/kg	Zn mg/kg	Cu mg/kg	Ni mg/kg
BH1	0.0-0.3	0	175.99	0	30.43	4.27	2.27	5.10
	0.3-0.6	6.65	278.50	0	23.32	10.17	4.84	6.36
	0.6-1.2	1.98	229.96	0	24.8	3.87	1.59	6.94
BH2	0.0-0.3	3.92	486.65	0.10	26.88	14.30	4.10	8.68
	0.3-0.6	0	422.44	0.19	10.06	0.70	2.13	7.16
	0.6-1.2	4.3	406.39	0.29	31.7	2.42	2.76	7.35
BH3	0.0-0.3	0.96	310.66	0.29	19.83	1.81	1.72	8.30
	0.3-0.6	5.78	482.85	0.58	27.75	0	2.12	8.86
	0.6-1.2	1.97	369.29	0.39	27.36	0.59	2.17	10.43
BH4	0.0-0.3	0	250.93	0.39	31.47	1.14	2.71	11.81
	0.3-0.6	1.94	391.28	0.39	34.69	0.04	3.49	12.79
	0.6-1.2	2.83	271.51	0.28	29.03	3.27	3.12	11.41
BH5	0.0-0.3	1.89	297.16	0.38	39.96	0.76	3.98	13.83
	0.3-0.6	9.52	368.38	0.19	28.57	15.01	6.48	10.86
	0.6-1.2	5.83	474.95	0.10	38.26	1.98	3.70	12.75
BH6	0.0-0.3	1.91	243.57	0.29	28.32	2.54	1.72	12.33
Control	0.3-0.6	0	322.58	0	36.62	1.61	1.14	12.14
	0.6-1.2	3.81	200.19	0.19	52.38	1.92	0.95	12.19
Limit USA		190	-	20	1540	1500	775	230
Limit EC		50-300	-	1-3	-	150-300	50-140	30-75

4 IMPACTS ON GROUNDWATER QUALITY

4.1 *Methodology*

Groundwater samples were collected from boreholes using a plastic bailer. The depth to water table, sampling depth and depth of the borehole, were noted during sampling. Grab samples were collected at 1.0 m below the water table (1.5 m from the ground surface). In between sampling, the bailer was rigorously cleaned 5 times with distilled water. Borehole purging volumes were limited to 1000 ml to avoid additional borehole development and increased turbidity of water samples. The first 500 ml of the sample were used to rinse the sampling equipment and the sample bottle. Samples were collected in sterile bottles clearly marked with a borehole number, parameters to be analyzed and preservative added. Samples for microbiology and macro pollutants were preserved at 4°C in a cooler box and sampled within 6 hours of sampling. Temperature and pH were measured in the field. Sampling was performed on three sampling occasions.

The parameters tested and the analytical procedures were the same as the ones described in section 3.1.1, except that FC and TC were determined in addition, by ELE Paqualab equipment, applying the membrane filtration method (Standard methods 1989). Duplicate samples were included in all analyses for quality control.

Table 9.4. Ground water quality characteristics – general parameters and nutrients.

Parameter	Sampling point							Maximum recommended levels for different water uses			
	BH1	BH2	BH3	BH4	BH5	Average wells	BH6 Control	Domestic*	Domestic**	Live-stock	Irrigation
pH	7.5 (0.07)	7.2 (0.14)	6.9 (0.23)	7.0 (0.19)	6.7 (0.25)	7.1 (0.28)	11.6 (0.10)	--	6 – 9	--	6.5-8.4
T °C	23.2 (1.51)	23.2 (1.04)	23.2 (1.50)	23.1 (1.08)	23.2 (0.07)	23.2 (0.07)	23.2 (0.21)	--	--	--	--
TDS mg/l	1009 (291)	310 (125)	861 (387)	650 (235)	679 (93)	886 (199)	855 (37)	--	450	3000	260
TSS (mg/l)	4678 (4014)	150 (50)	1061 (1228)	35 (33)	217 (100)	1230 (1152)	103 (135)				
Nitrate mg/l	1.78 (2.51)	1.67 (2.61)	2.49 (4.14)	2.74 (4.38)	3.17 (5.11)	2.37 (0.44)	1.52 (2.23)	50	6	100	--
TP mg/l	0.15 (0.021)	0.03 (0.021)	0.13 (0.121)	0.14 (0.073)	0.54 (0.701)	0.20 (0.203)	0.04 (0.032)	--	--	--	--
Ortho P mg/l	0.10 (0.011)	0.01 (0.014)	0.04 (0.042)	0.11 (0.012)	0.64 (0.902)	0.18 (0.193)	0.02 (0.022)	--	--	--	--
FC cfu/100 ml	295 (91)	188 (110)	812 (166)	960 (259)	1170 (624)	685 (426)	0	0	0	200	1
TC cfu/100 ml	481 (125)	377 (133)	840 (387)	1800 (1311)	2383 (1975)	1176 (878)	133 (23)				

* WHO guidelines (WHO 1996);
** South African guidelines;
Values in parenthesis reflect standard deviations of the mean values

4.2 *Assessing groundwater quality*

4.2.1 General parameters and nutrients

The groundwater within the farm area is not used for any beneficial purpose at present, but it recharges the two rivers, surrounding the pasture. Results with respect to ground water quality are presented in Table 9.4.

Unfortunately, it was no possible to obtain data regarding the aquifer characteristics and to use it with respect to the pathways of pollution transport. A comparison with water quality guidelines for different types of beneficial uses (Tables 9.4 and 9.5) has been made, based on WHO guidelines (WHO 1996) and the South African guidelines for domestic use, livestock watering and irrigation (DWAF 1996 a, b, c). The evaluation of risks to environment has been done based on the Zimbabwe regulations for effluent, runoff and seepage discharging into the environment (WWEDR 2000).

The pasture groundwater is neutral with respect to pH values with no significant spatial variations. It should be noted that the increased pH and TDS values at the control point could be explained with the fact that during the sampling period, the site has been affected by an agricultural lime application, leading to higher values of both parameters. TDS values of farm groundwater at all points show higher concentrations compared to surface water quality. They fall in the green classification posing a low hazard to the environment. TSS concentrations reflect the soils conditions. At the most frequently irrigated site (BH4), which is located at the highest part of the pasture, along the road to CWTW, the TSS is relatively low, indicating that the soil is well structured and the release of soil particles is relatively low. At BH1 and BH3, which are wetlands, and marshy conditions are typical throughout the year, the soil has changed its structure and a considerable amount of particulate material is released to the groundwater.

Nitrate concentrations are generally low at all points, which indicates that the nitrate leaching from the farm soil is limited despite the high pollution load (Table 9.1). Spatial variations show that at the BH3, BH 4 and BH5 the nitrogen concentrations are twice higher compared to the control point and the areas, which are not irrigated. A possible explanation of the relatively low transport of nitrates to the groundwater could be the fact that due to the sludge application, the larger fraction of nitrogen is in the form of organic nitrogen and ammonia, which are more prone to fixation and are retained in the top soil layer. This is confirmed by the soil characteristics (Table 9.2), which show high nitrogen concentrations at BH4 and BH5. Due to the high hydraulic and pollution loading rates, anaerobic conditions prevail and do not allow the biological transformation of these fractions to nitrates, which are more mobile. Also, the high hydraulic loading rate, especially during the rain season, leads to surface runoff, which transports part of the material, accumulated on the surface, to the River. This scenario was confirmed by the results in Chapter 8. In addition, the grass has a relatively high nitrogen uptake, and is growing throughout the year, its height being controlled by the grazing cattle.

TP values fall under the blue classification (safe to the environment), except for BH5, where the values are exceeding the safe limit slightly. However, the comparison with the control point shows that the farm groundwater quality considerably exceeds the control point values, which are similar to background quality (Hranova et.al. 2002), indicating that there is a trend of pollution from the farm with respect to phosphates. Results show that phosphorous in groundwater is mainly in the form of ortho-P and that despite the high loading rates, the groundwater has not yet been polluted at a level to pose an environmental risk. At present the phosphorous is immobilized in soil, which is confirmed by the relatively high TP concentrations in soil (Table 9.2). This could be explained with the high organic content of the top layer and the Fe concentrations in the soil, which contribute to the phosphorous immobilization. At BH5, the values of ortho-P are higher than the TP values, but this is due to the high standard deviations of the data set.

FC values indicate that groundwater is biologically contaminated and therefore is not fit for direct human consumption. In terms of environmental protection the counts are within the safe limit. Again the most contaminated sites are the ones more frequently irrigated and the results are consistent in this direction. On average the fraction of FC to TC is about 60%.

4.2.2 Metals

Results regarding metal concentrations in groundwater are presented in Table 9.5. The values of metal concentrations in ground water show that only Cu and Zn concentrations are within the prescribed environmentally safe limit. Comparing with the application rates regarding these parameters (Table 9.1), show that the rates are far below the maximum permissible limit. This explains the low concentrations of these metals in ground water.

Cd and Fe values fall in the yellow classification, indicating a medium hazard to the environment, while Cr, Pb and Ni are exceeding the red classification limit (high risk to the environment). The Cr concentrations in particular are two times higher than the red limit. These findings could be well explained with the high application rates regarding these metals (Table 9.1), which are close or exceed the maximum permissible limits. Also, these results confirm the low soil concentrations of these metals (except for Fe, which forms part of the soil background quality) and the fact that the conditions at the pasture are conducive for the release of metals in solution and leaching to the groundwater.

A point of serious concern is related to the spatial variation of the metals. Results show that no significant variation was found with respect to the most frequently irrigated areas, the areas, which have not been irrigated and the control point. The only possible explanation of this fact could be the assumption that the metals pollution plume from the pasture has been extended beyond the stream, separating the pasture from the control point and has affected the aquifer quality at BH6, due to a dispersive flow of contaminants from the pasture aquifer. This assumption could be supported by the fact that TDS concentrations at BH6 are also high. Little data on groundwater quality in the City of Harare is available, and none from anywhere near the study area, in order to confirm this assumption.

The levels of Cr, Ni and Pb at the control point are much higher than the levels of these metals in the Marimba River and tributaries (Chapters 5 and 8). The interaction and recharge mechanisms between the Marimba River water and the aquifer are not clear. It is expected, that during the dry season the aquifer supplies part of the base flow of the River and during this period, a higher influence with respect to metals concentrations could be expected. Additional investigation would be needed to evaluate in more details the impact of the pasture on the aquifer, and the surface water – groundwater interactions.

With respect to the possible beneficial uses of the groundwater (Tables 9.4 and 9.5), the high metal concentrations in respect to Cr, Ni and Pb makes it unsuitable for any of the categories listed. The use for potable purposes should not be recommended even after conventional treatment because it does not provide for the removal of dissolved metals and a very complex and costly treatment should be applied in order to achieve the recommended criteria.

5 CONCLUSIONS

Sewage sludge disposal on land could be considered as an extensive, low capital and operational cost technology, suitable for application in the countries of the Southern African region, because of the favorable climatic conditions and land availability. However, if not applied adequately it could be viewed as a diffuse source of pollution to groundwater and a possible reason for soils' degradation. In order to avoid such undesirable effects and to improve the operational practice, some general recommendations were made in Chapter 8. They are applicable for this case too. In addition, the following could be stated:

Table 9.5. Ground water quality characteristics – metals.

Parameter	Sampling point							Maximum recommended levels for different water uses			
	1	2	3	4	5	Average wells	6 Control	Domestic*	Domestic**	Live-stock	Irrigation
Cd mg/l	0.09 (0.082)	0.07 (0.011)	0.08 (0.033)	0.08 (0.035)	0.07 (0.023)	0.08 (0.013)	0.08 (0.053)	0.003	5	10	10
Cr mg/l	3.62 (3.072)	5.00 (1.713)	4.05 (1.142)	4.61 (0.483)	3.59 (1.421)	4.17 (1.375)	3.99 (0.723)	0.05	0.05	1	0.1
Cu mg/l	0.04 (0.042)	0.04 (0.032)	0.09 (0.031)	0.03 (0.023)	0.05 (0.034)	0.05 (0.014)	0.05 (0.042)	2	1	5	0.2
Fe mg/l	4.16 (2.773)	4.16 (3.714)	11.09 (7.721)	2.70 (1.413)	3.60 (1.711)	5.14 (3.234)	1.38 (0.921)	0.3	0.1	10	5
Ni mg/l	1.31 (0.902)	1.83 (0.532)	1.91 (0.522)	1.78 (0.154)	1.96 (0.453)	1.76 (0.264)	1.66 (0.332)	0.02	--	1	0.2
Pb mg/l	0.41 (0.142)	0.49 (0.192)	0.70 (0.741)	0.65 (0.452)	1.14 (0.382)	0.68 (0.523)	0.65 (0.141)	0.01	0.01	0.5	0.2
Zn mg/l	0.12 (0.071)	0.16 (0.042)	0.18 (0.051)	0.17 (0.132)	0.23 (0.164)	0.17 (0.074)	0.15 (0.062)	--	3	20	1

* WHO guidelines;
** South African guidelines;
Values in parenthesis reflect standard deviations of the mean values

- Sewage sludge land disposal, if applied adequately, leads to the improvement of soil quality in terms of organic content and nutrients.
- In cases of sludge land disposal and beneficial sludge reuse in agriculture, the enforcement of municipal by-laws limiting the discharge of wastes with high metals concentrations into the sewer system is essential in order to limit the metals concentrations in sewage sludge. The wastewater treatment process does not provide for reduction of metals, thus the only way to reduce these concentrations is the source control.
- Proper monitoring and control of pollutant concentrations in soils is essential, as soils saturated with pollutants might act as a time bomb and release them into groundwater after saturation limits are reached. Thus, the operational period of the dedicated site would depend on such monitoring practice and must stop before these limiting concentrations are reached.

In relation to the specific case of SEM disposal on Crowborough pastures, the following conclusions could be made:

- The soils' quality has been improved in relation to organic carbon and nutrients content;
- The levels of metals concentrations in the soil are within prescribed limits, but the soil has been acidified in the most often irrigated areas.
- Long term irrigation, high hydraulic and pollution loads and improper irrigation practice have resulted in elevated concentrations of metals in groundwater, with Pb, Cr and Ni being of greatest concern. This makes the aquifer water quality unsuitable for beneficial use and could be treated as a serious hazard for the environment.

The further use of the pasture at Crowborough farm as a land disposal site for SEM could be recommended only if proper measures are implemented, such as the:

- Improvement of pH of the soil and additional investigation in respect to soil characteristics;
- Reduction of the present hydraulic load. This could be achieved by:
  - Upgrading the BNR plant to prevent the overloading of the trickling filters line (already under consideration by the municipal authorities);
  - Reduction of the incoming effluent by diverting part of the sewage to one or more new local treatment plants;
  - Pumping part of SEM to new disposal sites.
  - The choice of an optimal solution should be based on detailed technical and economic evaluation.
- Enforcement of by-law regulations regarding limits of metal concentrations of effluents discharged into the sewer system.
- Strict application of a proper irrigation practice, based on a rotating principle of irrigation of selected plots and a regular loading of the whole active area.

Acknowledgements – This study was executed as part of the Harare Urban Groundwater Project, which was funded by the Water Research Fund for Southern Africa (WARFSA). The authors would like to thank the sponsors for the financial aid offered and the City of Harare authorities for their co-operation.

REFERENCES

Ayers, R. S. & Westcot, D.W. 1985. Water quality for irrigation. *F A O, Irrigation and Drainage Paper,* 29.

Barry, G.A., Chudek, P.K., Best, E.K., & Moody, P.W. 1995. Estimating sludge application rates to land based on heavy metal and phosphorous sorption characteristics of soil. *Journal of the International Association on Water Quality,* 29, 1945-1949.

Degremont 1991. *Water Treatment handbook, 6th edition,* Vol.1, Paris: Lavoisier Publishing.

Domenico, A.P. & Schwartz, F. W. 1997. *Physical and Chemical Hydrogeology.* 2nd edition. New York: Wiley & Sons.

DWAF (Department of Water Affairs and Forestry, South Africa). 1996a. *South African Water Quality Guidelines, 1: Domestic Use.* 2nd edition. Pretoria, Government Printer.

DWAF (Department of Water Affairs and Forestry, South Africa). 1996b. *South African Water Quality Guidelines, 5: Livestock Watering.* Pretoria, Government Printer.

DWAF (Department of Water Affairs and Forestry, South Africa). 1996c. *South African Water Quality Guidelines, 4: Irrigation.* Pretoria, Government Printer.

Fetter, C.W. 1984. *Applied Hydrogeology.* 3rd Edition. Englewood Cliffs: Prentice Hall.

Foth, H.D. 1984. *Fundamentals of Soil Science.* 7th Edition. Chichester: Wiley.

Freeze, R. A & Cherry, J. A. 1979. *Groundwater.* Englewood Cliffs: Prentice Hall.

Harter, T. Davis, H., Mathews, M.C. & Meyer, R.D. 2002. Shallow groundwater quality on dairy farms with irrigated forage crops. *Journal of Contaminant Hydrology, 55,* 287-315.

Hranova, R., Gumbo, B., Klein, J. & van der Zaag, P.2002. Aspects of the water resource management practice with emphasis on nutrients control in the Chivero Basin, Zimbabwe. *Physics and Chemistry of the Earth* 27, 875-886.

Lotter, L.H. & Pitman, A.R. 1997. Aspects of Sewage Sludge Handling and Disposal. WRC Report 316/1/97. Pretoria: WRC.

Madyiwa S., Chimbari M., Nyamangara J. & Bangira D. 2002. Cumulative Effects of Sewage Sludge and Effluent Mixture Application on Soil Properties of a Sandy Soil Under a Mixture of Star and Kikuyu grasses in Zimbabwe, *Physics and Chemistry of the Earth* 27, 747-753.

McLay, C.D.A., Dragten, R., Sparling, G. & Selvarajah, N. 2001. Predicting groundwater nitrate concentrations in a region of mixed agricultural land use: a comparison of three approaches. *Environmental Pollution,* 115, 191-204.

Metcalf & Eddy Inc. 1991. *Wastewater Engineering Treatment, Disposal and Reuse.* New York: McGraw-Hill.

Novotny, V. 2003. *Water Quality: Diffuse Pollution and watershed management.* New Jersey: John Willey & Sons, Inc.

Nyamangara J. & Mzezewa J. 2001 Effect of Long-term Application of Sewage to a Grazed Grass Pasture on Organic Carbon and Nutrients of a Clay Soil in Zimbabwe, *Nutrient Cycling in Agroecosystems* 59, 13-18.

Page A. L., Miller, R. H., & Keeney, D. R (Eds.) 1982. *Methods of Soil Analysis. Part 2.* Chemical and Microbiological Properties. Madison: ASA, Inc.

Pescod, M. B. 1992. *Wastewater Treatment and Use in Agriculture.* FAO Irrigation and Drainage Paper No. 47. Rome: FAO, UN

Polprasert, C. 1996. *Organic Waste Recycling Technology and Management.* 2nd Edition. Chichester: Wiley.

Reed S., Middlebrooks E., & Crites R. 1988. *Natural Systems for Waste Management & Treatment, USA:* McGraw-Hill Book Company.

Standard Methods for the Examination of Water and Wastewater 1989. 17th ed. American Public Health association/ American Water Works Association/Water Environment federation. Washington DC, USA.

USEPA (United States Environmental Protection Agency). 1992. Design manual: wastewater treatment/disposal for small communities. *USEPA Report* EPA\625\R-92\005.

USEPA (United States Environmental Protection Agency). 1995. Process design manual: surface disposal of sewage sludge and domestic septage. USEPA Report EPA/625/R-95/002.

WHO (World Health Organisation), 1996. *Guidelines for Drinking Water Quality, Volume 2: Health criteria and other supporting information.* 2nd Edition. Geneva: World Health Organisation Press.

WRC (Water Research Commission) 1997 Permissible utilization and disposal of sewage sludge *Report TT85/97,* Pretoria: WRC.

Wright, A. 1999. Groundwater contamination as a result of developing urban settlements. *Water Research Commission Report 514/1/99,* Pretoria: WRC

WWEDR 2000. *Water (Waste and Effluent Disposal) Regulations,* Statutory Instrument 274 of 2000, Republic of Zimbabwe.

Yadav, R.K., Goyal, B., Sharma, R.K., Dubey, S.K. & Minhas, P.S. 2002. Post-irrigation impact of domestic sewage effluent on composition of soils, crops and ground water—A case study. *Environment International,* 28, 481-4.

CHAPTER 10

Irrigation with ponds effluent – impacts on soils and groundwater

R. Hranova & W. Gwenzi

ABSTRACT: Waste stabilization ponds – the source of effluents to be reused beneficially for irrigation purposes, have been discussed together with specific guidelines and criteria regulating the practice of wastewater irrigation. The impacts of long-term irrigation with ponds effluent on soils and ground water quality have been assessed by means of a specific case study of pasture irrigation at Imbwa farm in Chitungwiza, Zimbabwe during the period 2000-2001. Results show that the quality of naturally acidic soils has been improved in terms of increase of the pH and the essential nutrients content. The metals content of soils were far below the recommended maximum values. Adverse impacts on ground water were found with respect to metals – Cd, Zn and specifically Cr, which showed high background values increased by the irrigation practice. High contamination with respect to ammonia, and pronounced pollution with respect to EC, nitrate, phosphate and FC was found as well. Recommendations to improve the present practice have been made.

1 INTRODUCTION

Wastewater stabilization pond systems (WSPS) are a reliable treatment technology, widely applied in countries with temperate climates and where land is available. At low construction and maintenance costs and with no need of expert labor for their operation, they achieve considerable treatment efficiency regarding organic material and pathogens. For these reasons, they are a preferred option of sewage treatment in the Southern African region. In Zimbabwe, from a total number of 139 off site wastewater treatment plants, 110 are WSPS. The treated effluent from these plants is reused for irrigation or disposed on land, as the regulatory instruments do not allow their discharge into surface water bodies. WSPS are mainly applied to serve small towns and rural settlements, and in some cases, combined domestic and industrial wastewater is treated. However, it has been reported (Mtetwa 1998) that the vast majority of the plants do not meet the design treatment efficiency mainly due to operational problems and overloading.

WSPS effluent contains a relatively high concentration of nutrients and organic material, it is safer microbiologically, compared to conventional and BNR processes, and is easy to pump due to relatively low suspended solids concentrations. Therefore, it should be viewed as a valuable source of water for irrigation purposes, especially in cases where the origin of the wastewater is purely domestic. In other cases, where combined and industrial wastewater is treated together, or in cases of purely industrial wastewater, a more cautious approach should be applied in terms of its beneficial reuse. Such applications would require a special investigation of the effluent quality and a corresponding choice of beneficial reuse, which would not lead to environmental or public health hazards.

A wise approach to water resources exploitation and a water demand orientated practice is an international trend followed in developed and in developing countries, but it is of highest priority and importance in countries where water resources are limited. Thus, considering wastewater reuse as an alternative source of water in general, and the specific case of ponds effluents reuse, are of significant importance for the water resources management practice in the region, especially considering the fact that the beneficial reuse of such effluents is not widely applied. This Chapter aims to help in this direction by:

- Presenting the most common design and operational aspects of WSPS, which would ensure a safe effluent quality, appropriate for further beneficial use;
- Revising guidelines and criteria for safe irrigation practice in cases of different types of agricultural applications, as well as regulatory aspects of such practice;
- Discussing the impacts on soils and ground water by presenting a specific case study of long-term irrigation of pastures with ponds effluent in Chitungwiza, Harare.

2 WASTE STABILIZATION PONDS – DESIGN AND OPERATION

2.1 *Waste stabilization pond systems*

Conventional sewage treatment systems, discussed in Chapter 8, are designed to speed up the process of natural aeration and bio-oxidation of organic material by forced aeration. They require a constant and usually high input of energy together with regular maintenance in order to achieve the required reduction of organic pollution, suspended material, nutrients and pathogens. In contrast, WSPS provides a cheap and easy to maintain alternative, which has considerable advantages especially when a high level of removal of organic material and nutrients is not required, as it is the case of beneficial reuse of the effluent. Another specific advantage of WSPS is their significant buffer capacity with respect to shock hydraulic, organic and metal loads. This means that the quality of the effluent would not be affected adversely due to such loads, but they will be absorbed in the system, due to the long retention time. The disadvantages of the application of WSPS could be summarized as follows:

- Large area requirement, correspondingly the land to place the treatment plant should be available and the cost should not be prohibitive;
- Release of malodors to the vicinity and possible breeding site for insects and mosquitoes;
- Possible pollution of ground water, in cases where the water table is high;
- Last but not least – high evaporation rates leading to a considerable loss of the volume of water available for reuse.

WSPS consists of series of artificial ponds, which are subject to a continuous flow and are arranged in such a way that successive ponds receive their flow from the previous one. The degree of treatment achieved is a function of the number and types of the ponds in the system and their configuration. Thus the performance of each type of pond would influence the performance of the subsequent one and the output of the system as a whole. The most commonly applied types of ponds are the anaerobic (AP), facultative and maturation ponds (M). Facultative ponds could be classified as primary (PF), which receive directly screened sewage and secondary (SF), which receive treated effluent in an AP or PF. The basic principles of the functioning of these three types of ponds have been described in Chapter 2. In general, the main reduction of pollutants occurs in the facultative stage. The maturation ponds have the primary objective of pathogen removal, with some additional polishing functions with respect to nutrients and organic matter removal. Some more contemporary modifications of the classic types of ponds are the high rate algal pond (HRAP) and the macrophytes pond.

The HRAP is a modification of the facultative ponds, where conditions for accelerated algae growth have been created in order to intensify the organic matter assimilation and removal. These ponds are designed as very shallow (up to 0.5 m depth) long channels to allow for light penetration. Furthermore, conditions for regular mixing have been provided to prevent sludge layer formation. HRAP ponds could produce a high volume of algae mass, which could be used as a protein source. It should be noted, however, that these treatment units require a constant removal of the excess algae growth from the suspension. The new alga mass generated should be removed regularly from the system with the effluent. If not removed from the pond, the excess algae mass would settle and decay in the pond, thus contributing an additional pollution load and jeopardizing the treatment process. In general, HRAP have been developed as experimental units with the primary objective of protein generation. However, the thickening and dewatering of the algae mass would be a complex and expensive process, which would influence the economic effectiveness of such a system. For this reason, HRAP have not found wide application internationally.

Macrophyte ponds could be considered as an alternative of facultative ponds, which have the primary objective of polishing the effluents by removing algae from the pond effluent and nutrients reduction. Macrophytes, also known as water hyacinths, are aquatic plants, which grow on the pond surface, forming a dense mat and preventing sunlight from penetrating the water. Algae are also removed by biological means. A micro-invertebrate, known also as *Daphnia* or water flea, breeds on macrophytes, and consume algae in large quantity through grazing, thus helping in the algae removal process. It should be emphasized that macrophyte ponds do not provide for pathogen removal. The water hyacinth reproduces rapidly in tropical conditions and could double its mass in 6 days. Thus, such types of treatment units require a frequent harvesting of the biomass generated in order to achieve their treatment effect, which in most cases should be done mechanically or by hand. The harvested material could be used as a protein source to complement animal feeding or as a substrate for anaerobic digestion and biogas production. However, such applications have not been widely applied in practice. One of the main disadvantages of macrophyte ponds is the fact that they provide a breeding environment not only to the water flea but to other insects as well, including mosquitoes.

An illustration of the spreading of macrophytes is shown in Figure 10.1, which shows Lake Chivero shores, covered with aquatic plants due to excessive nutrients concentrations. The upper reaches of the Lake, as well as Manyame River after the Lake, are densely covered with this aquatic plant. It could be observed from the picture that the harvesting process, by mechanical means or by hand, would require considerable efforts, with a low economic effect, and due to this fact, macrophyte ponds have not found wide application at present.

The choice of the WSPS configuration would be based on the quality of the sewage and the required effluent quality. Some common arrangements are shown in Figure 10.2. Configuration 1 is the arrangement, which is most commonly used in the region. Different alternatives of this configuration, substituting the anaerobic stage with an Imhoff tank or an Up-flow Anaerobic Biological Reactor (USBR), would be an attractive configuration as well that could provide conditions for sludge reuse in the form of dry fertilizer (Hranova 2003).

Figure 10.1. Water hyacinths in Lake Chivero.

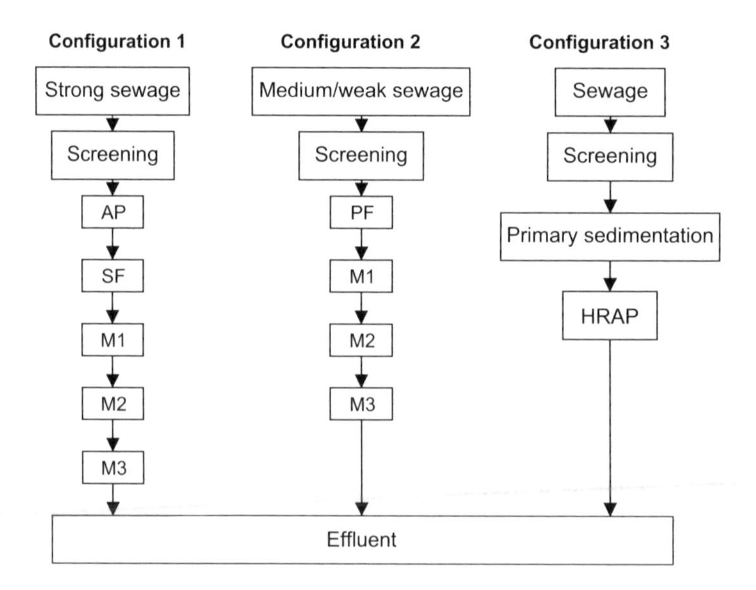

Figure 10.2. Alternative configurations of pond systems.

For medium to large size treatment plants, it is advisable to configure the number of ponds in the different stages in the form of two or more parallel units in order to provide for desludging without the need of bypassing the whole stage. This is most valid for the AP or PF stage, where most of the sludge is accumulated. Based on practical experience, it is recommended to configure several M in a series for a higher rate of pathogen removal (Horan 1990).

Configuration 2 is applicable for weak sewage and could be applied as a diffuse pollution abatement measure in cases of polluted storm water runoff. In such cases, provision for regular desludging is very important. Configuration 3 has the primary objective for algae production and use as a protein source, parallel to sewage treatment.

Another widely applied configuration consists in the combination of conventional treatment systems and maturation ponds. In such cases, the ponds have a polishing function, and if properly operated, could improve significantly the effluent quality of a conventional treatment system with respect to organic pollution and pathogens removal. Such configuration was discussed in Chapter 8 (Fig. 8.5). In cases where wastewater would be used for irrigation purposes, configurations 1 or 2 would be suitable, depending on the strength of the incoming sewage.

2.2 *Design approaches*

In general, ponds are considered as completely mixed biological reactors and their design is based on empirical equations or recommended design parameters, obtained on the basis of investigation of existing facilities under continuous operation. WSPS are biological systems, which resemble closely the biological processes in a natural water environment. The reactions involved are complex and mutually interrelated, and for this reason the most commonly applied design techniques are based on empirical data. The design parameters of highest importance are the organic loading rate, the water depth in the pond and the hydraulic retention time, which would allow for the treatment processes to take place in an appropriate environment.

The main processes taking place in an AP are the sedimentation of suspended material and the anaerobic digestion of accumulated sludge, together with partial bio-degradation of the suspended organic material in the wastewater. In general, these are the reactions involved in a septic tank, thus AP could be considered as large open septic tanks. In order to provide anaerobic conditions within the whole volume of the pond, AP are designed for high organic loading rates and as relatively deep structures, with a recommended depth between 2 and 5 m. The organic loading is given as a volumetric loading rate, e.g. kg BOD_5/m^3, which would depend on the average annual air temperatures. The required retention time is 3-5 days (Mara & Cairncross 1989). A well functioning anaerobic pond would be characterized by a vigorous release of biogas, as the end product of the anaerobic process. The biogas bubbles, formed at the bottom of the pond, would rise and will be released in the atmosphere. They would contribute to the mixing of the pond and would seed the upper layers with active microorganisms, which would contribute to the anaerobic degradation of the wastewater with corresponding partial reduction of BOD_5.

In facultative ponds, both aerobic and anaerobic biological processes take place in addition to the sedimentation of decaying material. The major process of importance during the wastewater treatment is the biodegradation of organic material (reduction of BOD_5). The vast majority is decomposed by aerobic microorganisms, present in the facultative layer, with the active involvement of algae, as shown in Figure 2.4 (Chapter 2). Therefore, the presence of DO and sunlight are crucial for a successful treatment process together with adequate temperature and pH conditions. DO is supplied by natural means through reaeration and through the algae respiration. Based on this, the pond is usually designed to have a relatively low depth (between 1 and 2 m) and a large area, to ensure a good reaeration rate due to a large contact area between the water and the air surface. The low depth also allows the sunlight to penetrate through a large portion of the reactors' volume. Organic loading rates are expressed as surface loading rates, e.g. kg BOD_5/m^2, which would depend on the average annual air temperatures. The biological degradation processes show a distinctive diurnal variation due to the photosynthetic activity of algae. During the day, the upper 50 cm layer of the pond is rich in DO due to the considerable contribution of oxygen as a product of the algae respiration. The oxygen rich conditions have their peak concentrations during midday, and this would be the time when the rate of the biological oxidation would be at its maximum. Due to the high intensity of the rate of photosynthesis, and the utilization of carbon dioxide by algae, the pH in the pond could rise to values of 8 to 9. During the night, algae switch from photosynthesis to respiration and start utilizing the available DO, thus reducing the reaction rate of the aerobic biological process. In the same pond, the sludge accumulated at the bottom would undergo anaerobic degradation, leading to the mineralization of the organic fractions. During the night, the anaerobic portion of the ponds' volume would increase. This type of fluctuations would be more pronounced in PF, which usually are subject to considerably higher organic loads, compared to SF. In general, due to the relatively low organic loading, the amount of biogas released in this type of ponds is not substantial.

Maturation ponds have the primary function to provide adequate conditions for the pathogen removal process. The mechanism of this process is not fully understood, but empirical data shows that stabilization ponds have high pathogens' removal efficiency, much higher compared to conventional treatment systems. This is associated with the long retention time, with the bactericide effect of the ultraviolet rays in the sunlight and with the higher values of pH due to algae activity. The design of M ponds is based on empirical equations, which assume a first order kinetic reaction to describe the removal of fecal coliforms, where the kinetic rate constant is determined based on empirical data and the retention time and temperature would determine the removal effect. The equations consider the retention time and removal effects at the previous treatment stages in AP and F ponds.

During the design process, the configuration, size, treatment efficiency and the layout of the ponds on the specific terrain, are determined. Another important consideration during this phase is to provide proper hydraulic conditions in the pond, in order to achieve a full use of the available volumes and avoid short-circuits of the flow in the pond. The latter could happen due to the improper design of the inlet and outlet facilities, due to high volumes of sludge accumulation, as well as irregular sludge deposition patterns. Short-circuiting leads to a considerable reduction of the actual retention time, compared to the originally anticipated one during the process of design, thus leading to much lower treatment efficiency than the anticipated ones. It could be avoided by a proper design of inlet and outlet facilities to each pond, together with adequate geometrical proportions of the ponds layout. In some cases, the provision of baffles in F and M ponds could improve the ponds hydraulics.

2.3 *Operation and maintenance requirements*

It has already been emphasized, that one of the major advantages of WSPS is the easiness of operation. It does not require high skills personnel, there is no mechanical equipment, or it is limited to screening and pumping facilities. An energy input would be required only for pumping purposes, and if the topography allows, could be avoided by careful design and the provision of gravity flow throughout the system. All these make them a very attractive treatment alternative for developing countries and for population centers or institutional/industrial centers in remote areas. Unfortunately, in many cases of application of WSPS in the region, they have been completely neglected, resulting in a very poor effluent quality. One example of such a case is described in Chapter 11. It should be well understood by the managing authorities that the easiness of operation does not mean lack of operation. A good operation and maintenance practice could achieve a long life of the treatment facility, during which effluent with a good quality could be provided. It should be remembered that if the ponds effluent would be used for irrigation, a high level of operation and maintenance should be provided in order to maintain a high and reliable quality of the effluents.

The most common WSPS operation requirements are:

- Regular desludging in order to provide for the minimum retention time of the wastewater in the pond;
- Provision of a simple monitoring program to control actual hydraulic and organic loading rates, treatment efficiencies, and other necessary environmental factors as pH, T, DO and the sludge dept at selected locations.
- Maintaining in good condition the following basic elements:
    - The hydraulic conduits and corresponding shutoff devices;
    - The inlet/outlet structures of the ponds;
    - Screening devices and pumping stations;
- Cutting the grass along the banks of the ponds and between the different treatment units;
- Periodical evaluation of WSPS performance and preparation of recommendations for the future operational plan of the system.

One of the most common reasons for the failure of WSPS to meet the expected treatment efficiency is the lack of regular desludging and the filling of the volume available for treatment with sediments. It must be admitted, that in many cases this could not only be a problem created by negligence during the operational stage, but a problem induced during the design stage. The design team should prescribe clear directions about the method of desludging and the period of desludging at all stages, but most importantly, with respect to AP or PF stages. Also, they should indicate appropriate methods for sludge disposal. If these provisions have not been included during the design stage, it would be more difficult to do it after the ponds are operational.

3 REGULATING THE PRACTICE OF IRRIGATION WITH SEWAGE EFFLUENTS

Wastewater contains physical, chemical and biological constituents that affect its suitability for reuse in irrigation. Some constituents have beneficial effects in terms of crop requirements and the improvement of soil characteristics. Others have adverse effects on soils and crops. Therefore, different regulatory instruments and literature sources provide guidelines and recommendations with respect to the required range of variation of concentrations with respect to different water quality constituents, which could help in the implementation of a safe practice of wastewater reuse in agriculture. Three categories of constituents are discussed in terms of their impact on crops and soils: salinity and related parameters, specific ions' impacts, and trace metals and related toxic effects. In addition, indications for the suitability of water to be used for irrigation purposes have been made.

3.1 *Salinity and soil permeability*

Accumulation of salts in the upper soil layers (the root zone) reduces considerably the ability of plants to extract water from the soil. The effects of soil salinity on plants depend on the type of soil, irrigation practices, crop type and the growing stage of the crop. Soils' salinity is measured by the electrical conductivity of the soil. Salinity of wastewater is expressed by TDS and the correlated parameter EC, or by the concentration of Na and Cl ions. Wastewater contains high TDS concentrations, which are not removed during the treatment process. Consequently, prolonged use of wastewater for irrigation would lead to salinization of the soil. This is one of the most common adverse effects of wastewater reuse for irrigation. High soil salinity results in high Na concentration in soils, leading to poor soil structure and dispersion of clay particles. As consequence, the soil permeability is reduced, leading to water logging, weed growth and erosion. On drying, the soil forms hard crusts that limit root penetration. Restrictive guidelines with respect to EC and TDS in wastewater reused for irrigation are presented in Table 10.1, which has been compiled based on data from Degremont (1991) and Ayers & Westcot (1985).

The parameter used to evaluate and measure the potential of wastewater to cause salinity problems is the sodium adsorption ratio (SAR). It is represented by equation (10.1) and shows the potential of wastewater to supply Na^+, which would replace Ca^{2+} and Mg^{2+} in soils and would lead to a change in the soils' structures.

$$SAR = Na^+ \;/\; \{\sqrt{(0.5\ Ca^{2+} + 0.5\ Mg^{2+})}\} \qquad (10.1)$$

where, the concentrations of the different ions are expressed in meq/l.

A low SAR (2 to 10) indicates little danger, medium (7 to 18) indicates medium hazards and high hazards are indicated within the range of 11 to 25 (Fetter 1984). High SAR can destroy the soil structure owing to the dispersion of the clay particles, and may result in hard setting, reduced penetration and infiltration rates and reduced crop yield (DWAF 1996b).

3.2 *Specific ions*

Nutrients in their different forms, present in wastewater used for irrigation are the major beneficial aspect of this practice, as they provide the quantity necessary for plant uptake without the need to use artificial fertilizers. Khouri et al. (1994) have performed comparative studies on the yields of selected crops irrigated with different types of wastewater and with fresh water combined with the application of a commercial fertilizer. Results show that stabilization ponds' effluents provide a comparative crop yield to other forms of wastewater and it is higher than the yield achieved when commercial fertilizers are applied. It has already been emphasized in previous chapters that the nutrients loadings should be comparable to the specific plants nutrients uptake rates to avoid excessive amounts accumulated in soils, leached to ground water or washed up by runoff. Some adverse effects of nutrients would be associated with the presence of ammonia, which could cause crop burn. Also, excessive nitrogen in the form of nitrates and ammonium causes excessive vegetative growth resulting in delayed crop maturity. High bicarbonates in irrigation water can unsightly deposits on crop leaves that are sprinkler-irrigated. Table 10.1 presents recommended guidelines with respect to these elements.

The adverse effect of Na and Cl ions on soils has been discussed, but it should be noted that these ions, together with boron, could have a toxic effect on some plants as well. Table 10.1 presents guidelines regarding the recommended concentrations of these metals in wastewater reused for irrigation purposes.

Phosphorous (P) applied on cropland could be taken up by plants, incorporated into soils as organic P, or could become weakly or strongly adsorbed onto clay particles and fixed by Al, Fe and Ca ions, depending on the soils' pH. After initial adsorption, there is a gradual fixation of added P that renders part of it unavailable to plants. With regular P application, the importance of the fixation process is diminished, as the soil P-sorption capacity becomes slowly saturated, and a larger P concentration remains in soils in soluble

Table 10.1. Recommended concentrations of selected parameters in wastewater reused for irrigation.

Parameter	Level of restriction for irrigation reuse		
	None	Slight to Moderate	Severe
EC (deciSiemens/m at 25°C)-in respect to salinity	< 0.7	0.7-3.0	> 3.0
EC (deciSiemens/m at 25°C)-in respect to soil permeability			
SAR[2] = 0-3	< 0.2	0.2-3.0	> 3.0
SAR[2] = 3-6	< 0.3	0.3-1.2	> 1.2
SAR[2] = 6-12	< 0.5	0.5-1.9	> 1.9
SAR[2] = > 12	< 2.9	2.9-5.0	> 5.0
TDS (mg/l)	< 450	450-2000	> 2000
Sodium			
Surface irrigation (meq/l)	< 3.0	3.0-9.0	> 9.0
Sprinkler irrigation (meq/l)	< 3.0	> 3.0	-
Sprinkler irrigation (mg/l)	< 70	> 70	
Chloride			
Surface irrigation (meq/l)	< 4.0	4.0-10.0	> 10.0
Surface irrigation (mg/l)	< 140.0	140.0-350.0	> 350.0
Sprinkler irrigation (meq/l)	< 3.0	> 3.0	
Sprinkler irrigation (mg/l)	< 100.0	> 100.0	
Boron (mg/l)	< 0.7	0.7-3.0	> 3.0
Nitrate nitrogen (mg/l)	< 5.0	5.0-30.0	> 30.0
Bicarbonate (meq/l)	< 1.5	1.5-8.5	> 8.5

form. In a study of the behavior of P in soils, it was noted that after 6 months of P application, the plant available P decreased from 47 to 27 % due to the fixation by clay, organic carbon, iron (Fe), aluminum (Al) and calcium carbonate (Muchovej & Rechgil 1995, Wild 1995). The assessment of the plant available P should consider soluble P concentrations in soil.

Recommendations with respect to different types of crops and corresponding nutrients' uptake rate is presented in Table 10.2. These could serve as a guideline during the design of the irrigation field and the determination of the nutrients loadings.

The continuous long-term application of P at levels exceeding crop requirements (Table 10.2) increases the potential for P loss through runoff and drainage water leading to the eutrophication of surface water

Table 10.2. Nutrients' uptake rate by selected crops.

Crop	Nutrients' uptake rate (kg/ha.yr)		
	Nitrogen	Phosphorous	Potassium
Forage crops			
Alfalfa	225-540	22-35	175-225
Bermuda grass	400-675	35-45	225
Reed canary grass	335-540	40-45	315
Field crops			
Barley	125	15	20
Corn	175-200	20-30	110
Grain sorghum	135	15	70
Cotton	75-110	15	40
Potatoes	250	10-20	30-55

Selected from Metcalf & Eddy (1991)

bodies. However, excessive P has not been a problem for irrigation purposes and no guideline value is given for its evaluation.

3.3 *Toxic metals*

The concern with respect to the impacts of toxic metals on public health, soils, ground and surface water has been discussed in previous chapters. From the point of view of crops irrigation, the most hazardous trace elements are Cd, Zn and Pb (Wild 1995). This is due to three factors:

- They accumulate in the edible portions of crops;
- They easily enter the food chain;
- They form part of many products, used widely in the every-day life, and consequently, their concentrations in wastewater are relatively high.

Although a high trace metal concentration poses a health hazard, few fatalities due to the ingestion of trace metals have been traced to the contamination of soils (Wild 1995). Of highest concern is the mobility of the toxic metals to ground and surface water. Table 10.3 shows the maximum permissible values of selected metals in wastewater reused for irrigation with corresponding adverse impacts. The recommended values with respect to long-term and short-term applications are presented in table 10.4.

Cd is a relatively rare metal and natural concentrations in soils are very low. Man-made sources of Cd are usually associated with industrial activities. It forms part of automotive batteries and pigments, and could be released to the environment due to the improper management of waste products. It forms part of some P fertilizers as well. The uptake of Cd by plants is greatest in acid soils of high Cd concentration.

Table 10.3. Threshold values of trace elements in water for crop production.

Element	Recommended maximum concentration (mg/l)	Remarks
Al	5.0	Causes non-productivity in acid soils (pH<5.5) but no problems at alkaline pH>7
Cd	0.01	Toxic to beans, beets, turnips in a concentration as low as 0.1 mg/l in nutrient solution. Conservative limits recommended due to the potential to accumulate in plant tissues to concentration that may be harmful to humans.
Cr	0.1	Not an essential element. The conservative limit is due to lack of knowledge on toxicity.
Fe	-	Not toxic to plants in aerated soils but can increase acidity and reduce the availability of phosphorus and molybdenum. Sprinkling can cause unsightly deposits on plants, buildings and equipment
Mn	0.2	Toxic to a number of crops at a few tenths to a few mg/l only in acid soils pH<5.5
Ni	0.2	Toxic to a number of plants at 0.5 mg/l to 1 mg/l. Reduced toxicity at neutral or alkaline pH.
Pb	5.0	Can inhibit plant cell growth at a high concentration
Zn	2.0	Toxic to many plants at widely varying concentration. Reduced toxicity at pH>6.0 and in fine textured and organic soils

After Pescod (1992)

Table 10.4. Concentration of trace elements in wastewater suitable for irrigation (mg/l).

Element	Long-term*	Short-term*	Long-term**	Short-term**
Aluminum	-	-	5	20
Boron	0.75	2.0	0.75	2.0
Cadmium	0.01	0.05	0.01	0.05
Chromium	0.1	20.0	0.1	1.0
Copper	0.2	5.0	0.2	5.0
Nickel	0.2	2.0	0.01	0.05
Lead	5.0	20.0	5.0	20
Zinc	-	-	2,0	10.0
Manganese	-	-	2.5	2.5

After: * - WWEDR 2000
** - Papadopoulus (1985)

Cadmium is very mobile and is readily available to crops. It accumulates in crops without showing signs of phytotoxicity.

Pb exists in three major oxidation states (Pb^{4+}, Pb^{2+}, Pb^{0}). Solubility of Pb compounds is strongly dependent on pH and redox potential. The main sources of Pb in wastewater are batteries, cable sheathing, paints, and plastics. Uptake of Pb by plants is low even under conditions of a high Pb in soil solution and most of it is retained in the roots (Wild 1995).

Sources of Zn in wastewater include metal coating, alloys, batteries and pigment. A high Zn content in soil solution can be phytotoxic and reduces growth (Landon 1991, Henning et al. 2001). The remedy to Zn toxicity is to lime the soil to pH 6.5. Zn is most readily absorbed than any other trace metal and has the potential to move below the depth of incorporation (Oloya & Tagwira 1996, Rechgil 1995, Polprasert 1989). Crops sensitive to Zn deficiency include maize, citrus, alfalfa and cotton. Soil parent material, organic matter content and soil pH affect Zn availability. Zn deficiency is associated with a high pH, coarse-textured and highly leached soils and soils with low inherent zinc concentration (Mikkelsen & Camberato 1995).

In soils, Cu exists mainly as Cu^{2+} adsorbed to clay minerals or organic matter. Crops sensitive to copper deficiency isnclude cereals and vegetables. Copper toxicity is not very widespread but concentrations greater than 150 ppm are toxic to citrus (Landon 1991). Like other trace metals, solubility of copper is highest at pH less than 5.

4 IMPACTS ON SOILS

4.1 *The study area*

The study was carried out at Imbwa Farm in Beatrice with a total area of 1444 ha of which about 481 ha were under irrigation. Before pasture irrigation, the farm was used for the production of tobacco under dry land conditions. Since 1979, the farm has been converted to wastewater-irrigated pasture, which is the most common wastewater-irrigated crop in Zimbabwe. Grass species grown on the farm include a mixture of Kikuyu (*Pennisetum clandestinum, Chiov.)* and star grass (*Cynodon dactylon*, L) under sprinkler irrigation and furrow irrigation. About 80 ha were grown to Bana grass (*Pennisetum typhoides*, L.*)* under furrow irrigation.

Imbwa Farm is located at about 22 km from Harare. It has grid co-ordinates 30°55^{1}E, 18 ° 05^{1}N. During the period of study, it was surrounded by a large-scale commercial farming area, which grows tobacco as the major crop. Thus soils are referred to as the 'tobacco soils' of Zimbabwe. It is in the Natural Region

II and is part of the Manyame Catchment with a mean annual rainfall of 850 mm and a mean annual temperature of 23 °C. The study area drains into the Mtsike River to the Southwest.

The area is gently sloping with an average slope of 1:60. The natural vegetation in the area consists of sparse "miombo" woodland with msasa (*Julbernadia globiflora*) and mnondo (*Brachstegia sperciformis*) being the dominant tree species. Other isolated species include muhacha (*Parinari curatelifolia*).

The geology consists of intrusive igneous rocks with granite as the dominant rock. Soils are predominantly sands and loamy sands typical of granitic origin. Nyamapfene (1991) classified the soils at the site as: " Harare 6 G.3", according to the Zimbabwe classification, as *Gleyic Luxisol,* according to the FAO soil classification, and as *Udic kandiustalf,* according to the USDA soil classification. The soils are 1 m deep on average and show clay movement down the profile as observed in the field. The soils have high K (about 164 mg/kg) and crops grown on such soils do not respond to the K fertilizer application. Nitrogen and phosphorus are low, thus A and P fertilizers are needed in most cases. The soils have low phosphate sorption capacity. The pH of the soils is naturally acidic (about 4.3). The reason advanced by Nyamapfene (1991) for the low pH was that the soils are prone to the leaching of basic cations such as potassium, calcium, sodium, and magnesium. Additional physico-chemical characteristics of the soils including water retention at different soil sections could be found in Purves (1976) and Nyamapfene (1991).

A scheme of the farm is shown in Figure 10.3. The hydraulic structures on the site, consisting of 6 ponds, which treat wastewater pumped from Zengeza Sewage Treatment Works (ZSTW), 2 reservoirs and irrigation installations are owned by the Department of Water Resources, which sells the wastewater to the Ministry of Lands and Agriculture. The Ministry of Lands and Agriculture in turn leases the farm to individuals who are compelled to operate the pasture.

Five pumps are used for irrigation but most of the times, only four are working. The pumps discharge a total of about 30×10^3 m^3 per day and each pump commands about 80 ha. The pumps operate 24 hours a day throughout the year and the six laterals are moved every 12 hours with an irrigation interval of 3 days. During the rainy season, excess effluent is discharged through runoff channels into natural grassland. Gauging devices at the farm include flumes at the main canal, sub-canals and at the outlet to Boronia Farm, but during the period of study, they were not functioning properly. Also, during the study, it was observed that part of the equipment is not functional and could not be used for operation, which actually means that only the blocks that are irrigated have adequate pumping and irrigation facility. Block 1 was chosen as a representative for the sprinkler irrigated blocks, because it was the most often irrigated block.

A pipeline from ZSTW, located about 15 km to the north east of the farm, supplies the wastewater used for irrigation. The plant treats wastewater from the satellite City of Chitungwiza by a conventional trickling filter scheme. After that, it is pumped to six stabilization ponds, which can be operated in a series or in parallel. Five of the ponds have a depth of 2 m and a size in plan of 75 m by 75 m (ponds number 2, 3, 4, 5 and 6). The first pond is designed as anaerobic pond with a depth of 5 m. The ponds were designed with a total mean retention time of nine days. However, due to rapid urbanization and the increase in the population of Chitungwiza, as well as due to irregular desludging, the ponds have been filled with sediments and the retention time has been reduced considerably. The primary function of the pasture and irrigation facilities has been the disposal of the effluents, rather than intentional reuse. Based on data provided by the operators, during the period of study, the ponds received 36×10^3 m^3/day of wastewater against a design flow of 21.75×10^3 m^3/day.

Most of the effluent has been used on the farm, whilst a relatively small portion goes to Boronia Farm and Ellerton Farm to the north and east respectively. Farm records and JICA (1997) showed that the effluent had TKN of 30-40 mg/l and 60 mg/l for the rain and the dry seasons respectively. Recently, after this study was performed, a new BNR plant was commissioned, parallel to the trickling filter line, in order to alleviate the hydraulic and pollution loads at ZSTW and the pastures.

Farm records during the initial stages of the pasture operation, showed that the cattle stocking rate for the irrigated pasture was seven livestock units (LU) per hectare compared to one livestock unit per 3 hectares for veld grazing. The total irrigated area under furrow and sprinkler irrigation was 390 ha giving a total of 1820 LU grazing on the pasture for 8 months. Approximately 15×10^3 m^3/ha yr of effluent was applied during this time. This hydraulic load, together with the average annual rainfall of approximately

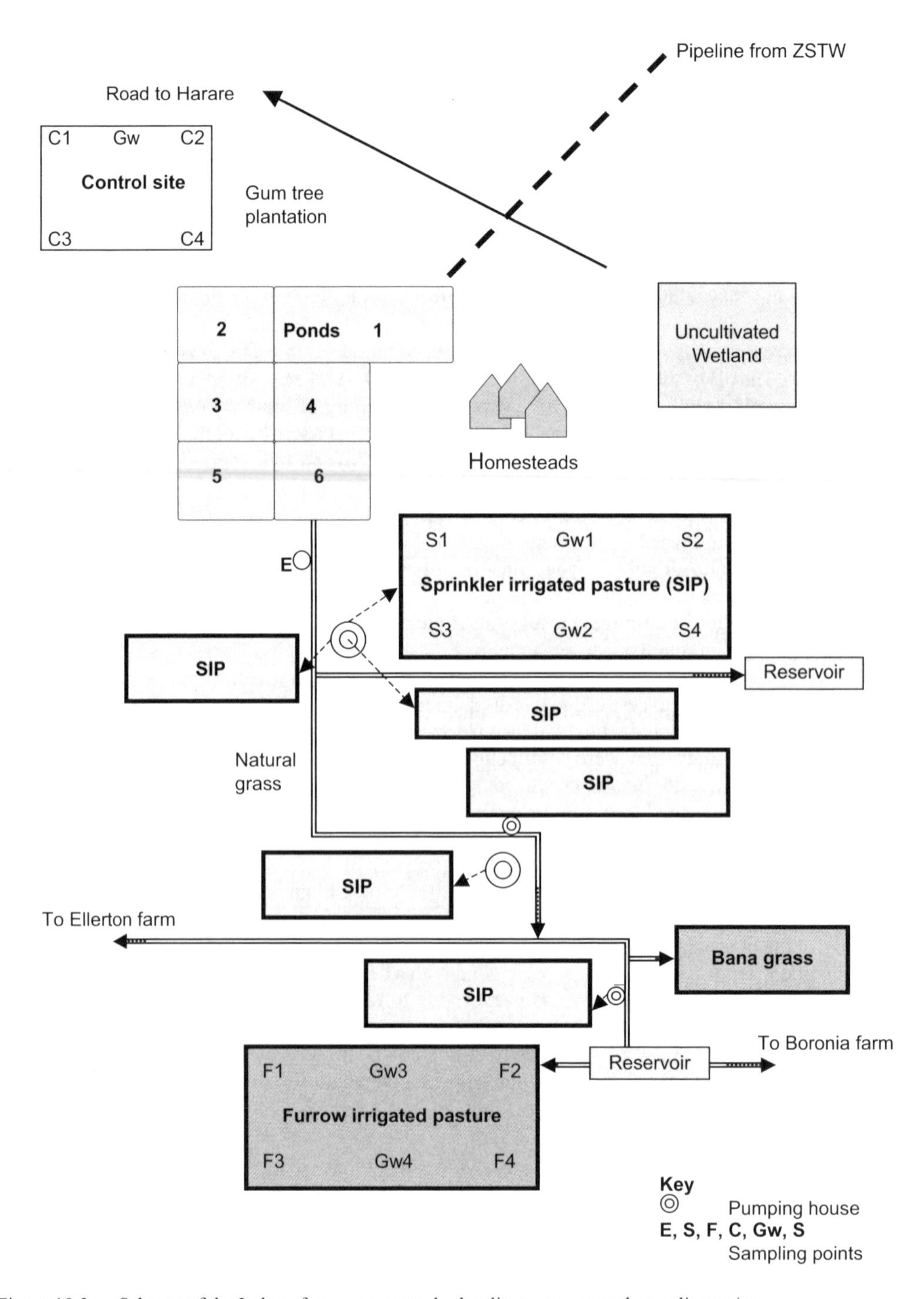

Figure 10.3. Scheme of the Imbwa farm – pastures, hydraulic structures and sampling points.

850 mm gives a total hydraulic load of 23.5×10^3 m^3/ha yr, which is about double the requirement of the pasture grass.

4.2 *Methodology*

Two pasture blocks that have been under irrigation for the past 22 years were selected for sampling purposes. One is the sprinkler-irrigated block close to the pump house and adjacent to the stabilization ponds, which is the one most frequently irrigated. The other block is under furrow irrigation and is the least frequently irrigated one. The control site is an area, located at the highest position, and reflecting the upper part of the aquifer flow gradient. It is located at 1 km from the irrigated area. It has never been irrigated and previous land use for the control site was maize production without the addition of fertilizers. Other factors that were considered in site selection include accessibility, similar soils and the availability of wells for groundwater monitoring.

Sampling was done from November 2000 to April 2001. For both the irrigated and control sites, 4 replicates of soil samples were collected from each site, following recommendations of Henning et al. (2000). As shown in Figure 10.3, points C, S and F indicate sampling locations for soil samples from the control site, sprinkler and furrow irrigated sites respectively, and Gw indicate sampling locations of ground water samples. During the sampling procedure, areas that showed evidence of localized ponding, cow dung, erosion and other visible disturbances were avoided. Samples were collected at 0-30 cm, 30-60 and 60-90 cm depths using a 50 mm diameter steel hand auger. For the control sites, sampling was done on land that had the minimum disturbance from cultivation. To avoid contamination, soil sampling started from the control to the irrigated sites following the recommendations outlined in Rubio & Vidal (1998). When sampling the irrigated sites, the auger was washed with distilled water to minimize cross-contamination. Samples were collected in clean polythene bags and labeled with prefixes "C" to indicate control, "S", sprinkler and "F", furrow irrigated sites. The following parameters were tested: pH, EC, nitrates, ammonia, phosphates, K, Ca, Cd, Cr, Cu, and Zn. Samples for nutrients were analyzed in the Analytical Services Laboratory (ASL), the Department of Soil Science and Agricultural Engineering, University of Zimbabwe, and for metals in the Institute of Mining Research (IMR) laboratory.

Parameters that are sensitive to pre-treatment such as ammonia, nitrate and pH were analyzed first before air-drying the samples. For parameters that are relatively insensitive to drying, soil samples were air-dried in a well-ventilated room in wooden boxes. The soils were ground, using a mortar and pestle, passed through a 2 mm diameter steel sieve and kept at 4 ^{o}C until analyzed, following recommendations of Page et al. (1982). All analyses were done within 72 hours of sampling.

The soil characteristic parameters were tested according to Page et al. (1982). Ammonium and nitrates were extracted from soil using 2N potassium sulfate (KSO_4) followed by steam distillation. Ammonia and nitrate were then determined by the Salicylate method using an ultra violet/visible (uv/vis) spectrophotometer (model: PYE UNICAM PU 8600, Philips Pvt. Ltd). Phosphorus as ortho-P was determined by the Bray 1 method followed by ultra violet/visible (uv/vis) spectrophotometer (model: PYE UNICAM PU 8600, Philips Pvt. Ltd). Exchangeable bases (Ca, Mg, K and Na) were extracted by acidified ammonium acetate. Ca and Mg were then determined by an atomic absorption spectrophotometer (AAS) (model: PYE UNICAM SP9) employing nitrous/acetylene as fuel. Na and K were determined by a flame photometer (model: Corning 400) employing a low-pressure butane gas as fuel. EC was measured from a soil: water (1:5) suspension using an electrical conductivity meter (model Jenway 4020) after allowing settling. The determination of pH was done in a 1:5 soil: 0.01$CaCl_2$ suspension using a pH meter (model: Corning 220).

Trace nutrients B, Cu and Zn and total trace metals were extracted using the USEPA 3050 method of double acid (nitric-perchloric ($HNO_3/HClO_4$)) digestion following the recommendations of Henning *et al.* (2000) and Page *et al.* (1982). The method has been adapted by USEPA as a standard method of extraction and recovers almost 100% of the total metals in the samples (Henning et al. 2000). Soil samples were weighed, ashed and taken up in HNO_3 /$HClO_4$ and made up to a known volume. The trace metals in the extract were determined with an atomic absorption spectrophotometer (AAS) (model: VARIAN TECHTRON SPECTRAA 50B [110 SOFTWARE]) employing an air-acetylene flame.

The determination of pollution loads to the farm was not possible because water quantities were not measured during the period of study, due to non-functioning measurement devices, but effluent characteristics were determined. Time-integrated effluent samples were collected every 20 minutes from 9 a.m. to 12 p.m. from the location E as shown in Figure 10.3. The parameters tested and testing procedures were the same as the ones used for groundwater determination (section 5.1).

4.3 *Ponds effluent quality*

Results with respect to ponds effluents used for irrigation are presented in Table 10.5. The SAR[2] value, calculated based on the mean concentrations of Na, Ca and Mg, is 4. In general, this is a low value, which would not require restrictions to use this effluent for irrigation purposes. However, if it is considered in conjunction with the EC value, it shows that this effluent would require a moderate restriction for reuse (Table 10.2). With respect to Na ions, the effluent could be used with no restrictions. The same applies to Zn and Cu. (Tables 10.3). However, the Cd concentrations exceed twice the maximum permissible limit with respect to Cd and about 10 times the limit with respect to Cr, which is a point of serious concern. Another observation is connected to the very high values of TKN and ammonia, typical for raw sewage, which indicates the very poor performance of the treatment facilities. These two parameters are usually not restricted by regulations, because it is expected that they would be transformed and absorbed by the soil, and special attention is given to nitrates, with respect to their mobility. However, high loads with respect to ammonia, could lead to leaching of this element to ground water.

4.4 *General parameters and nutrients*

The results with respect to general parameters and nutrients characteristic for this specific site, are presented in Table 10.6. With respect to the pH of the control sites, the results show natural acidic soils. This agrees well with similar results in respect to sandy soils in Mhondoro and Chinhamora Communal lands studied by Nyamangara et al. (2000). Nyamapfene (1991) explains the low pH by leaching of basic cations (Ca^{2+}, Mg^{2+}, Na^{+} and K^{+}). However, the pH values of the irrigated sites were about 1.0 unit higher than that of the control.

As supported by Nyamangara & Mzezewa (1999), the higher soil pH at the irrigated sites is due to the "alkalinizing" effect of effluent used for irrigation. Thus the pH values of the irrigated sites were significantly higher than that of the control. In all the sites, there were no marked pH variations along the soil profile.

Table 10.5. Ponds effluent characteristics during the study period.

Parameter	Units	07/03/01	03/04/01	10/04/01	Mean	Standard deviation
TKN	mg/l	35.60	34.80	35.00	35.10	0.42
NH_4	mg/l	13.20	12.80	15.20	13.70	1.29
NO_3	mg/l	0.90	1.20	1.00	1.03	0.15
P	mg/l	3.02	3.00	2.80	2.94	0.12
K	mg/l	29.70	30.40	29.90	30.00	0.36
Na	mg/l	46.00	46.60	47.70	46.00	0.86
Mg	mg/l	8.90	9.70	11.50	10.00	0.45
Ca	mg/l	22.20	26.60	23.10	24.00	2.32
pH	[-]	7.23	7.18	7.10	7.17	0.07
EC	μS/cm	950	1021	978	983	35
Zn	mg/l	0.06	0.05	0.06	0.06	0.01
Cu	mg/l	0.00	0.01	0.01	0.01	0.01
Cd	mg/l	0.00	0.05	0.02	0.02	0.03
Cr	mg/l	0.02	2.18	2.01	1.40	1.20

Table 10.6. Soil characteristics – general parameters and nutrients.

Depth [cm]	Site	pH [-]	EC [μS/cm]	Ortho-P [mg/l]	Ammonia [mg/l]	Nitrate [mg/l]
0-30	C1	4.2	27	21	3.8	1
0-30	C2	4	22	77	3.06	0
0-30	C3	3.5	588	15	2.5	0
0-30	C4	3.8	26	104	2.8	1
	Average	3.88	165.75	54.25	2.31	0.50
0-30	F1	5.5	223	172	3	15
0-30	F2	5.3	103	288	3	3
0-30	F3	4	132	282	3	4
0-30	F4	4.7	428	224	3	1
	Average	4.88	221.50	241.50	3.00	5.75
0-30	S1	5.2	134	365	3	5
0-30	S2	4.7	83	172	5	2
0-30	S3	4.9	109	178	3	2
0-30	S4	5.2	96	383	4	0
	Average	5.00	105.50	274.50	3.75	2.25
30-60	C1	3.3	4	18	1	1
30-60	C2	5.6	41	260	1	4
30-60	C3	3.8	22	25	5	0
30-60	C4	3	18	120	0	2
	Average	3.93	21.25	105.75	1.75	1.75
30-60	F1	5.3	211	193	0	0
30-60	F2	4.5	122	288	4	3
30-60	F3	4.1	101	509	2	0
30-60	F4	5.4	94	254	2	0
	Average	4.83	132.00	311.00	2.00	0.75
30-60	S1	5	88	334	2	4
30-60	S2	4.8	56	172	3	0
30-60	S3	4.5	55	190	2	0
30-60	S4	5.5	72	346	2	0
	Average	4.95	67.75	260.50	2.25	1.00
60-90	C1	3.6	30	3	1	0
60-90	C2	3.8	56	260	1.02	7
60-90	C3	3.7	203	3	1.8	0
60-90	C4	4.1	61	89	2	0
	Average	3.80	87.50	88.75	1.21	1.75
60-90	F1	5.9	176	257	2	0
60-90	F2	5.5	95	101	1.5	0
60-90	F3	4.5	198	607	1.6	1
60-90	F4	4	102	401	2	0
	Average	4.98	142.75	341.50	1.78	0.25
60-90	S1	4.7	110	313	2.7	0
60-90	S2	5.9	88	636	3.2	0
60-90	S3	5.5	138	496	3.4	0
60-90	S4	5.6	119	447	3	0
	Average	5.43	113.75	473.00	3.08	0.00

Ammonium concentrations in soils at the irrigated sites were higher than those at the control. There was a significant difference between the control and the irrigated sites at all depths. This can be attributed to anaerobic conditions that inhibit the nitrification process. For each site, highest concentrations were found in the topsoil (0-30 cm). This could be expected because according to Reemtsma et al. (2000) the ammonium cation is adsorbed on the negatively charged sites of the soil by cation exchange reactions.

Ammonium concentrations in the sprinkler-irrigated site were significantly higher than those of the furrow-irrigated site at all depths. However, in the sprinkler irrigated site there was a higher level of ammonium at 60-90 cm depth than at 30-60 cm depth, evidencing the movement of the NH_4 down the soil profile.

In general, the nitrate concentrations are low. Only the furrow-irrigated site had a significantly higher nitrate concentration than the control and the sprinkler-irrigated sites at 0-30 cm depth. In general, this could be explained with the low rate of the process of nitrification, together with the intensification of the denitrification processes, due to anaerobic conditions in the soil created by a high hydraulic load, together with high organic and nutrients loadings. It could be argued that the frequency of irrigation and its attendant effects on aeration of the soil had a more profound effect than the method of application (sprinkler or furrow). For all the sites and soil depths, partial co-relation between the ammonium and nitrate levels in the soil could be observed. For example at the 60-90 cm-depth profile in the sprinkler-irrigated site, the low levels of nitrates in the soil are supported by a corresponding higher concentration of ammonia (Table 10.6). Although conditions are mainly aerobic at the control site, the low concentration of nitrates was due to limited amounts of ammonia and organic nitrogen.

The control point showed very low concentrations of ortho-P. Similar results have been reported by Grant (1995), who noted that the sandy soils of Zimbabwe are generally deficient in P. A marked difference in ortho-P concentrations between the control and the irrigated sites was found. At the 30-60 and 60-90 cm depth profiles, there was a significant difference among the three sites. This accumulation of P in the soil was consistent with data obtained by Oloya & Tagwira (1996) and Thompson (1968) in their separate studies on the effect of effluent irrigation on soil properties at Aiselby and Goodhope farms in Bulawayo. For the irrigated sites, there was evidence of P movement down the profile. This can be explained by the saturation of the P adsorption capacity of the soil over time. This downward movement was enhanced by over-irrigation, which was reflected in the significantly higher ortho-P concentration in the sprinkler-irrigated site at 60-90 cm than the furrow irrigated. Nyamangara & Mzezewa (1999) also observed P movement down the profile on sandy soils at Crowborough Farm. This was highly expected because according to Nyamapfene (1991), such sandy soils have very low P sorption capacity. Muchovej and Rechgil (1995) noted that the P sorption capacity of sandy soils with low clay content, Fe/Al oxides and organic matter, was limited. Despite being less mobile compared to other nutrients, Withers & Sharpley (1995) quantified that the P movement in the soil becomes significant once 25 % of the P sorption capacity of the soil is saturated. According to Reemtsma et al. (2000), the movement of P down the profile was also enhanced by the anaerobic conditions that favour the reduction of iron (III) to iron (II) and the release of phosphates and Fe^{2+} in solution.

4.5 *Metals*

Results with respect to metals concentrations in the soil are presented in Table 10.7. With respect to K, the metal essential to plant growth, and the levels measured were comparable at all sites. The statistical test for a significant difference showed that there was no significant difference between the control and the sprinkler-irrigated sites. However, potassium levels in the furrow-irrigated site were significantly higher than that of the other sites at all depth profiles. Muchovej & Rechgil (1995) observed that after 40 years of K application, no K accumulated in the top 75 cm of a sandy soil due to leaching. This explains the low K levels in the sprinkler-irrigated site, which receives the highest hydraulic load. The relatively high K concentrations at the control site could be attributed to a number of reasons. Firstly, granitic soils have high concentrations of K and no K deficiencies have been reported on sandy soils (Nyamangara et al. 2000, Nyamapfene 1991). Thus the high native K from the weathering of K-rich felspars in granites masked the effects of wastewater-derived K. Secondly, compared to other basic cations such as Mg and Ca, K is required by plants in high amounts (Landon 1991), thus it could also be argued that the K added through wastewater irrigation ($\approx$30 mg/l) and crop uptake have reached an equilibrium. The high K in the topsoil (0-30 cm) was partly due to adsorption of K on the cation exchange sites. For all the sites, there was a decrease in K concentration at the 30-60 cm depth and a rise in concentration at 60-90 cm depth. This showed that K was leached down the soil profile as supported by Muchovej & Rechgil (1995), who observed that leaching in sandy soils could be as high as 90 % of the K input.

A major characteristic behavior of trace metals is their ability to accumulate in the topsoil. Contrary to this, there was no distinct evidence of accumulation. In general, the levels of trace metals in all the sites were comparable. This showed that the native concentrations in the soil were relatively high. In general, high Cd concentrations were obtained in all the sites including the control. Henning et al. (2000) even reported higher concentrations in the control compared to sludge-amended soils. They argued that such results were due to non-homogeneity of the sandy soils. In the topsoil (0-30 cm) depth profile, the highest concentration was obtained at the control site and the lowest was observed at the sprinkler-irrigated site. According to Pennsylvania State University (1985), Cd uptake by plants can be very high especially in acid soils of high Cd concentration and has the potential to accumulate in the plant without showing phytotoxicity effects.

Henning et al. (2000), in a study of the behavior of heavy metals in sludge-amended soils on a sandy soil in South Africa, observed that Cd was very mobile especially in acid soils. This indicates a high level of plant uptake and leaching at the irrigated sites, leading to a reduction in the Cd concentration at these sites, compared to the control. Although Zimbabwe does not have legislated guidelines for a maximum allowable concentration of Cd in the soil, all concentrations exceeded the FAO guideline limit of 3 mg/kg. The Cd concentration in the control was significantly higher than that of the sprinkler-irrigated site reflecting the effect of over-irrigation on the mobility of Cd.

Unlike Cd, Zn is an essential plant nutrient. The EDTA-extractable Zn in the major soils of Zimbabwe ranged from 1.8 to 3.8 mg/kg. Since the double acid ($HNO_3/HClO_4$) is a stronger extractant than EDTA, higher values than those of Nyamangara & Mzezewa (1999) were expected. The results fall within the ranges of 10-300 mg/kg for most soils, but were about 10 times lower than those of Reemtsma et al. (2000), who reported a Zn concentration of 400 mg/kg due to wastewater application on a sandy soil. Similar to Cd variations, the highest concentrations were obtained for the control site. Rechgil (1995) noted that zinc was most readily absorbed by plants than any other trace metal and has a high potential to move beyond the depth of incorporation. Oloya & Tagwira (1996) reported similar low zinc concentrations in effluent-irrigated clay soils at Good hope and Aiselby farms, which did not correspond to the loading rates of Zn. They attributed this to the leaching of Zn by large volumes of irrigation water. During this study, the highest Zn concentrations were about 3-fold lower than the FAO cumulative maximum allowable guideline limit of 150 mg/kg. The leaching of Zn to the ground water could be expected and is supported by the clearly defined trend of increased concentrations with an increasing depth at the sprinkler-irrigated site.

For all the soil depths, the highest concentrations of Cr were obtained in the furrow-irrigated site with an average of around 18 mg/kg with no major variation with depth. The values obtained are in line with the native concentration of Cr in a natural soil (Landon 1991). There was a significant difference between the control and the furrow-irrigated site at all depths. Unlike other trace metals discussed before, Cr is relatively immobile. In the furrow-irrigated site, there was evidence of a slight accumulation at the 0-60 cm depth. Higher concentrations in the irrigated sites than the control were attributed to wastewater irrigation and to the high concentrations of Cr in the ponds effluent. However, since the application rates at the two sites could not be identified, the higher Cr level in the furrow-irrigated site compared to the sprinkler-irrigated cannot be fully accounted for. According to WRC (1997), the recommended maximum permissible value is 80 mg/kg, which is much higher than the observed values in this study.

Table 10.7. Soil characteristics-metals.

Depth [cm]	Site	K [mg/kg]	Ca [mg/kg]	Cd [mg/kg]	Cr [mg/kg]	Cu [mg/kg]	Zn [mg/kg]
0-30	C1	191.1	40	7	12	6	50
0-30	C2	78	94	6	12	0.01	55
0-30	C3	78	18	6	14	0.01	40
0-30	C4	128.7	72	5	10	0.01	57
		118.95	56.00	6.00	12.00	1.51	50.50
0-30	F1	183.3	60	6	19	0.01	34
0-30	F2	191.1	130	5	20	0.01	30
0-30	F3	159.9	98	6	16	0.01	31
0-30	F4	214.5	138	5	16	0.01	36
		187.20	106.50	5.50	17.75	0.01	32.75
0-30	S1	163.8	64	6	13	0.01	35
0-30	S2	113.1	76	5	13	0.01	33
0-30	S3	78	82	5.5	16	0.01	38
0-30	S4	93.6	70	3	12	0.01	48
		112.13	73.00	4.88	13.50	0.01	38.50
30-60	C1	105.3	56	6	12	6	52
30-60	C2	81.9	102	6	11	0.01	50
30-60	C3	78	46	5	13	0.01	37
30-60	C4	93.6	50	6	11	0.01	56
		89.70	63.50	5.75	11.75	1.51	48.75
30-60	F1	120.9	66	7	20	0.01	47
30-60	F2	140.4	96	6	21	0.01	31
30-60	F3	136.5	104	7	17	0.01	27
30-60	F4	117	72	6	15	0.01	34
		127.28	84.50	6.50	18.25	0.01	34.75
30-60	S1	109.2	56	6	14	0.01	43
30-60	S2	66.3	44	6	14	0.01	47
30-60	S3	66.3	94	4	15	0.01	48
30-60	S4	78	44	3	12	0.01	47
		79.95	59.50	4.75	13.75	0.01	46.25
60-90	C1	93.60	72	7	11	13	58
60-90	C2	105.30	140	7	11	0.01	58
60-90	C3	70.20	56	6	14	0.01	65
60-90	C4	214.50	92	6	12	0.01	65
		120.90	90.00	6.50	12.00	3.26	61.50
60-90	F1	167.70	102	6	19	0.01	63
60-90	F2	179.40	82	7	18	0.01	51
60-90	F3	198.90	154	7	16	0.01	54
60-90	F4	175.50	102	7	14	0.01	47
		180.38	110.00	6.75	16.75	0.01	53.75
60-90	S1	156.00	92	6	12	0.01	66
60-90	S2	140.40	82	6	13	0.01	52
60-90	S3	140.40	122	6	16	0.01	70
60-90	S4	140.40	86	5	13	0.01	66
	Average	144.30	95.50	5.75	13.50	0.01	63.50

5 IMPACTS ON GROUNDWATER

5.1 *Methodology*

The farm did not have suitably located boreholes for monitoring the effect of wastewater irrigation on groundwater quality. For this reason, well drilling was carried out following the procedures in Chapman (1998) and Todd (1980). A 50 mm diameter steel hand auger was used to drill holes to a depth of about 3 m below ground level at five locations within the study area (Fig. 10.3). About 10 cm of the pipe was left above the ground to prevent runoff getting into the well. The wells were then capped to avoid direct irrigation water getting in. Two wells were drilled at a site grown to a sprinkler-irrigated pasture of Kikuyu and Star grass, one at the upper aquifer gradient part of the block and the other at the down aquifer gradient part of the block. The same was done on a furrow-irrigated pasture of the same grass species. Another well was drilled at the control site at the upper aquifer gradient of the area and taken as the control well. Soil samples collected at 0-30, 30-60 and 60-90 cm during drilling were sent to laboratories for trace metal and nutrient analysis. The auger holes were fitted with perforated 32 mm diameter polyvinyl chloride (PVC) pipes. Well logging was done during drilling at each site by 'finger' assessment. To determine the direction of groundwater flow and variation in the water level over time, groundwater levels were measured using a steel tape. On average, the groundwater water table was 1.5 m from the ground surface.

Groundwater samples were collected from monitoring wells using a suction plastic hand pump (Nalgene pump, Newark Enterprises). The pump has a flexible polythene extension tube, which was lowered to a depth 2.5 m into the PVC tube. In between sampling, the suction pump was rigorously cleaned 5 times with distilled water following the recommendations of Chapman (1998). Monitoring wells were purged for 10 minutes before sampling. The first 20 ml of the sample were used to rinse the sampling equipment and the sample bottle following recommendations of Quevauviller (1995).

The effluent and ground water samples were analyzed for Cd, Cu, Zn, and Cr (as total metals) at the IMR laboratory and for ortho P, nitrate, ammonia, FC, TC and TKN at the Water Quality Laboratory, Civil Engineering Department, UZ. Ca, Mg, Na and K at the ASL laboratory. EC and pH were determined in the field. During the preliminary survey of ground water samples Ni, Pb, Mn and Hg were tested as well, but concentrations were below the detection limit and these metals were not included in the study.

The effluent and groundwater characteristics were determined following Standard methods (1989). Total trace metals were determined using the USEPA 3050 method of acid ($HNO_3/HClO_4$) extraction followed by AAS (model: VARIAN TECHTRON SPECTRAA 50B [110 SOFTWARE]) employing an air-acetylene fuel. Electrical conductivity, nitrate and pH were determined in the field, using a filed unit (model:ELE Paqualab, 1996). Ammonium was determined by the direct Nesslerisation method, followed by a photometric determination at a wavelength of 425 nm (model: Spectronic 21D, SPECTRONIC Instruments). Ortho – P, FC and TC were determined by the same procedures, described in Chapter 9. TKN was determined by the macro-Kjeldahl method.

5.2 *Groundwater characteristics*

The ground water quality results are shown in Table 10.8. The comparison with water quality guidelines for different types of beneficial uses has been based on WHO (1996) and the South African guidelines for domestic use, livestock watering and irrigation (DWAF 1996 a, b, c). Risks to the environment have been evaluated, based on the Zimbabwe regulations (WWEDR 2000).

With respect to metals characteristics, the only problematic metal, which exceeds the recommended limits for all listed beneficial uses, is Cr. This element could be classified as posing a high environmental risk, as the measured concentrations exceed considerably the maximum permissible values. However, the concentrations at the control site are high as well, which could be attributed to natural soil conditions. There is a significant increase in the concentrations at the furrow-irrigated sites, which shows that the irrigation practice has an additional adverse effect, especially considering the high Cr concentrations in the effluent. Cd exceeds the stipulated WHO guideline for potable purposes, but is lower than the SA requirement. In general, all metals show a distinct spatial variation with a significant increase in the concentrations at the furrow-irrigated site. This should not be attributed to the difference in the irrigation methods, but with the transport of pollutants down-stream the main gradient of the aquifer and the fact that this area, which is the lowest of all irrigated blocks, receives pollution transported from up-gradient areas in addition to the one released at this specific block. Observed concentrations of Cd and Zn pose a low to medium environmental risk.

In respect to pH there is no significant difference but in general the ground water is in the lower range of the neutral zone. EC values show a distinct difference between the control and irrigated blocks. However, the measured values are lower, when compared to the results in Chapter 9, where a sludge and effluent mixture is applied. According to this parameter, the aquifer water could be used for livestock watering only.

The nutrients variations in ground water quality show clear signs of pollution from the irrigated sites. These concentrations at the control site are very low. Ortho-P shows a maximum value at GW1, which is the point, receiving the highest hydraulic load, with decreasing concentrations along the gradient. Ammonia shows a similar distribution, with well-pronounced higher concentrations at the sprinkler-irrigated site. This could be explained by the creation of anaerobic conditions at this block due to high hydraulic and pollution loads, which contribute to the mobilization of these elements at these specific sites, but their movement within the aquifer is relatively low. In contrast, nitrate variations show a gradual increase from G1 to G3, where the highest concentrations were measured and a considerable decrease at GW4, which could be explained by the higher mobility of this element within the aquifer and in the direction of the main flow. The observed TKN concentrations are very high, and even at the control point, a concentration of 2.5 mg/l was measured, which could not be explained and might be due to a systematic error during the testing procedures, but the pollution pattern with respect to this element is pronounced as well. The environmental risks with respect to nitrate and phosphorous could be evaluated as low to medium, while ammonia concentrations exceed the high hazard values up to two times.

The ground water at Imbwa farm shows an indication of microbiological pollution as well, with FC in the range of 20-50 counts in the irrigated sites. Again, the sprinkler irrigated site showed higher counts. The TC showed the same trend. A point to be noted is the relatively high difference between FC and TC, which could be associated with the presence of animal excreta on the pastures. Westcot (1997) reported that cattle feces contain 0.23×10^6 indicator micro-organisms/g and that a single cow can contribute 5400×10^6 of indicator organisms/24 hours. This could have had an effect on groundwater close to the watering points. Under a system of rotational grazing in open pastures, such a type of contribution could be minimal. At present, the ground water has not being used for beneficial purposes, but the evidence of fecal contamination should be recorded in terms of future use. Under normal uniform conditions, Papadopoulos (1995) reported that the horizontal travel of microorganisms rarely exceed 20 m except in karstic formations. Todd (1980) observed that most pollutants (except those which are conservative) tend to be reduced in concentration with time and distance traveled, through attenuation processes such as infiltration, sorption and chemical processes, microbiological decomposition and dilution with the native groundwater.

Thus, it could be considered that the microbiological contamination observed during this study, does not pose a significant public health hazard, but it does not allow the ground water to be used for direct potable consumption. The level of contamination in terms of fecal coliforms observed was much lower compared to Crowborough farm (Chapter 9). From an environmental perspective the ground water is considered safe with respect to FC.

6 CONCLUSIONS

WSPS proves to be a viable option for wastewater treatment in the region and the effluent produced is suitable for irrigation. Under proper conditions of design and operation, this treatment option could cause problems with respect to an increased soils' salinity and, in some cases, ground water pollution and toxic metals contamination. However, in order to reduce at a minimum level such risks, the beneficial reuse of the effluent should be considered during the design stage and the WSPS should be designed in conjunction with the corresponding irrigation facility, after a thorough examination and careful consideration of the geological and climatic conditions, available and future quantities and qualities of effluents and types of crops to be irrigated. The design should include all transport and storage facilities, as well as, instructions

Table 10.8. Ground water characteristics.

Parameter		Control	Gw1	Gw2	Gw3	Gw4
Metals [mg/l]						
Cd	1	0	0.020	0.020	0.030	0:070
	2	0.002	0.030	0.040	0.040	0.060
	3	0	0.010	0.010	0.040	0.060
	Mean	0.001	0.020	0.023	0.037	0.063
Cr	1	1:020	1:180	1:380	1.800	3.020
	2	0.970	1:020	1:300	2.100	2.800
	3	1:060	1:230	1:240	1.600	3.000
	Mean	1.020	1.140	1.310	1.800	2.940
Cu	1	0	0:030	0:010	0.020	0.440
	2	0	0.010	0:050	0.040	0.480
	3	0	0.010	0:040	0.070	0.400
	Mean	0	0.020	0.030	0.043	0.440
Zn	1	0	0.700	1.200	0.800	1.600
	2	0.005	1.000	1.230	1.300	1.450
	3	0	0.800	1.400	1.080	1.800
	Mean	0.002	0.830	1.280	1.060	1.620
Nutrients [mg/l]						
P	1	0	1.200	0.900	0.500	0.600
	2	0	1.800	1.000	0.600	1.200
	3	0	2.000	0.900	0.300	0.800
	Mean	0	1.670	0.930	0.470	0.870
NO_3	1	0	2.800	3.500	5.600	2.800
	2	0.010	3.000	4.400	4.700	3.200
	3	0	2.000	4.600	8.800	3.000
	Mean	0	2.600	4.170	6.370	3.000
NH_4	1	0.200	4.500	2.500	0.200	1.300
	2	0	6.800	3.200	0.600	2.100
	3	0.020	3.700	3.000	0.500	1.700
	Mean	0.017	5.00	2.900	0.430	1.700
TKN	1	2.300	19.600	21.400	10.600	11.500
	2	3.200	21.000	18.500	9.500	12.000
	3	1.900	16.900	20.000	11.200	9.000
	Mean	2.470	19.170	19.970	10.430	10.830
PH [-]	1	5.650	5.950	6.030	5.880	5.880
	2	6.000	6.120	6.300	6.000	6.400
	3	5.430	5.980	5.990	6.230	5.880
	Mean	5.690	6.020	6.110	6.040	6.050
EC [μS/cm]	1	540	800.00	670.00	680.00	645.00
	2	546	820.00	678.00	693.00	678.00
	3	500	815.00	654.00	675.00	652.00
	Mean	528.67	811.67	667.33	682.67	658.33
Indicator organisms [counts/100 ml]						
Faecal coliforms	1	2	44	31	20	28
	2	0	31	19	17	24
	3	0	28	22	14	16
	Mean	1	35	24	17	23
Total coliforms	1	8	450	378	168	210
	2	3	420	400	180	189
	3	5	412	300	128	216
	Mean	6	428	360	159	205

for operation and maintenance of both the WSPS and the irrigation system. The same applies for cases, when the irrigation system has to be designed for a beneficial reuse of the effluent from an existing WSPS.

The recommendations made in Chapters 8 and 9 are applicable for the case of irrigation with ponds effluent as well, especially in relation to the monitoring requirements and source control of metals in wastewater. A specific point with respect to irrigation with effluents, which should be considered, is related to the regulatory instruments and guidelines, which usually do not limit the application loads with respect to nutrients, which are considered to be beneficial for the crops or the irrigated plants. However, this specific case study, as well as the studies described previously, show that excessive nutrients concentrations could pose environmental hazards and are a potential source of diffuse pollution.

With respect to the impacts of long term irrigation with ponds effluent at the Imbwa pasture, it could be mentioned that:

- The effluent quality was substandard, showing an inadequate functioning of the WSPS and leading to high pollution loading. In addition, the irregular practice of irrigation was observed with selected portions of the pasture irrigated more intensively, compared to others, due to equipment breakdown.
- The major benefits of a 25-years period of irrigation with ponds effluent were the enrichment of the soil with essential plant nutrients especially nitrogen, phosphorus and potassium. It also raised the pH of naturally acid soils, thus reducing the need for lime application and the risks of trace metal phyto-toxicity.
- The major drawback observed during this study was associated with the transport of contaminants of wastewater-origin down the soil profile, resulting in elevated levels of trace metals, nutrients and coliform bacteria in the underlying groundwater. The major pollutant constituents of concern were ammonia, ortho-P, Cr and Cd. With respect to metals, FC and phosphorous, the contamination levels were lower, compared to Crowborough farm, which was irrigated with an effluent and sludge mixture.
- In order to prevent groundwater pollution in the future, the farm and treatment plant management should consider the changed hydraulic loads due to the new treatment plant of the town, together with rehabilitation of the WSPS, proper maintenance of the irrigation system, adequate monitoring program and the enforcement of municipal by-laws in terms of the prevention of discharges of high metals concentrations to the sewer system.

Acknowledgements – The authors would like to thank the management of the WREM program, through the "Collaborative Program for Capacity Building in the Water Sector in Zimbabwe and the Southern Africa Region", jointly executed by the Civil Engineering Department-UZ, IWSD and IHE-Delft, for the financial support offered during this study. To the technical staff and the management of all laboratories involved – thanks for their support and understanding.

REFERENCES

Ayers, R. S. & Westcot, D.W. 1985. Water quality for irrigation. *F A O, Irrigation and Drainage Paper*, 29.

Chapman, D. 1998. *Water Quality Assessments A Guide to Use of Biota, Sediments and Water in Environmental Monitoring*. Second Edition. London: Spon Press.

Degremont 1991 *Water Treatment handbook, 6th edition*, Vol.1, Paris: Lavoisier Publishing.

DWAF (Department of Water Affairs and Forestry, South Africa). 1996a. *South African Water Quality Guidelines, 1: Domestic Use*. 2nd edition. Pretoria, Government Printer.

DWAF (Department of Water Affairs and Forestry, South Africa). 1996b. *South African Water Quality Guidelines, 5: Livestock Watering*. Pretoria, Government Printer.

DWAF (Department of Water Affairs and Forestry, South Africa). 1996c. *South African Water Quality Guidelines, 4: Irrigation*. Pretoria, Government Printer.

Fetter, C.W. 1984. *Applied Hydrogeology*. 3rd Edition. Englewood Cliffs: Prentice-Hall.

Henning, B. J.; Cinnamon, H. G. & Averring, T. A. S. 2001. Plant-Soil Interactions of Sludge-Borne Heavy Metals and the Effect on Maize (*Sea Mays* L.) Seedling Growth. *Water SA* Vol. 27,. 1. Pretoria: Water Research Commission.

Horan, N.J. 1990 *Biological wastewater treatment systems – Theory and operation*. Chichester: John Wiley & Sons.

Hranova, R. 2003. Treatment of Municipal Wastewater in Anaerobic Ponds – Seasonal Performance Evaluation and Design Alternatives" *Proceedings of the 10th World Congress AD 2004 – Anaerobic bioconversion...answer for sustainability, 29th August-2nd September 2004, Montreal, Canada.* Vol.1 212-218.

JICA, 1997. The Study of Water Pollution Control in the Upper Manyame River Basin in the Republic of Zimbabwe, *Final Report Vol.2. MLGRUD,&, Nippon Jogesuido Sekkei Co. Ltd/Nippon Koei Co.Ltd.* Harare: City of Harare.

Khouri, N., Karlbermatten, J. M., Bartone, C. R. 1994. Reuse of Wastewater in Agriculture: A Guide to Planners. *Water and Sanitation Report No. 6-UNDP-World Bank Water and Sanitation Program.* New York: UNDP.

Landon, J. R. (Ed). 1991. *Booker Tropical Soil Manual A Handbook of Soil Survey and Agricultural Land Evaluation in the Tropics and Subtropics.* Oxon: Booker Tate Ltd.

Mara, D. & Cairncross, S. 1989. *Guidelines for the Safe Use of Wastewater and Excreta in Agriculture and Aquaculture. Measures for Public Health Protection.* Geneva: WHO.

Metcalf and Eddy Inc. 1991. *Wastewater Engineering Treatment, Disposal and Reuse.* New York: McGraw-Hill.

Mikkelsen, R. L, & Camberato, J. J. 1995. Potassium, Sulphur, Lime and Micronutrient Fertilisers. In: Rechgil J. E (Ed). *Soil Amendments and Environmental Quality, Agriculture and Environment Series.* Boca Raton, Flowda: CRC Press, Lewis Publishers.

Mtetwa S 1998 Waste stabilisation ponds and their problems in Zimbabwe in Workshop *Biological Nutrient Removal In Wastewater Using Aquatic Plants*, Harare, Zimbabwe.

Muchovej, R. M.C. & Rechgil, R. E.1995 Nitrogen Fertilisers. In: Rechgil J. E. (Ed). *Soil Amendments and Environmental Quality, Agriculture and Environment Series.* Boca Raton, Flowda: CRC Press, Lewis Publishers.

Nyamangara, J. & Mzezewa, J. 1999. The Effects of Long-term Sewage Sludge Application on Zn, Cu, Ni and Pb Levels in a Clay Soil under Pasture Grass in Zimbabwe. *Agriculture, Ecosystems and Environment* UK: Elsevier Science Ltd.

Nyamangara, J. & Mzezewa, J. 2000. Effect of Long-term Application of Sewage Sludge to a Grazed Pasture Grass on Organic Carbon and Nutrients of a Clay Soil in Zimbabwe. *Nutrient Cycling in Agrosystems*, Dordrecht: Kluwer Academic Publishers.

Nyamapfene, K. 1991. *The soils of Zimbabwe*, Harare: Nehanda publishers.

Oloya, T. & Tagwira, F. 1996a. Land Disposal of Sewage Sludge and Effluent in Zimbabwe. 1. Effects of Applying Sewage Sludge and Effluent on Elemental Accumulation and Distribution in the Soil Profile. *The Zimbabwe Journal of Agricultural Research*, 34.

Page A. L., Miller, R. H., & Keeney, D. R (Eds.) 1982. *Methods of Soil Analysis. Part 2. Chemical and Microbiological Properties.* Madison: ASA, Inc.

Papadopoulos, I. 1985. *Wastewater Management for Agricultural Production and Environmental Protection in the Near East Region*, Technical Paper, Cairo: FAO.

Pennsylvania State University 1985. *Criteria and Recommendations for Land Application of Sludge in the North East.* Bulletin 851, Pennsylvania Agricultural Experiment Station, USA.

Pescod, M. B. & Arar, A. (Eds.). 1985. *Treatment and Use of Sewage Effluent for Irrigation*, Proceedings of the FAO Regional Seminar on the Treatment and Use of Sewage Effluent for Irrigation held in Nicosia, Cyprus 7-9 October 1985. Nicosia, Cyprus.

Pescod, M. B. 1992. *Wastewater Treatment and Use in Agriculture.* FAO Irrigation and Drainage Paper No. 47. Rome: FAO, UN.

Polprasert, C. 1996. *Organic Waste Recycling Technology and Management.* 2nd Edition. Chichester: Wiley.

Purves, W. D. 1976. Soil Physical Properties of Samples from Imbwa Farm. Chemistry and Soil Research Institute, Department of Research and Specialist Services, Causeway, Harare.

Quevauviller. P. 1995. *Quality Assurance in Environmental Monitoring: Sampling and Sample Pretreatment.* Weinheim: VCH.

Rechgil J. E (Ed). 1995. *Soil Amendments and Environmental Quality.* Boca Raton, Flowda: CRC Press, Lewis Publishers.

Reed S., Middlebrooks E., & Crites R. 1988. *Natural Systems for Waste Management & Treatment,* New York: McGraw-Hill.

Reemtsma. T, Gnirβ, R. and Jekel, M. 2000. Infiltration of Combined Sewer Overflow and Tertiary Municipal Wastewater-An Integrated Laboratory and Field Study on Nutrients and Dissolved Organics. *Wat. Res.* Vol. 34, 4, 1179-1186.

Rubio, R. & Vidal, M. 1998. Quality Assurance of Sampling and Sample Pretreatment for Trace Metal Determination. In: Chapman, D. *Water Quality Assessments A Guide to Use of Biota, Sediments and Water in Environmental Monitoring.* 2nd ed. London: Spon Press.

Standard Methods for the Examination of Water and Wastewater 1989. 17th edn, American Public Health association/ American Water Works Association/Water Environment federation, Washington DC, USA.

Todd, D. K. 1980. *Groundwater Hydrology*. 2nd Edition. New York: John Wiley and Sons.

Thompson, J. G. 1968. Report on the Use of Municipal Sewage Effluent for Irrigation on Aiselby and Goodhope, Bulawayo. In: *The Department of Conservation and Extension/Southern African Regional Commission for the Conservation and Utilisation of Soil Symposium on the Reuse of Sewage Effluent.* Bulawayo, 23-24 July 1968, Harare: Government Printer.

Westcot, D. W. 1997. Quality Control of Wastewater for Irrigated Crop Production. *FAO Water Reports No. 10.* ROME: FAO.

WHO (World Health Organisation), 1996. *Guidelines for Drinking Water Quality, Volume 2: Health criteria and other supporting information.* 2nd Edition. Geneva: World Health Organisation Press.

Wild, A. 1995. S*oils and the Environment. An Introduction.* Cambridge: Cambridge University Press.

Withers, P. J. & Sharpley, A. N. 1995. Phosphate Fertilizers. In: Rechgil J. E (Ed). *Soil Amendments and Environmental Quality, Agriculture and Environment Series.* Boca Raton, Flowda: CRC Press, Lewis Publishers.

WRC (Water Research Commission) 1997 Permissible utilization and disposal of sewage sludge *Report TT85/97,* Pretoria: WRC.

WWEDR 2000. *Water (Waste and Effluent Disposal) Regulations*, Statutory Instrument 274 of 2000, Republic of Zimbabwe.

CHAPTER 11

Diffuse pollution of urban rivers – case studies in Malawi and Swaziland

R. Hranova, S. Nkambule & S. Mwandira

ABSTRACT: Diffuse pollution of two urban rivers has been presented in the light of two independent case studies executed during the rainy season of 2001. The first examines and evaluates the spatial river water quality variation (pH, EC, TC and TDS, BOD and COD) of Lilongwe River and tributaries, in its upper reaches, before it enters the capital city of Malawi – Lilongwe. The river is the major source of potable water supply of the city. Different types of land use patterns have been associated with the sampling locations chosen. Results show a considerable level of pollution with respect to TC only, associated mainly with informal settlements along the riverbanks and, possibly, with effluents from food industries. The second case study examines the spatial water quality variation (BOD_5, ortho-P, nitrates, sulfates, ammonia, chlorides, DO, EC, pH, T, TSS, TC and FC) of Mbabane River and tributaries, which flows along the capital city of Swaziland – Mbabane. It collects the runoff and municipal effluents of the City, and provides water for direct domestic and agricultural use of the rural population, living downstream of the city. Results show that the major source of diffuse pollution is associated with municipal effluent discharges due to the malfunctioning of the treatment plant. The major parameters of concern are nutrients, organic and bacteriological pollution. However, 20 km downstream the city, the water quality improves due to additional discharges by tributaries mainly, and partial self-purification.

1 INTRODUCTION

Urban development patterns in the Southern African region, and in many other developing countries, are characterized by rapid population growth and the formation of informal or semi-formal settlements without the necessary provisions for adequate infrastructure, and often without a basic water supply and sanitation structures. In numerous cases the existing wastewater treatment facilities are neglected and suffer from chronic deficiencies in the operation and maintenance practice, due to insufficient funding and lack of trained personnel. All these, create conditions for severe public health hazards, which are usually recognized by official authorities. However, the hazards associated with the pollution of the environment and the available water resources are not widely recognized and addressed. In numerous cases, informal settlements are formed in the vicinity of rivers and streams, which serve as sources for a potable water supply of urban population centers. In other cases, the downstream users of polluted rivers and streams, flowing through urban population centers are rural communities, who use the polluted waters directly, without any treatment. Thus, the uncontrolled urban development has negative impacts, not only in terms of social and economic conditions, but in terms of environmental and public health protection as well.

This chapter presents two specific case studies of spatial variation of river water quality and the impact of land use patterns in urban environments, typical for the region. The first case study investigates the water quality variation of Lilongwe River, which passes through the city of Lilongwe, the capital city of Malawi. The River serves as the main source of potable water supply to the city. The impact of different land use patterns along the upper reaches of the river has been investigated up to the abstraction point for the water treatment plant.

The second case study focuses on the spatial water quality variation of the Mbabane River, passing through the city of Mbabane, the capital city of the kingdom of Swaziland. The water quality of the river, passing throughout the city has been investigated at several locations, reflecting specific land use patterns, and a point downstream the city, where the river water is used for direct domestic and agricultural use.

Both studies have been conducted in an environment where no regular water quality monitoring of the river water quality has been performed previously. Also, in general, the existing water quality monitoring laboratories deal with selected parameters only and are orientated mainly to the control of potable water quality. Any other related data, concerning natural or effluents' water quality and quantity, is missing as well. Considering these conditions, and the fact that the studies have been performed within a limited time frame, they could be viewed as a preliminary investigation for the development of a more substantial monitoring program, and a water quality management policy for both cases.

2 LILONGWE RIVER, MALAWI

2.1 *About the country and the catchment area*

Malawi is a republic in Southern Africa, formerly the British protectorate of Nyasaland, bounded on the North by Tanzania, on the east by Lake Malawi (formerly Lake Nyasa). Malawi extends about 835 km to south and varies in width from about 80 km to 160 km. The total area of the country is 118,484 km^2; nearly one fourth of it is water surface, mainly Lake Malawi and three smaller lakes. Malawi is a land locked country located between 16° S, and 9° N and between 36° W and 32° E. A schematic map is shown in Figure 11.1.

Lilongwe is the capital city located in the central part of Malawi. Lilongwe district is located between longitudes 14.5^o and 13.5^o, and latitudes 33.5^o and 34.5^o. The district has the altitude ranging from 1000 m to 1500 m above sea level.

The Dzalanyama catchment is located on the south-west of Lilongwe district. The rivers of Lilongwe, Lisungwi and Likuni originate from this catchment. The catchment area for Lilongwe River is about 1800 km^2.

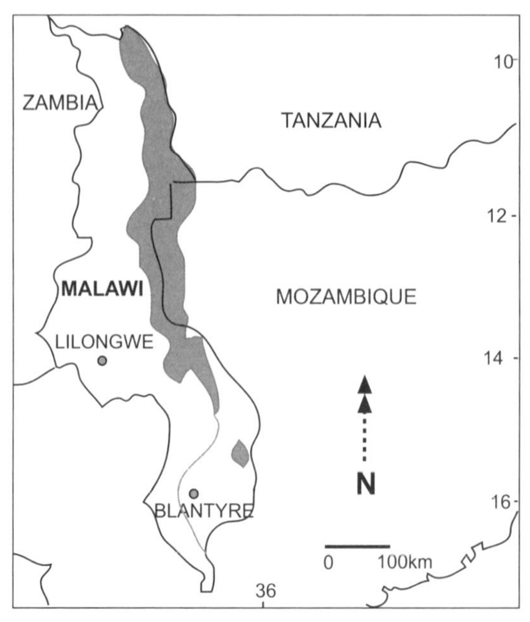

Figure 11.1. A schematic map of Malawi and bordering countries.

2.2 *Water quality management – institutional aspects*

The water sector in Malawi comprises several levels of responsibility that range from national management policy to the construction, operation and the maintenance of water supply structures and sanitation facilities. These levels of responsibility are assigned to different government institutions and parastatal organizations. The Ministry of Water Development is the key government institution responsible for water management services and consists of four major sections – the Hydrology, Hydrogeology, Rural Water Supply and Sanitation, and the Water Quality Sections. Other important organizations in the management of water resources in Malawi are the National Water Resources Board and the local Boards, namely the Blantyre and Lilongwe Water Boards, and the Northern, Central and Southern Regions Water Boards.

The Hydrology Section is responsible for the collection, processing, analysis, archiving and dissemination of hydrometric data. The section operates 171 gauging stations throughout the country. Data collection has been going on since 1896. The major constraint in the collection of hydrometric data is the poor state of equipment used at the gauging stations.

The Hydrogeology Section is responsible for the monitoring, assessment, management and exploration of groundwater resources. However, the Section is almost exclusively engaged in borehole construction and borehole maintenance activities. Groundwater monitoring, in terms of aquifer zoning and water quality assessment/management, are therefore basically non-existent. By the year 2000, there were more than 20,000 boreholes countrywide.

The Water Quality Section is responsible for the monitoring and assessment of the physical, chemical and biological aspects of the water quality. The Section also conducts inspections on pollution levels in effluent and wastes discharged in public waters. The major setback to the assessment of water quality is the inadequacy of laboratory equipment for conducting the analyses, the insufficient training of the laboratory staff, and an acute shortage of equipment for use in water sampling programmes. The analyses of various water quality parameters are therefore carried out using the little equipment available and in conjunction with other laboratories.

The Water Resources Boards oversee policies as outlined in the Water Resources Act and other regulations. Furthermore, the Boards collaborate with other institutions responsible for the environmental monitoring and control on water resource management issues. The National Water Resources Board comprises of the following members:

- The chairperson;
- The Secretary for the Ministry of Water Development;
- The Secretary to the Office of the President and Cabinet;
- A representative from the Water Resources Division of the Ministry of Water Development;
- The secretary responsible for Commerce, Trade and Industry;
- The Chief executive of the Electricity Supply Commission of Malawi, and;
- Four other members representing the public.

Apart from the Water Quality Section of the Ministry of Water Development, the five water Boards have set up their own Water Quality sections, which look into the day-to-day monitoring activities of the water from its source to the consumer. The Lilongwe Water Board has a well-developed water quality laboratory and is able to carry out both chemical and biological analyses including trace elements.

The enforcement of laws governing water resources in Malawi rests with the Water Resources Boards. These laws are outlined in the Water Resources Act along with Pollution Control and other regulations. The Act provides for the granting of water rights such as diversion, storage, abstraction and use of public water in specified quantities. Under the Act, the Minister responsible for water resources is empowered to punish all users of water that contravene the regulations. To ensure that the Water Resources Act is complied with by all users of public water resources, the Board has been entrusted with the authority to monitor the adherence to water regulations without discrimination. Although the Act calls for the implementation of punitive measures against users that contravene the regulations, the Board has been seen to be ineffective in punishing offenders and conducting monitoring works. For example, no punitive measures have been taken against people who have opened maize gardens in the water catchment areas. The local Water Boards help in the enforcement of laws and regulations through bylaws. Each water board enforces its own by-laws that cover the whole range of water resources management practices among which is the management of water quality.

2.3 *The study area*

The study area is located in the upper reaches of Lilongwe River, between the Dzalamayama Forest Reserve and the water abstraction point for the city of Lilongwe. The area has got three main rivers that flow through out the year. The Lilongwe River is the main river and the other two – Lisungwi and Likuni Rivers – are its tributaries (Fig. 11.2).

Other smaller streams only flow later in the year when the rainfall has started. There is no recent data, with respect to water flows, recorded by the Hydrology section of the Ministry of Water Development for neither the Lilongwe River nor its tributaries. SAFEGE (2001) gives the average flows for the three rivers as being 0.995 m^3/s for the Lilongwe River after the Likuni confluence and 0.24 m^3/s for Likuni tributary.

Most of the catchment area is gently or very gently sloping, and is intensively cultivated. Almost all the natural forest has been cleared for agriculture. The only significant areas of forest are found in the Dzalanyama Forest Reserve towards the international border with Mozambique, upstream of the study area.

Lilongwe district like all other areas in the central region of Malawi has a dry season and a rainy season. The rainy season lasts no more than six months from November to April. There is no rainfall in the last half of the year. Data from the department of Metrological Services shows that the average rainfall in the district is of the order of 600-1200 mm per year. Average temperatures range from a minimum of 6°C in winter to a maximum of 30°C in summer.

The land use patterns in the study area and the location of the sampling points, denominated with capital letters from A to F, are presented in Figure 11.2. The Lilongwe River drains the Dzalanyama catchment. The drainage area is about 1800 km^2. The Katete, Lisungwi, and Likuni Rivers are tributaries of the Lilongwe River and originate within the catchment. A large portion of the land has been used for agriculture with different types of crops grown.

Along the Lilongwe River, two dams have been constructed namely Kamuzu Dams I and II (Dam I and Dam II). Dam I was constructed in 1966 and Dam II was constructed in 1999. The primary purpose of these dams is to provide the water supply of the city of Lilongwe. The flow downstream of Dam II is regulated to meet the demands of the downstream users. The capacity of Dam I was initially 4.5 Mm^3, but has been reduced significantly due to deforestation and reduced land cover that has exposed the soil to erosion. The Dam II capacity is 19.8 Mm^3 with a 1 in 200-year return period. Downstream of the dams, the river passes through an area of rural pattern settlements. Most people in the area rely for their survival on subsistence agriculture. Few families are completely self-sufficient in food production. Crops grown in the area include: tobacco, paprika, vegetables, maize and groundnuts. Other typical economic activities in the area include brick making and the brewing of beer in the form of informal sector economic activities, as described in Chapter 5.

Informal villages and squatter settlements have emerged along the riverbanks during the last decade, and their presence is pronounced between Dam II and the abstraction point (E) of the Lilongwe Water Board Water Treatment Works (LWBWTW), which is the main source of the potable water supply for the City. The settlements are not provided with basic sanitary facilities, and their liquid and solid wastes are discharged straight into the river or its tributaries. The settlements have been developed along the riverbanks, in most places not more than 500 m away from the rivers, and this is understandable, considering the lack of basic water supply structures. The most notable ones are Likuni and Chinsapo settlements. These areas are heavily populated with much of the housing being unplanned.

Further downstream of Dam II, on the eastern riverbanks Katete livestock farm is located, with the land sloping towards the river. There used to be a good ground cover because of the forest in the 100 m-buffer zone to the river, but it is no longer there, due to deforestation. This area also has a few village settlements that are located very close to the river. The western side is an area that is developing fast with unplanned

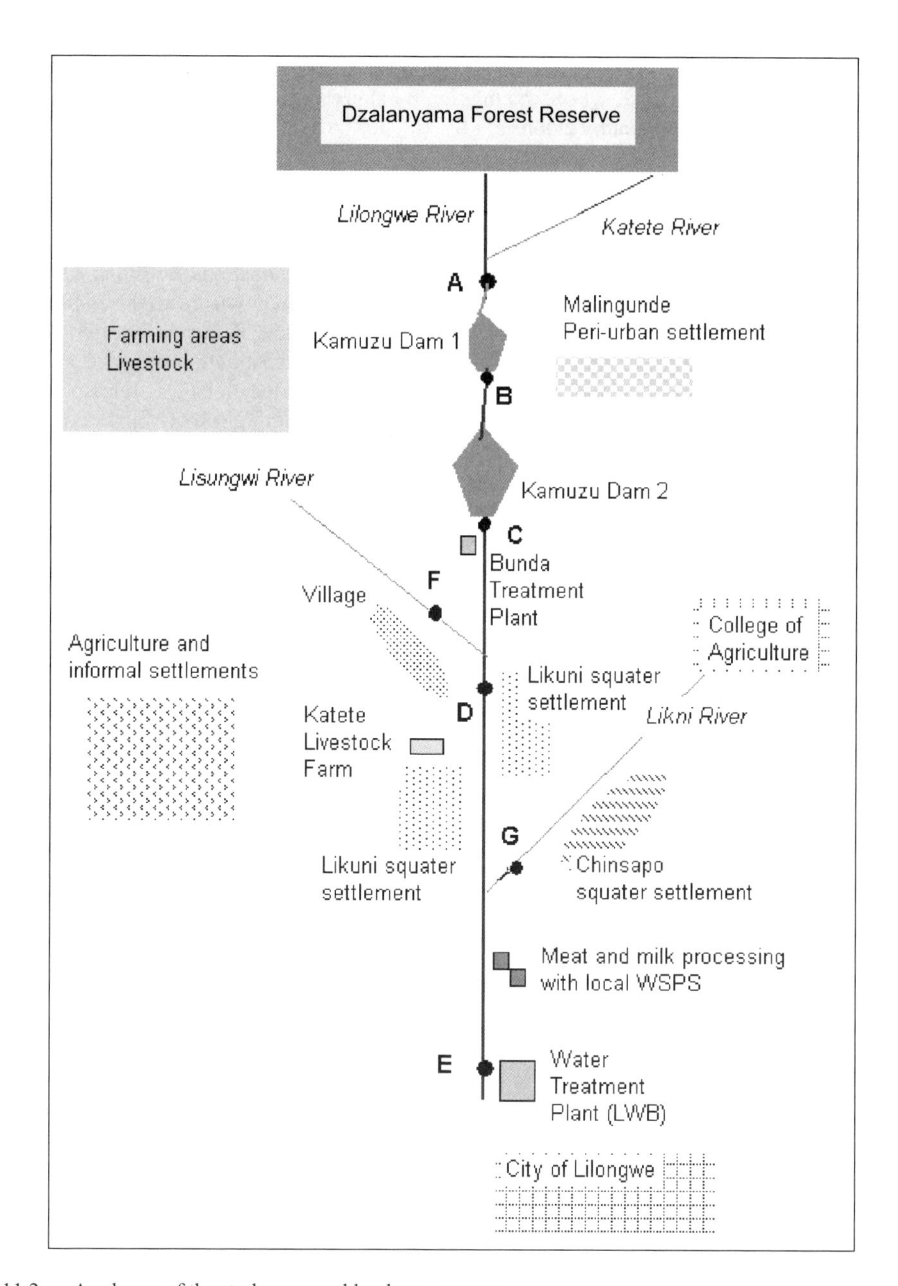

Figure 11.2. A scheme of the study area and land use patterns.

(squatter) settlements with marginal sanitary facilities. The widely used sanitation facilities are pit latrines that are overloaded, each pit latrine serving 15 to 20 persons. The preferred sanitation technique, applied in institutional developments of rural type (clinics and schools), and in selected houses, is the septic tank.

Just upstream of point E, a few small-scale industries are located, including a dairy processing plant and a meat cold-storage facility. The last two have their local wastewater treatment plant consisting of waste stabilization ponds system (WSPS). It is poorly maintained, the screens and grit chambers are clogged, and aquatic vegetation is invading the ponds, in addition to a large amount of sludge accumulated.

The abstraction point for the LWBWSTP is located about 1 km downstream of this area. The plant consists of a conventional treatment scheme, where coagulation by alum is provided, followed by sedimentation, rapid sand filtration and disinfection by chlorine.

2.4 *Methodology*

Sampling points were established at several locations along the Lilongwe River and its tributaries in places where the conditions allowed accessibility for sample collection. The locations are shown in Figure 11.3 and were given identification symbols from A to G. Pont A was located downstream the confluence of the Lilongwe and Katete Rivers and was selected as a control point. Points F, D and G were located at the bridges on the corresponding rivers, and point E was located near the abstraction structure, collecting the river water to be treated in the LWBWTW. The choice of sampling locations was related to suspected sources of pollution. Sampling was carried out twice a month on a fortnight basis for four months from the month of November to February 2001. Several additional samples were collected during January 2002. Three grab samples were collected at different depths to form a depth-integrated representative sample at each sampling point. Samples were collected and preserved as required and recommended in the Standard Methods (1994). During sampling, efforts were made to avoid areas that showed evidence of localized stagnation, erosion and other visible disturbances. Lilongwe Water Board provided equipment for the

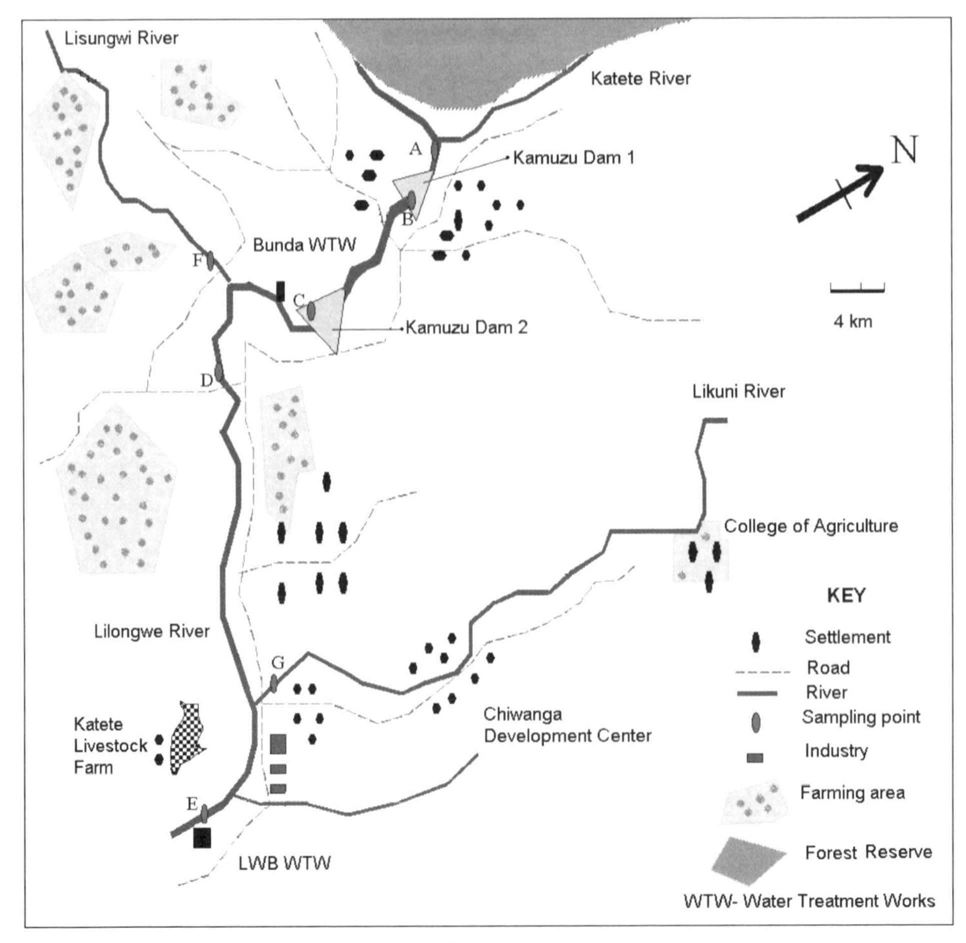

Figure 11.3. Map of the study area and sampling points' location.

sampling and the transportation to all the sampling points. Samples were preserved in a cooler box with ice cubes to maintain a low temperature during the transportation to the laboratories.

Samples were analyzed at the LWB laboratory and the Central Water Laboratory. The LWB laboratory carried tests for pH, EC, TC and TDS. The Central Water Laboratory carried out the tests for nitrates, phosphates, BOD, and COD, according to the Standard Methods (1994).

The results were compared with the water quality standards in Malawi and the WHO guidelines on drinking water quality, because many of the people in the study area use the water in the rivers directly without any treatment. Samples for nutrient analyses were filtered through a 0.45 nm filter paper to allow the determination of only dissolved nutrients and were analyzed within 24 hours of sampling. Samples for TC were collected in sterile bottles.

The analysis of samples was done within 6 hours of sampling. Nitrates and phosphates were measured using the Ultra Violet Direct Reading method by a multipurpose HACH DR/4000 UV – VIS Spectrophotometer, with a range of 0 – 10.2 mg/l. The "Potassium permanganate in sulfuric acid titration method" was used to determine COD. EC and pH were measured in the field using GIL International (model 33) meters.

The land use patterns and new developments were established through field reconnaissance surveys in the area and the information obtained was compared with existing maps. In addition, interviews were conducted with officials that are responsible for such developments and practices. They were requested to cite problems they are facing in managing their respective departments. These include the Lilongwe City Council, the Ministry of Lands and Housing, the Department of Forestry, the Ministry of Agriculture and Natural resources. The information with respect to existing land patterns was obtained in the form of maps or any other documentation that was in existence. These were obtained from the Department of Surveys and Ministry of Lands and Housing.

Statistical data analysis was performed by the use of standard EXCEL statistical tools.

2.5 *Results*

Results obtained during the period of study show significant pollution in terms of EC and TC only (Fig. 11.4 and 11.5). The values obtained with respect to pH are within the normal range of 7 to 7.7 (variation of SD from 0.1 to 0.7) showing no signs of pollution and no spatial variation.

The results with respect to dissolved nutrients (both phosphate and nitrate) vary within the range of 0.01 mg/l to 0.04 mg/l. No significant spatial variation was found. The magnitude of the dissolved nutrients concentrations observed is very low; showing no impact of the different land use patterns on the river water quality in the study area. It could be accepted that the measured concentrations are close to the background pollution values. This suggests that the informal agriculture in the region does not have a considerable impact on the river water quality in respect to nutrients.

The spatial variation of EC and TC concentrations along Lilongwe River is presented on Figures 11.4 and 11.5 respectively. It shows a considerable increase in both parameters after the confluence of the Lisungwi River and the Likuni River.

The EC values increase from about 100 mg/l at the control point, which could be considered to be the background pollution value, to about 500 mg/l at point G. Although this value of EC does not pose any imminent hazard to public health in terms of drinking water quality requirements, it indicates a considerable pollution and the possible presence of other pollutant constituents. At point E the measured concentration was lower due to the dilution of the constituent. The same trend of spatial variation was shown with respect to TDS values, confirming the indication of pollution with respect to dissolved ions.

The TC variation shows a very high level of pollution in terms of bacteriological contamination and a high risk of diseases' spreading, if the water is consumed without treatment. The trend of the spatial variation is similar to the EC variation, with the highest mean concentration of 5812 mg/l at site E. In this case, it is possible that the additional influx of coliforms from the informal settlements, the industrial site and the livestock farm is considerable and could not be attenuated by the increased flow rates. Similar results of high bacteriological contamination have been reported for other rural areas in Malawi (Ndolo et al. 2002).

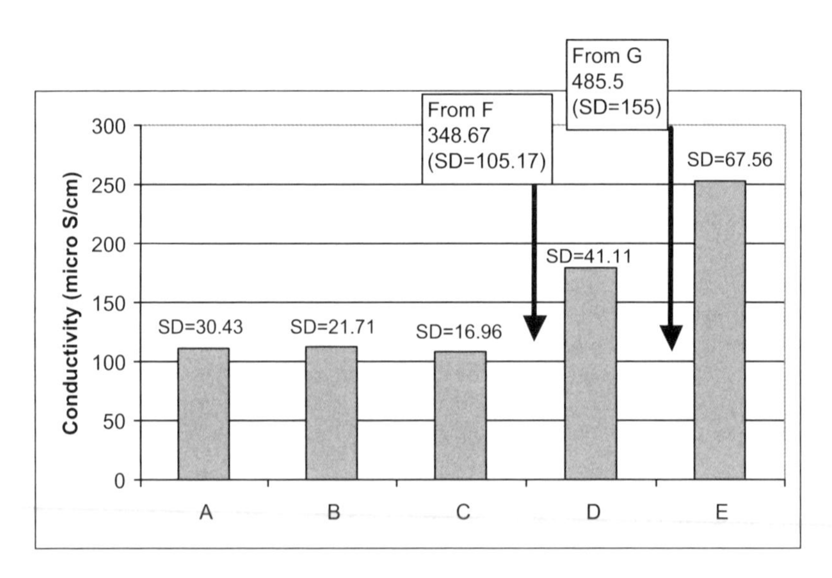

Figure 11.4. Spatial variation of EC along the Lilongwe River.

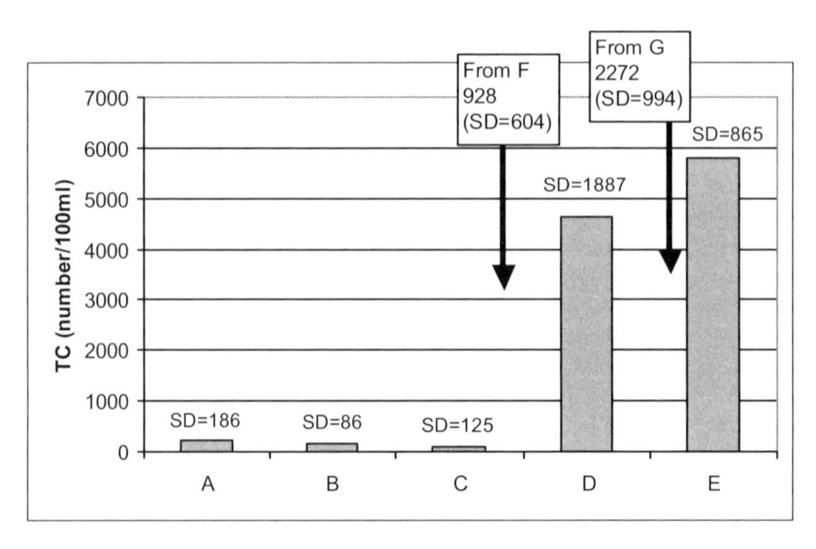

Figure 11.5. Spatial variation of TC along the Lilongwe River.

The results with respect to COD and BOD are relatively low and do not support the findings with respect to TC. COD values along the river vary within a range of 16-31 mg/l (variation of SD = 6-25 mg/l) and BOD values vary from 2 mg/l to 5 mg/l. No significant spatial variation was found among the different sampling locations. Considering the high TC and EC values measured, it could be expected that they would be supplemented by concentrations higher than the measured ones. This discrepancy needs additional investigation. It might be due to a consistent error in the laboratory analytical procedures. Such a possible scenario would require a thorough examination of the existing laboratory procedures and the implementation of a reliable quality assurance program in the corresponding laboratories. UN/ECE (1996) could provide useful information in this direction.

2.6 *Discussion*

The land use patterns along the stretch of the river vary, where the conditions along the upper reaches of the river, near the Dzalanyama Forest Reserve and the Kamuzu Dams, could be evaluated as pristine. The agricultural areas downstream of the Dams do not have an adverse impact on the river water quality and this could be expected, as informal and small-scale agriculture usually does not apply large amounts of fertilizers and pesticides, which could have an adverse effect on the river water quality. A high level of pollution found during this specific study, was significant with respect to TC only, and could be associated mainly with the pattern of informal and quasi-formal settlements along the riverbanks. It was not possible to account for the contribution of each specific area, but this rapidly spreading form of uncontrolled urban development could be a serious threat and if not controlled, inevitably, would lead to larger pollution loads in future. The industrial enterprises located just upstream the abstraction point of LWBWTW, together with the denser informal settlements development in this area contribute to the increase in the pollution constituents of the river water.

A point of serious concern is the recent trend of uncontrolled urban development along the upper reaches of Lilongwe River. Such a development is a direct contradiction to the recommendations for source control measures, aiming at the control of diffuse pollution problems. Informal settlements with no basic sanitation facilities create conditions for diseases' spreading and environmental pollution. Instead of creating new environmental buffer zones between settlements and riverbanks, the existing ones have been destroyed to create living space for the new-coming population.

The location of industrial enterprises in such proximity to the water supply source of the city is highly unacceptable and is another point of concern, which needs to be addressed with urgency. The recent study, due to its limitations, shows only a limited part of the whole picture of the water quality status of the river. The parameters studied, do not show a serious danger with respect to the water supplied to the city, as during the treatment process and after disinfection, the produced water should correspond to the requirements for potable use. However, a most thorough investigation and a continuous monitoring program could show the presence of pollutants, such as higher BOD or COD values, or some trace toxic elements released from the industrial enterprises, which could not be eliminated by the conventional process of the treatment plant. High organic pollution loads of natural water are of serious concern, when such water is used as a source of potable water, even after conventional water treatment. Such loads could be the bases for THMs formation after disinfection, which have carcinogenic and mutagenic effects, when present in excessive concentrations in potable water.

The results of this study show that the adverse effects of the pollution found in the river are affecting mostly the same people from the informal settlements, which generate it, as they are consuming the polluted water directly, without treatment. This finding should be accepted as a red signal for the managing authorities, which should expect serious public health hazards and the spreading of waterborne diseases in the areas. Despite the limitations of this investigation, several recommendations with respect to the managing authorities could be made, as listed below:

- Considering the understandable difficulties in the control of informal settlements establishing and spreading, the adverse effects of such a type of urban development could be alleviated by:
  - The provision of basic water and sanitation structures, such as ablution blocks, at convenient locations within the informal settlements;
  - The strict prohibition of settlement formation near the riverbanks;
  - The creation of environmental buffer zones between the settlements and the river;
  - In conjunction to the above-mentioned activities, awareness programs and educational activities should be implemented.
  - The development of Malingunde settlement in the vicinity to Kamuzu Dams is rather inappropriate and should be limited and gradually eliminated.
- The implementation of simple monitoring programs for the control of river water quality is also important, based on appropriate technical and laboratory facilities and robust and validated testing procedures, to provide a reliable and properly managed data.

- There is need to improve the flow of information amongst institutions that in one way or another can affect the water quality in the Lilongwe River. The information will assist in the development of an integrated water quality management strategy in the catchment of the Lilongwe River and will avert the possible adverse impacts emanating from the degraded water quality due to the improper land use development patterns. Institutions in the fields of Forestry, Land Husbandry, Agriculture, Town Planning and Social Welfare need to be involved in the management processes and the implementation of the different activities.
- The development and enforcement of water quality standards with respect to surface water quality could help in the implementation of activities to change the recent trend and to prevent a future increase in the pollution loads discharged to the river.

3 MBABANE RIVER, SWAZILAND

3.1 *About the country*

Swaziland is an independent monarchy in the Southern Africa, bounded on the east by Mozambique and in all other directions by South Africa, as shown on Figure 11.6. It has an area of 17,363 km^2. The western part is mountainous with an altitude reaching 1220 m, the central region is hilly with an average elevation of 610m, while the eastern region, known as the Low veld, has an altitude of 120-300 m and is bounded by the Lubombo Mountains. The climate ranges from tropical in the Low veld to temperate in the mountains, with the rainfall pattern typical for the region. The different ecological zones and corresponding rainfall figures are shown in Table 11.1.

The population is about 950,000 according to the 1997 census. The capital city is Mbabane, located in the High veld of the country and has an area of 8035 ha, with a population of about 60 000. Temperature

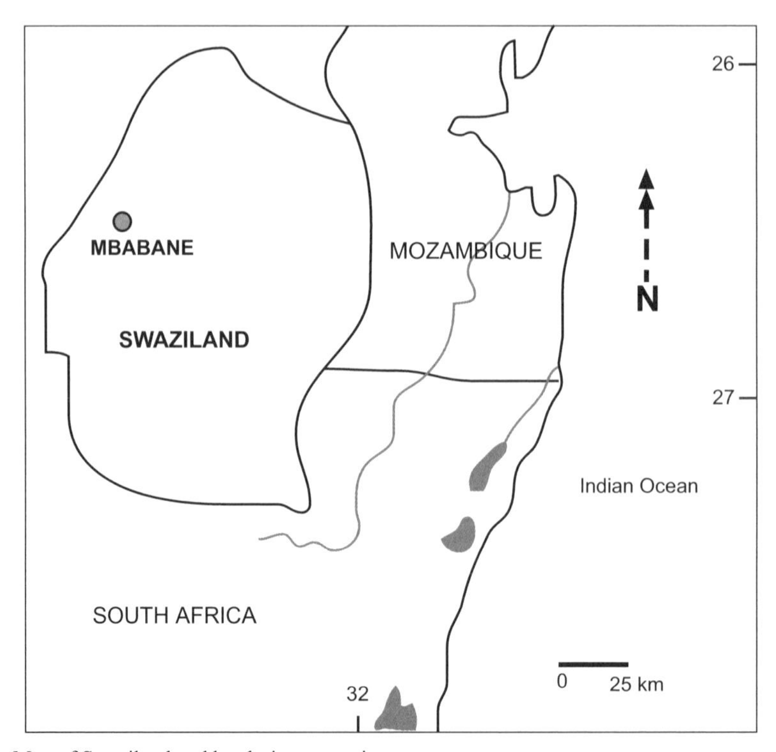

Figure 11.6. Map of Swaziland and bordering countries.

Table 11.1. Ecological zones and rainfoll data, Swaziland.

Zone	Altitude (m)	Average annual rainfall (mm)
High veld	1100-2000	1016-1524
Middle veld	300-1100	762-1016
Low veld	150-300	508-762
Lubombo	600-900	635-889

Source: Ministry of Natural Resources and Energy, Swaziland (Msibi 2001)

and rainfall vary markedly with season, as well as with altitude. In winter (May to August), rainfall is scarce and temperatures are low. The minimum temperature in the High veld could reach 4°C, although the average annual temperature is 10°C. The annual average temperature in the Low veld during the summer (November to February) is 29°C.

3.2 *The study area*

The Mbabane River originates at the eastern foot of the Nkoyoyo hills a few kilometres west of the city. It flows throughout the city (Fig. 11.7), collecting surface run off from residential areas such as Nkoyoyo, Sidwashini, Selection Park, Sandla, Manzana, Extension 3, and Mobeni flats. Other types of land use patterns include the commercial areas in the city centre and different type of small industries such as: Chemlog and Superchem (manufacturer of industrial and domestic cleaning, floor care material, chemical solutions etc); Rand Continental Tool (manufactures of paints); automotive service and filling stations; abattoir; and different types of food processing industries.

The river passes beside the very busy Mbabane bus terminus where people throw waste, fruit peels and bottles into the river, and along the Mbabane Market where traders wash their fruits and vegetables. After the market place and further downstream, the river flows through the industrial estate, where the main industrial activities are mechanical workshops, service stations and garages. Spilled used oil, grease and other chemicals from these small enterprises find their way into the river. From the eastern flank, there is Pholinjane stream – a tributary of the Mbabane River, which collects surface run-off from the Sidvwashini industrial estate. The river then flows south and links with Mvutjini River at the foot of Malagwane hill.

Gobholo treatment plant discharges into a small nearby stream, which joins Mbabane River a few kilometres downstream the discharge point. This study concentrates on about 30 km stretch from where the river originates (Nkoyoyo Hills) upto the Mvutjini area downstream of the city.

In the upper reaches of the Mbabane River, the sources of pollution are run-off from pastures and a few small areas of vegetable growing. The additional run-off from the areas of indigenous vegetation also enters the river. From general observation, the water quality upstream appears to be good.

In the vicinity of the city, there is a change of colour in the river. This is noticeable just after the river flows past the industrial and residential areas. In one of the residential areas (Sidvwashini), there is evidence of landslide, which eventually enters the river during rainy days. In general, the riverbanks and riverbed have been used for the disposal of solid wastes and garbage, which is visible at many locations and near the city centre. Further downstream, after the industrial site, there is evidence of foam, possibly from the chemical industries and also a film of oil and grease on the river from the mechanical workshops, service stations and automotive garages. Organic pollution from the abattoir could be an additional pollution source, which finds its way into the river. In addition, the sewage effluent from the Gobholo treatment plant also reaches the Mbabane River (Fig.11.7)

The river downstream is used for irrigation of crops (mainly vegetables), stock watering, domestic use (directly from the river without preliminary treatment) and recreation.

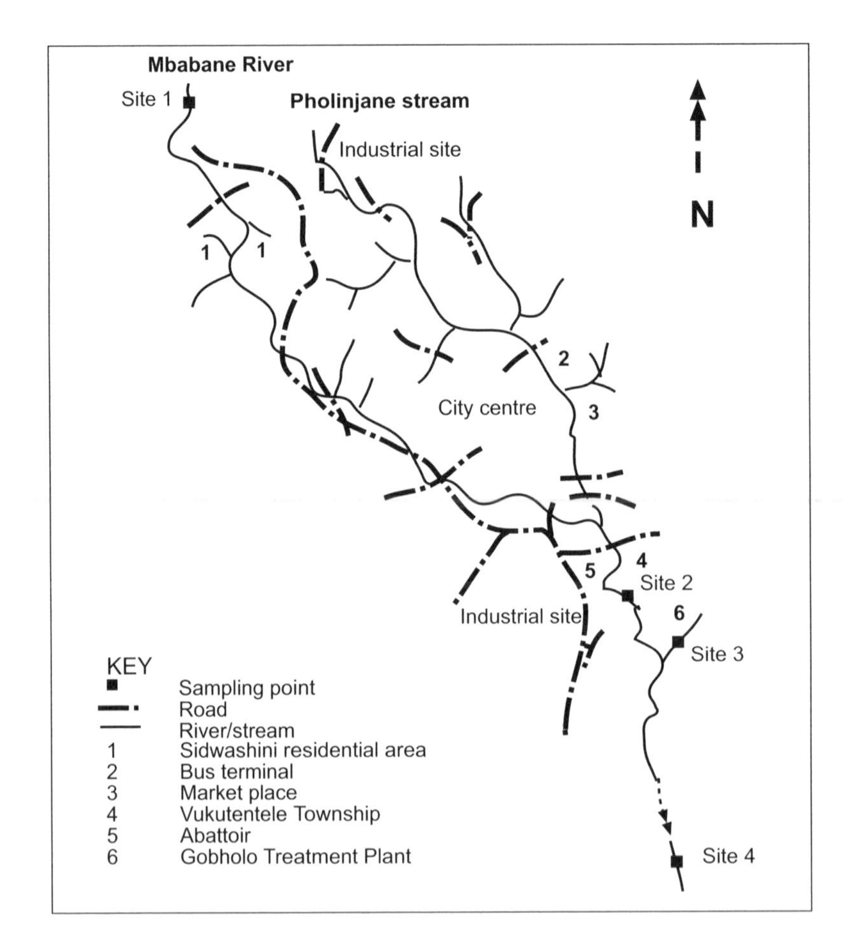

Figure 11.7. A schematic representation of Mbabane River and the study area.

3.3 *Water quality management – institutions and legislation*

The planning, implementation and operation of water resources development in Swaziland during 2001 was undertaken by five government departments and one parastatal organisation. Collectively, there are ten different divisions involved. The management of the water resource falls primarily under the Ministry of Natural Resources and Energy (MNRE) in which the water resources branch is responsible for monitoring stream flow, planning development activities, and controlling pollution. The Rural Water Supply Branch of the same institution is responsible for the rural water supply and sanitation. The Urban water supply and sanitation is looked after by a parastatal organization – the Swaziland Water Services Corporation (SWSC).

Several other Government Departments provide important input, in particular the Ministry of Agriculture and Co-operatives (MOAC), with the responsibility of assisting farmers, conservation works and small irrigation schemes. The Ministry of Health and Social Welfare recently established an Environmental Health Unit (EHU), with the role of assisting in controlling water pollution and minor spring protection schemes. The Ministry of Economic Planning and Development (MEPD) provides a channel for procuring external funding for rural water supply schemes. In addition, there are a large number of non-governmental organisations active in the water field. An important player in the water quality management practice is the Water Sector Committee, which was set up to oversee the drafting of the New Water Act.

Such a type of institutional arrangement could lead to a considerable overlapping and duplication of efforts. The institutional arrangements, with corresponding powers and responsibilities, which would be established under the new water legislation, would serve to overcome such a duplication of efforts. The new legislation would make provision for the establishment of a National Water Authority (NWA) and a Department of Water Affairs (DWA). The Director-General of the DWA will provide technical support to the NWA and help to co-ordinate the activities of the different boards, task forces, government departments and international water commissions.

The Swaziland legislation with respect to water quality control and regulation is included in the old Water Act, developed in 1967, which addresses point sources of pollution only. However, a new Water Bill has been developed in 1998, to amend this and to include a water quality objective approach, which during the period of study was in the process of approval by the Parliament. The water-quality monitoring program implemented to support the legislation has two different sections. The first one is envisaging the river water quality and one of its objectives is to determine the background pollution levels. The other section envisages effluent discharges and their control with respect to specified maximum pollution levels. The legislation focuses on industrial discharges and provides the possibility for a temporary exemption permit, which relaxes some of the requirements of the effluent regulations. Before such an exemption is granted, an environmental impact study should be carried out, which should be paid by the polluter. The legislation provides for offences and penalties in cases of non-compliance as well.

3.4 *Water supply and sewerage systems in Mbabane*

The City is provided with a reticulated water distribution system. Two water treatment plants provide about 15,000 m^3/day; the older one at Sidwashini is supplied from the upper reaches of Mbabane and Pholinjane Rivers and consists of slow sand filters and a chlorination unit. The new one (Woodlands treatment plant), is supplied from the Hawane Dam on Mbuluzi River, and consists of a classic treatment technology with a rapid sand filtration. Both water supply sources for the town are located up-stream the Mbabane City.

The wastewater system of Mbabane City is centralized. There are two main trunk sewers, located along the Mbabane and Pholinjane Rivers, which collect mainly domestic and industrial wastewater from a gradually expanding sewer network. The wastewater is discharged into the Gogholo treatment plant consisting of waste stabilization ponds. Considerable part of the population relies on septic tanks and soakways, which, except for Thembelihle district, is a fully acceptable and economical solution, given the favourable soil characteristic and the large plot size (low density). In the low-income areas and informal settlements, pit latrines and soak ways provide the services, the facilities generally being in a poor condition.

Table 11.2 shows the past, present and projected population of the city to be connected to the sewer system and represents the number of population, who have a connection to the reticulation water supply system. Public tap users have not been included because of the limited sanitary facilities that go with this level of service. It also shows the city's population growth trend, which is high, as emphasized in previous chapters.

The existing wastewater treatment plant of Gobholo, situated near Mbabane River, has been in operation since 1973. Some extensions and modifications were made in 1976 (Euro Consult 1994). The design capacity of the plant is approximately 3000 population equivalent. Considering the population today and the connections to the ponds, the plant is receiving over ten times the design load. The Gobholo Works comprises an inlet works followed by seven waste stabilisation ponds in series. The inlet works comprises of screens, grit channels and a flow measurement devise, which are not functional. All the ponds are full of sludge. Actually, they are acting as settling basins and are providing primary treatment only. The retention time is so short that there can be very little reduction of organic pollution and pathogens through the works and these pollutants are discharged into the stream, which discharges into Mbabane River as shown in Figure 11.7. Some 2 km down stream of the confluence of the stream into Mbabane River, the water is darker in color and bears a noticeable smell of sewage. An estimate from SWSC indicates that the discharge to the ponds is 1920 kg BOD/day, which corresponds to a population equivalent of about 38,400.

3.5 *Methodology*

The study was performed during the period February to April 2001. Heavy rains characterize this season of the year and the weather during this study was typical for the season. The sample sites' location is shown in Figure 11.7. Site 1 was chosen as a control point because it is located about 4 km upstream the Mbabane City. Along the upper reaches of the river, no major anthropogenic activities are taking place. Site 2 was located about 3 km downstream the industrial areas. At this location, the river and tributaries have collected runoff from different types of urban land use patterns, including residential areas, institutional/commercial areas, such as the city center, and several industrial sites. Thus the water quality at this location represents the combined effect of the urban runoff. Site 3 was located at a small stream, receiving the effluents from the Gobholo treatment plant, which discharges into Mbabane River at a distance of about 1 km downstream the sampling point. The water quality at this point reflects the impact of the treatment plant only. Site 4 was located at a distance of about 18 km downstream of the city, thus reflecting the River water quality, which will be used by downstream users. Along this stretch of the River, between site 3 and site 4, several tributaries discharge into it. Their contribution could be regarded as natural pristine water, as the area is sparsely populated.

Samples were collected on eight occasions during the study in the form of composite samples, collected as the integrated sample of three separate grab samples at 0.5 h interval. Samples were collected at a distance of 20 cm below the water surface and represent river water quality only. River sediment samples were not examined.

The following parameters were tested: BOD_5, ortho-P, nitrates, sulfates, ammonia, chlorides, DO, EC, pH, T, TSS, TC and FC. The tests were performed as follows: DO, EC, pH and T were determined on the field by portable conductivity, DO and pH meters. The National Water Laboratory performed the chemical parameters. The Analytical Services Laboratory (SWSC) performed BOD_5, TC and FC. The sampling procedure and the laboratory tests were performed according to the Standard methods (1994). Nitrates were determined by UV spectrophotometry at 220 nano-meters wavelength, ammonia was determined by the Nesslerization method (direct and following distillation), chlorides were determined by the Argentometric method, and sulfate by the Turbidimetric method.

Data with respect to the river water flows of Mbabane River and tributaries was not available, because there are no gauging stations and this parameter has not being monitored.

The statistical analysis of the data obtained was executed by ANOVA 2 package. The Duncan's multiple range test was applied for comparing the results and establishing a statistically significant difference and ranking of the measured mean values at the different sampling locations.

Table 11.2. Mbabane population figures served by sanitation facilities.

AREA	1986	1991	1996	2001	2006
Woodlands	1500	2500	4000	5000	6000
Sidvwashini	5500	8000	10,000	12,000	13,200
Lower Fonteyn	63,000	85,00	10,700	12,500	12,300
Thembelihle	2000	6000	7500	9000	15,300
Nkwalini	-	-	2000	4100	4100
Total	15,300	25,000	34,200	42,600	50,900

Source: Swaziland Water Services Corporation, 2001.

3.6 *Results*

The results with respect to the spatial water quality variation are shown on Figures 11.8 and 11.9. Site 3 was located on a small tributary of the River, and represents the impact of the wastewater treatment plant discharge. The magnitude of pollution at this point is considerably higher than in Mbabane River. This could be explained not only by the influence of the plant but also, because of the relatively low water quantity in this tributary, which does not allow considerable dilution of the effluent. The discharge from the plant could not be considered as a well-defined point source of pollution. The ponds are full with sediments and vegetation has grown at separated spots. Wastewater is flowing along naturally formed channels and small ponds, the banks of the ponds are not well defined, and the inlet and outlet facilities are clogged and do not function properly. The effluent finds its way to the stream at several, naturally formed locations, but not only at the discharge outlet. As a result, the treatment effect in the ponds is limited to partial sedimentation and retention of the course material and a fraction of the suspended solids. It could be expected that the washout of pollutants constituents during peak flow rates would be high.

The measured T values are within the range of 18°C to 20°C, with no significant spatial variation. The values of pH (Fig. 11.8a) vary within the normal range and the variation is not significant with respect to sites 1, 2 and 4, while at site 3, the mean value is significantly lower compared to the other sampling sites. This could be explained with the contribution from the ponds effluent, where anaerobic conditions exist, leading to a decrease in the pH of the effluent, which alters the river water quality with respect to this parameter. This explanation is supported by the high concentrations of ammonia, BOD_5, and the low DO values at this point. With respect to EC (Fig.11.8a), a significant increase was found, when comparing the control point to sites 2 and 4. At site 3, the EC value is significantly lower compared to all other points, which contradicts the rest of the results. Additional measurements with respect to this parameter, could clarify this contradiction. Most probably, it could be explained by a human error during the results' processing stage, because the chlorides and sulfates variation (Fig. 11.8c) show a completely different trend. Chlorides' concentrations show a considerable increase at sites 2 and 3, with a maximum concentration at site 3. Chloride, being a conservative constituent, could reduce its value from site 3 to site 4 due to dilution only. Sulfates show a similar trend, however the maximum concentration was found at site 2, which could be associated with industrial discharges.

The results of the Duncan's test for a statistically significant difference of the mean values among the different sampling locations are presented in Table 11.3. The test is performed based on 8 observations for all sampling locations and parameters, at an "α" value of 0.05. The ranking of the mean values is presented with capital letters from A to D in decreasing order. Mean values denominated with the same letter are not significantly different.

TSS and turbidity (Fig. 11.8b) show a good correlation, with a clear indication of pollution at sites 2 and 3. With respect to turbidity, all measured values were significantly different, with the highest concentration at site 3. This spatial variation underlines the same trend shown by the other parameters.

With respect to organic pollution Fig.11.9a the spatial variation is similar with a well-pronounced influence of the ponds effluent. The spatial variation of DO confirms this trend, with the lowest concentrations measured at site 3. In general, it was found that the urban runoff does not influence significantly the river water quality in terms of organic pollution. As the city is located in a hilly mountainous area, the river flow is relatively rapid, which allows for good aeration, and this explains the relatively high DO concentrations, and indicates to a potentially high self-purification capacity.

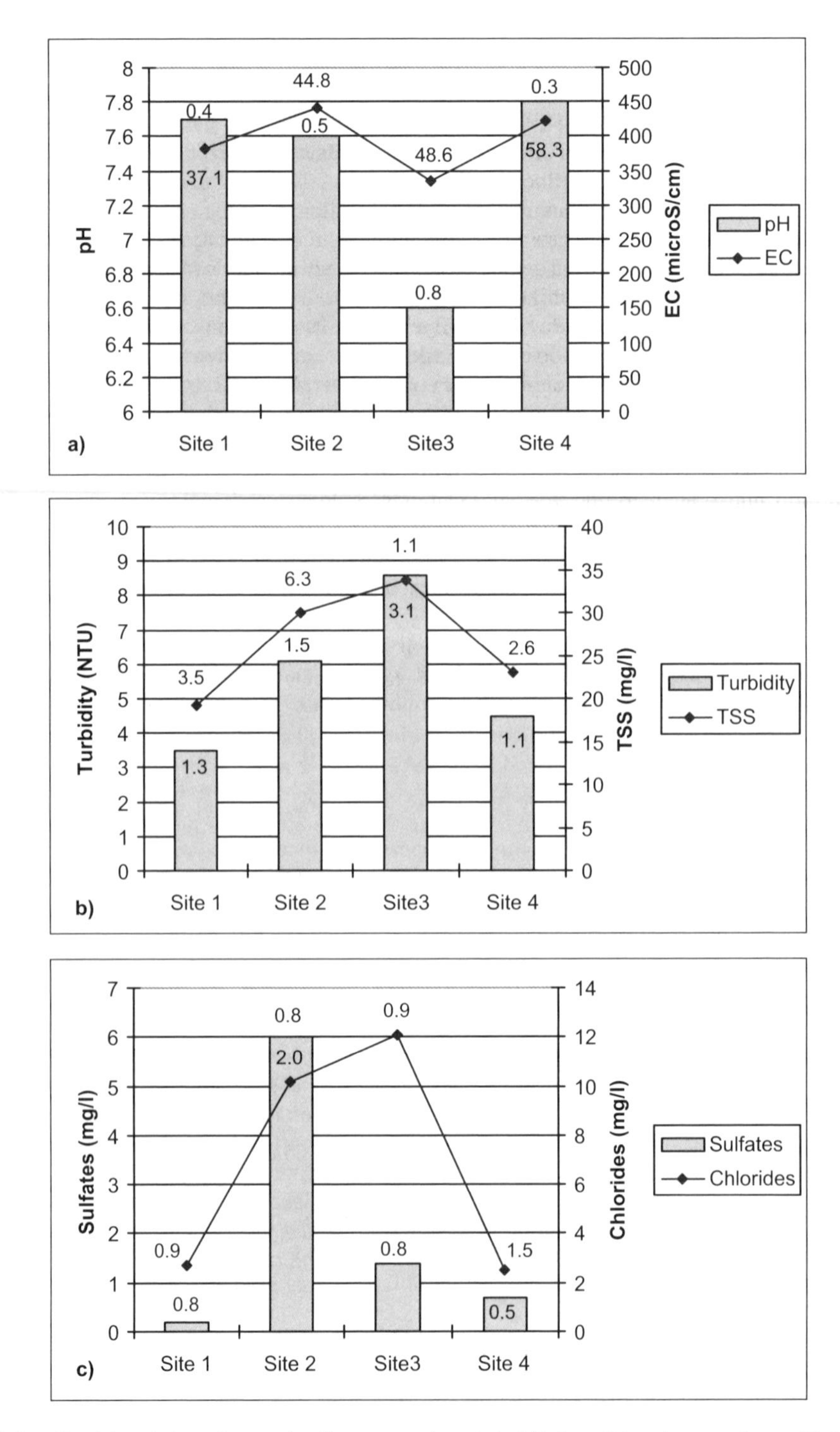

Figure 11.8. Spatial variation of general pollutant constituents in Mbabane River (mean values with the standard deviation shown as numerical value for n = 8).

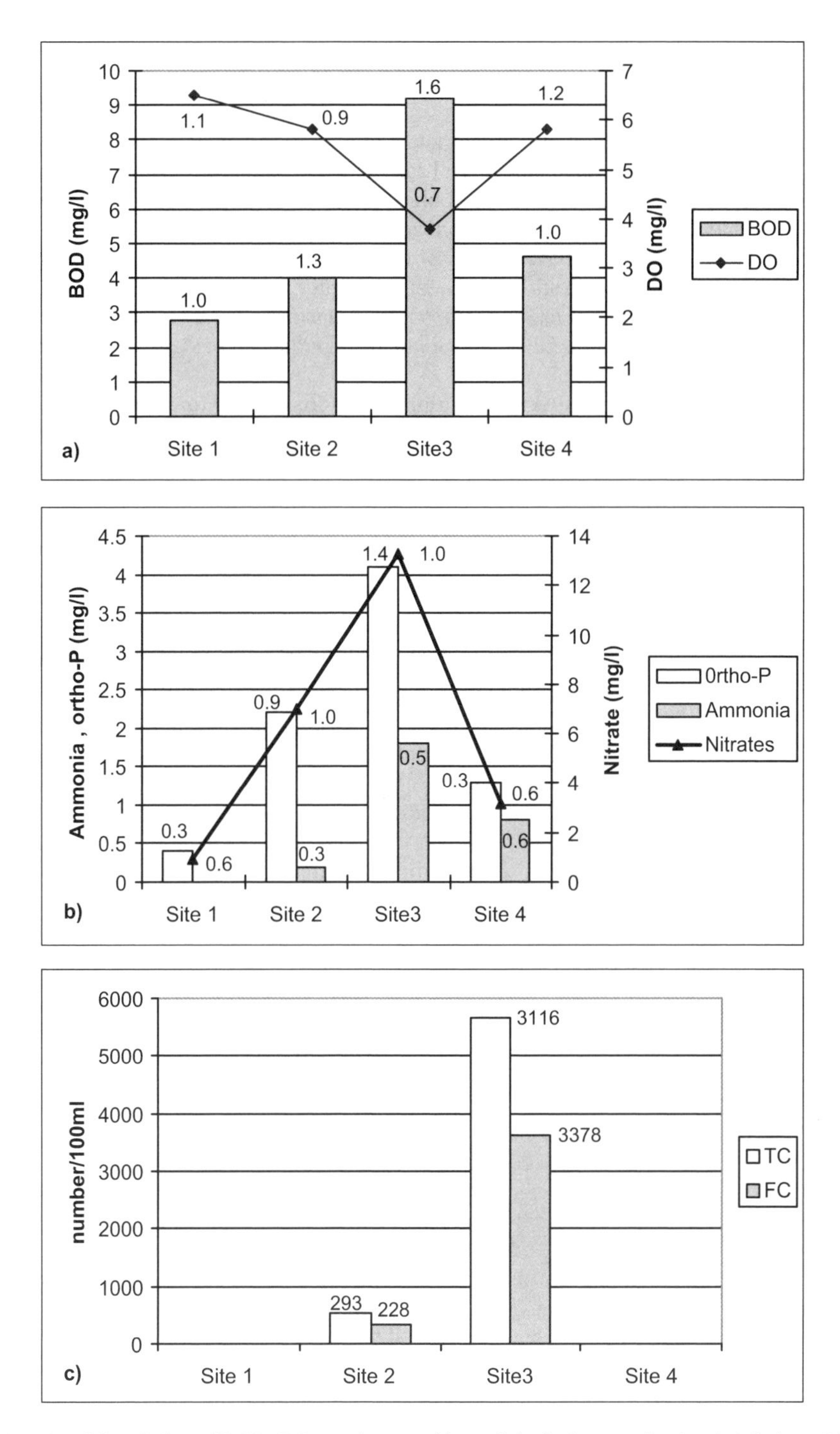

Figure 11.9. Spatial variation of BOD, DO, nutrients and bacteriological contamination in Mbabane River (mean values with the standard deviation shown as numerical value for n = 8).

Table 11.3. Results of the test for significant difference of mean water quality concentrations and ranking.

Parameter	Site 1	Site 2	Site 3	Site 4
T (°C)	A	A	A	A
EC (micro S/cm)	B	A	C	A
pH	A	A	B	A
Turbidity (NTU)	D	B	A	C
TSS (mg/l)	B	A	A	B
DO (mg/l)	AB	A	C	B
BOD_5 (mg/l)	C	BC	A	B
Nitrate (mg/l)	D	B	A	C
Ammonia (mg/l)	C	C	A	B
Ortho-P (mg/l)	D	B	A	C
TC (number/100 ml)	B	B	A	B
FC (number/100 ml)	B	B	A	B
Chlorides (mg/l)	B	A	A	B
Sulfates (mg/l)	C	A	B	C

Nutrients variations (Fig.11.9b) follow the trend with the highest concentrations of the three observed parameters at site 3. The significantly increased concentrations of nutrients at site 2 could be associated with informal discharges from blocked sewer systems. It is also possible that some of the industries are contributing as well. The highest concentration, observed at site 3, has an alarming value and is attributed to the malfunctioning of the treatment plant. The observed nitrate concentrations are very high, compared to other similar cases, presented in previous chapters. It could be expected that with prevailing anaerobic conditions at the treatment plant, nitrification processes would be suppressed, leading to lower nitrate concentrations. It is possible that the generally high DO concentrations due to the steep slop, could help in boosting this process. Additional investigations should be performed to determine the source of the nitrates observed during this study.

A very high bacteriological contamination has been observed at site 3 (Fig.11.9c), the major source being the treatment plant. At site 2, the presence of TC and FC could be associated with the solid wastes, spread along the riverbanks, and informal discharges from sewers and industries. At the control point, the TC and FC counts were negligible, and at site 4, the TC and FC counts were less than 10 numbers/100 ml.

In general, a significant increase in the water quality concentrations between the control point and sites 2 and 3 was found with respect to almost all measured parameters, showing the presence of diffuse pollution from urban runoff and from waste material dumped into the river. The relatively low pollutant constituents at site 4 could be associated with the dilution effect of tributaries and partial self purification due to the high DO values.

3.7 *Discussion*

During the period of study, the water quality legislation of the country was under revision. A comparison with the suggested draft river water quality standards shows a high level of pollution of the stream, which receives the effluents from the treatment plant. The limiting value for TSS was 25 mg/l and it was exceeded at sites 2, 3 and 4. The BOD_5 limiting value of 5 mg/l was exceeded at site 3 only. With respect to ammonia, the stipulated limit was 0.08 mg/l and it was exceeded at sites 3 and 4. The ortho-P values were exceeded at sites 2 and 3. The nitrate limiting value (50 mg/l) is higher compared to other regulatory instruments, which recommend a 10-mg/l threshold (WHO 1984, Viljoen 1992, WWEDR 2000), and was not exceeded at any of the tested locations. With respect to bacteriological contamination, the TC and FC indicators exceeded the safe limit at sites 2 and 3. It could be stated that at site 4, where river water could be used directly for domestic purposes, the water quality was not exceeding the maximum permissible values with respect to the tested parameters, except for ammonia.

The results of this study show that the major sources associated with the diffuse pollution of Mbabane River (site 2) are associated with urban runoff from industrial sites, residential areas (in terms of informal sewage discharges) and solid wastes' depositions along the riverbed and banks. The comparison of the observed pollutant constituents with regulatory documents indicates that the extent of the pollution is not alarming and is pronounced with respect to ortho-P, TC and FC only. A considerable pollution with respect to the organic and bacteriological constituents and nutrients has been contributed by the malfunctioning of the treatment plant (site 3).

The enforcement of the new regulatory instruments would require the implementation of a regular and continuous monitoring program in order to control the river water quality status and to enforce them. This study could be regarded as a preliminary survey for the establishment of such a program. In this aspect, the following recommendations could be made:

- Considering the fact that downstream of the City, the Mbabane River water is used for domestic purposes without pre-treatment, a stringent monitoring program should be implemented in order to control the river water quality in terms of seasonal variations and long-term trends. A careful consideration of different possible pollution sources from industrial enterprises could help to develop an optimal choice of parameters (Ongeley 1998). The information collected could serve to justify management decisions with respect to pollution prevention and provide a warning to downstream users in cases of high concentrations of selected pollutants.
- Monitoring network – in addition to the locations in this study, sampling sites could be established along the Pholinjane stream (after the industrial site and the City center) and on Mbabane River after the confluence of the stream, which collects the discharge of the treatment plant.
- Parameters tested – additional parameters with respect to toxic substances should be included in the monitoring program, such as toxic metals, grease and oil, selected synthetic organic compounds and trace elements.
- Laboratory backup – the proposed extended monitoring program would require a corresponding back up in terms of laboratory facilities, equipment and trained personnel.
- Quality assurance and records – these two aspects need to be given a specific emphasis in order to obtain reliable data and to store it in an easy to use and comprehensive way.

The management aspects related to the diffuse pollution problems of Mbabane River require a multi-disciplinary approach and the implementation of technical, social and educational activities. The following problems require specific and urgent attention:

- The status of the treatment plant – it should be given priority, as it is the major source of pollution of the Mbabane River. During the period of study the reconstruction of this facility was envisaged. The plant is located in a narrow valley, with very steep slopes of the surrounding hills. The original choice of a stabilization pond system could not be evaluated as very appropriate, as the place is difficult to access and in addition there is not enough space for expansion. The reconstruction of the plant should look at the provision of a more compact treatment scheme, where the existing ponds could be cleaned and transformed into a wetland system for effluent polishing. Special attention should be given to a proper access road in order to provide for the smooth operation and maintenance of the plant. The implementation of such a management decision would require the provision of the necessary funding, for the completion of the whole project, including the design, construction and initial operation phases, together with the proper training of the required human resources for the plant operation.
- The status of the solid waste management – it could be classified as a substantial source of diffuse pollution. The proper organization of the refuse collection, provisions for refuse containers at all important public places, and the proper disposal and treatment of the collected solid waste forms the technical base for a successful solution. However, this problem has a social aspect as well, and needs to be backed-up by a well-organized educational and public awareness program.

- The illegal discharges from industrial sites, residential, institutional and commercial areas – the technical solutions in this respect would include regular inspections and the control of water quality, which could help to identify the most important polluters and provide for improved sanitation facilities and preliminary treatment of effluents. The existing municipal by-laws should provide a provision and basis for such activities and should be supported by the necessary institutional set-up and laboratory facilities. However, no technical or managerial measure could succeed if there is no appropriate social behavior, awareness and understanding about the risks of illegal discharges and the possible effects on the whole society. Therefore, the development and implementation of such programs is an important requirement for successful solutions of the problems associated with the diffuse pollution of Mbabane River.

4 CONCLUSIONS

The two case studies presented show that the diffuse pollution problems of river water quality have common grounds and sources in the different countries of the region. Therefore, pollution management and abatement measures could have common bases for solutions as well, in terms of general approaches, methods and regulatory instruments.

The case study of the Lilongwe River shows bacteriological contamination due to informal settlements, and possibly from industrial discharges in the upper reaches of the river, which serves as the main water supply source of the Lilongwe City. This specific result should not be regarded as an alarming situation, because the treatment plant could provide for a safe disinfection of the potable water supplied to the city. However, it should serve as a warning to the managing authorities, and they should consider solutions for limiting the practice of development and formation of informal settlements in the future, specifically in this area. The creation of environmental buffer zones, free of any human interaction along the river, covering the whole study area, should be considered, and the necessary regulatory instruments developed.

The study along the Mbabane River focuses on the stretch of the river, which passes through the Mbabane City and receives its runoff and effluent discharges. The major pollution source was associated with the effluents from the malfunctioning treatment plant in terms of all tested nutrients, organic pollution and bacteriological contamination. Urban runoff, the spreading of solid wastes along the river banks and possible informal discharges from the city, all have an adverse impact of river water quality, which is less pronounced, but is noticeable in terms of ortho-P, nitrate and bacteriological contamination. About 20 km downstream the City, the river water quality did not show a considerable pollution. This could be explained by the dilution effect of tributaries and partial self-purification.

The results of both studies should be considered in the light of the mentioned limitations and as an evaluation of the spatial variation with respect to selected parameters only, given the available resource at the time of the study. The presence of industrial discharges (usually without treatment or with limited treatment) would require a more extensive and regular monitoring program with respect to some toxic elements, which might create health problems of downstream users.

Acknowledgements – The authors would like to thank the management of the WREM program, through the "Collaborative Program for Capacity Building in the Water Sector in Zimbabwe and the Southern African Region", jointly executed by the Civil Engineering Department – UZ, IWSD and IHE-Delft, for the financial support offered during this study. To the technical staff and the management of all laboratories involved – thanks for their support and understanding.

REFERENCES

Euro Consult. 1994. Siting and process options for the Mbabane sewerage treatment works Report by Euro Consult, Mbabane: SWSC.

Ndolo, V.U., Masamba, W.R.L., Binauli, L., & Kambewa P. 2002. Unsafe drinking water: examination of water quality in drought-prone areas in Balaka district, Malawi. . In Proc. *3th WaterNet/WARFSA Symposium-Integrating Water Supply & Water Demand for Sustainable Use of Water Resources",* 481-493, University of Dar es Salaam Publications, Dar es Salaam.

Msibi, K. 2001. Water resources Engineer, Ministry of Natural resources and Energy, Water resources Branch, Swaziland, personal communication, 8 April 2001.

Ongeley, E.D. 1998. Modernization of water quality programs in developing countries: issues of relevancy and cost efficiency. *Water Quality International*, Sept./Oct.: 37-42.

SAFEGE 2001. *Engineering Studies for Lilongwe Water Board* Report. SAFEGE Consulting Engineers, France, Lilongwe: LWB.

Standard methods for the examination of water and wastewater. 1994. 19th ed. *American Public Health Association/ American Water Works Association/ Water Environment federation*, Washington DC, USA

UN/ECE 1996. *State of the art on monitoring and assessment of rivers* UN/ECE Task Force on Monitoring & Assessment, RIZA report No 95.068 (5): 60-71.

WWEDR 2000. *Water (Waste and Effluent Disposal) Regulations*, Statutory Instrument 274 of 2000, Republic of Zimbabwe.

World Health Organization (WHO). 1984 *Guidelines for drinking water, health criteria and other supporting information*. Vol.2, Geneva: WHO.

Viljoen, F. C. 1992. Risks, Criteria and Water Quality. *Municipal Engineer* July 1992, SA: Rural Water Board.

CHAPTER 12

Integrated management of urban diffuse pollution in the Southern African region

R. Hranova

ABSTRACT: The multidisciplinary nature of diffuse pollution problems requires an integrated and systematic approach to their solution and the management of water resources. The integration with respect to regulatory instruments and corresponding monitoring practices for their enforcement, including biomonitoring, has been discussed. The spatial variation of storm and natural water quality at a small catchment level and interactions has been presented, including groundwater characteristics at selected locations. Management tools and practices, suitable for application in the region, have been presented in the light of the integrated cycled approach to the subject, together with social, economic and financial considerations. The most common sources of diffuse pollution have been discussed, and appropriate abatement techniques and methods suggested. Directions for future developments have been put forward.

1 SELECTED ASPECTS OF INTEGRATION

Diffuse pollution management should be viewed and implemented in the context of the integrated approach to the management of the whole water resources system, at catchment level. This is a multi-objective task, which could be approached from different perspectives – social, economic, political, technical, etc. Integration might be sought in different directions and at different levels. Some examples of different directions of the integration process might include:

- Consideration of impacts and interactions of the different elements of the water resources system e.g. man-made structures and natural water bodies;
- Spatial variations in terms of upstream – downstream uses;
- Changes within one element of the system (river or lake), e.g. impacts of erosion and sediment-water interactions;
- Land use patterns and their impacts on natural water resources;
- Surface – ground water interactions;
- Institutional arrangements and the need for coordination between different institutions involved in the process.

The levels of integration correspond to the scale at which the integration is sought, e.g. local, regional, national or international and corresponding links and interactions. Different scales of management would influence the objectives and goals to be achieved, e.g. a small sub-catchment could have different objectives, specific to the locality, in contrast to the objectives and goals of the whole river catchment. Integration in terms of hierarchy, and the incorporation of the sub-catchment objectives, within the goals and perspectives of the whole catchment are examples of integration at different levels and scales. Thus, the subject of integration covers a wide spectrum of activities and aspects, which involve a variety of professionals working in completely different areas, not only in the different fields of technical sciences, but also in politics, economics and social sciences. Problems discussed in this book have been addressed to different specialists from the field of engineering and natural sciences involved in the process, with specific references to management techniques, such as programs planning and development, public involvement, stakeholders participation and the need for a phased approach to the solution of diffuse pollution problems in particular and the water resources management in general. Two specific aspects of integration, as discussed below, need to be given more attention in the region, and require more efforts for their practical implementation.

1.1 *The multidisciplinary approach to diffuse pollution management*

Any management program related to the solution of one or several diffuse pollution problems, even at the lowest level in terms of scope and objectives, would require the involvement and integration of the efforts and expertise of different specialists working in the field. Therefore, the execution of the program would need the formation of multidisciplinary teams or task groups, in order to achieve the common objectives. Each member of the multy-disciplinary team could have different sub-objectives within the common objective, e.g. to solve a specific technical problem, to develop a policy in the field, to evaluate the current status of a natural water body, etc. The approach to the formation of such groups or teams in the developed and developing countries is different. Developed countries have well-developed human resources capacity and numerous specialists working in different areas of specialization, therefore the formation of such multi-disciplinary teams could involve a relatively large number of specialized professionals. Developing countries usually lack this capacity and the number of trained specialists is limited, both in numbers and in fields of specialization. Therefore, the approach to the formation of such multidisciplinary task groups should be different from the one applied in developed countries. The multi-disciplinary team would involve a smaller number of different specialists, who could cover larger, less specialized areas. Figure 12.1 shows a graphical representation of such a team, where the number of different specialists (n), involved in the team, would vary based on the local conditions. No matter how large "n" is, in order to obtain common understanding and an integrated result, all team-members should have basic knowledge with respect to common concepts and definitions, represented by the inner circle of the diagram. The larger the "n", the smaller the diameter of the inner circle representing the extent of the basic knowledge of each one of the team-members, as each one of them would be required to work mainly in his field of expertise. In the case of small number of specialists involved, the extent of basic concepts and principles (inner circle) should be grater, compared to the specialized knowledge of each team member, as one person would be required to solve several tasks from mutually related fields.

In all cases, the role of the team leader is crucial, as he would distribute the tasks and integrate the results, obtained from the different specialists. Consequently, he/she should be an experienced professional with broad expertise in the field.

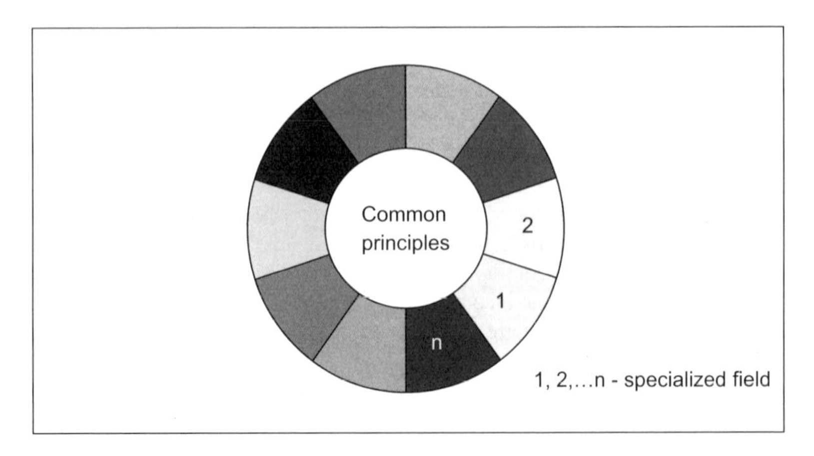

Figure 12.1. The multidisciplinary team – common principles versus specialized knowledge.

The balance between common principles and specialized knowledge is a delicate one and would depend on the specific conditions. This balance is very important with respect to the process of capacity building in general and academic institutions in particular, which would prepare the future specialists in the field. It should be reflected in the academic programs content and structure. However, in numerous cases, the conventional specialization structures are still in place and water resources managers have mainly a hydrological background. Academic programs in the field of natural sciences lack basic knowledge of the specifics of water quality and water resources management. In other cases, new programs, orientated specifically to water resources management, tend to cover extended fields of knowledge at the level of basic concepts and definitions. Such an approach would produce professionals with a broad but shallow base, who are not capable to solve practical problems, because they lack the specialized knowledge, methods and tools to do so. We should acknowledge, however, that this is a dynamic process and new developments and forms are to be sought in future.

The need for a multi-disciplinary approach to the solution of such types of problems is even more important in the field of research and investigation. The formation of multi-disciplinary teams is a rare case in the region; the usual practice being to designate a multidisciplinary task to one researcher or to a research group from one specific body – academic, parastatal, government or non-governmental institution. It works independently, and naturally, would tend to solve the problem from its own perspective and speciality only. Therefore, the need to enhance multi-disciplinary research teams, working on diffuse pollution topics, is important and would contribute considerably to the effectiveness of the management process.

1.2 *The cycled approach to diffuse pollution management*

This approach requires integration in terms of the steps and the sequence of implementation of the activities, which are incorporated in the management process. It has been emphasized previously that the process of diffuse pollution management should be based on the preparation and execution of programs, which are orientated to the solution of specific problems. Such programs could be developed at different levels, with higher hierarchy programs incorporating lower hierarchy programs at local level, and being designed on a cycling principal, each cycle covering a specified period of time. At the end of this period, the program implementation should be revised and evaluated, and the next cycle designed to incorporate corrections and changed conditions. This approach requires a sequence of activities, which are mutually interrelated and would logically conclude at the end of the cycle. The most important steps of the cycle include the monitoring stage, development and implementation of regulatory instruments, design, and implementation of abatement measures, evaluation of the implementation activities and the design of the next cycle. The abatement measures include not only technical solutions and structural measures, but also management practices, such as institutional arrangements, formation of governing water bodies, educational and public awareness programs, etc.

The practice of diffuse pollution management in the region often lacks this sequence and leaves the loop open. The most common example of a broken cycle is the case of the non-enforcement of regulatory instruments. Different reasons could be mentioned, such as: very high requirements, which are not practically achievable; insufficient funds to back up the planned activities' implementation, or lack of political will to punish individual polluters.

Another typical example of a broken cycle, which is common in the every-day practice, is the case when monitoring of water quality is executed on a regular basis, but the results are not processed properly and used during the management process.

In general, it should be well understood that the cost and efforts involved to manage diffuse pollution and water quality in general, would be lost for the society and for the funding agencies, if the integration of the activities to form a closed loop cycle is not implemented. Consequently, the anticipated objectives would not be achieved. The reasons behind such an undesirable practice is often related to the lack of awareness for the problems associated with diffuse pollution and water quality in general, among the public, and in some cases, among the managing authorities. Thus, relatively low priority is given to these problems, and very often, they are completely neglected. The following sections present directions for future improvements.

2 MONITORING DIFFUSE POLLUTION

2.1 *Why is it necessary to monitor natural water quality?*

This question might sound erroneous but in many cases, even in developed countries, the costs and efforts involved in the execution of a monitoring program are high, and the outputs often do not meet the expected goals. Usually, it is due to a wrongly designed monitoring program and is associated with the lack of understanding of the problems and consequently, wrongly formulated objectives. In other cases, it might be due to the unsatisfactory quality of the data obtained, as result of unadequate laboratory quality control. This question is much more valid for developing countries, which face acute social and economical problems and hardships. Therefore, they need to prioritize the problems and their solutions. Understandably, diffuse pollution problems lack priority.

The importance of water resources for the development and well being of any country need not be overemphasized, but for the countries in the region, water is a scarce resource and its importance is essential in terms of sustainable development. Numerous documents and publications deal with water-related issues from different perspectives. Only as an example, we could mention Turton et al. (2003), a publication discussing problems associated with the governance and management of the Okavango River basin. Even the best visions and intentions, reflected in different policies and strategies, could fail to achieve a sustainable use of water resources for the development of the country or the region, if sufficient and reliable information with respect to water quantity and quality is not available. Such type of information could not be borrowed or adapted from other sources, because of the specifics of each geographical region and the level of social and economic development. Therefore, it needs to be collected, stored and used in this specific location. Only a well-designed monitoring program could provide for such information. In addition, water quantity and quality parameters are randomly varying in terms of time and spatial frame; therefore, continuous and regular monitoring is essential.

In numerous cases in the region, information with respect to water quality is missing, and in such cases, decisions in respect to the use of water resources could be regarded as an "informed" guess, with corresponding economic and social consequences, not only for a specific locality, region or country, but in terms of international relations as well, considering upstream – downstream users. The need for an informed decision-making process is emphasized in Shultz (2003) as well.

2.2 *Water quality – a random variable*

River flow rates and water quality constituents are random probabilistic parameters, which vary with time, therefore the analysis of a data set should include statistical time-series tools. The information, which could be obtained from a data set, would depend heavily on the statistical data analysis tools adopted. The basic statistical parameters, which represent a water quality monitoring data set, are the mean and median values, the variance and the standard deviation. All measurements with respect to a given parameter could be fitted to a probability distribution. The most widely applied distribution is the "normal" or Gaussian distribution, which is represented by a typical bell-shape curve. If a data set complies with the normal probability distribution, the mean and median values are identical, or close to each other. In asymmetrical data sets, characterized by high standard deviation, the data set should be transformed to fit an asymmetrical probability distribution curve. The coefficient of skewness is the statistical parameter, characterizing the type of probability distribution of data sets. Values of this coefficient close to zero, show a normal distribution. In analyzing water quality monitoring data, it could happen often that the data sets are not distributed normally.

It is often necessary to find extreme values of a data set, which represent minimum or maximum concentrations, characteristic for a data set. To do this, we should answer the question: which one is the value representing the maximum of the data set? For a 95% confidence interval, this is a numerical figure, which is larger or equal, compared to 95% of the members of the data set, and would be exceeded by 5% of the data set values only. The same applies to minimum values. In order to determine such extremes, equation 12.1 could be applied:

$$X_{extr} = \mu + K\,\sigma \tag{12.1}$$

where: X_{extr} = extreme value; μ = mean data set value; K = frequency factor; σ = standard deviation of the data set.

The frequency factor is given in statistical tables and varies for different levels of the confidence interval and different types of distributions of the data sets. The determination of extreme values, some times known as percentiles, could be represented in a graphical form, showing the relation between actual concentrations and the confidence interval, also known as cumulative probability. For a normal distribution, the relation is linear within a logarithmic representation, and is represented schematically on Figure 12.2. The line shows the extreme values with respect to concentrations, for each specific level of confidence. For a 95% cumulative probability, the corresponding point on the line would show a concentration, which is specific for this data set only. It will be equal or higher than 95% of the data set observations, and will be exceeded by 5% of the observed values only. The statistical analysis of a data set would give a specific maximum value for each monitored parameter at each specific monitoring site.

The implications of the data sets statistical analysis with respect to diffuse pollution monitoring and control are related to the pollution assessment and applications of standards and regulatory instruments. Extreme (maximum) values of the observed data sets should be compared to specified criteria, opposed to the usual practice of comparing the average values, reflecting much lower cumulative probability. The practical meaning of this is that in cases of the comparison of average values with limiting criteria, a considerable number of observations, included in the data set, will exceed the recommended criteria, and would indicate a violation of the regulatory instrument.

Other implication is related to the frequency of occurrence of the observed extremes, reflecting the return period or the recurrence interval of the observation. This concept was explained in Chapter 1, related to hydrological data and is applicable to water quality data as well.

For each specific data set, the return period is directly related to the frequency of the observations within a given period of time, and could be determined by equation 12.2.

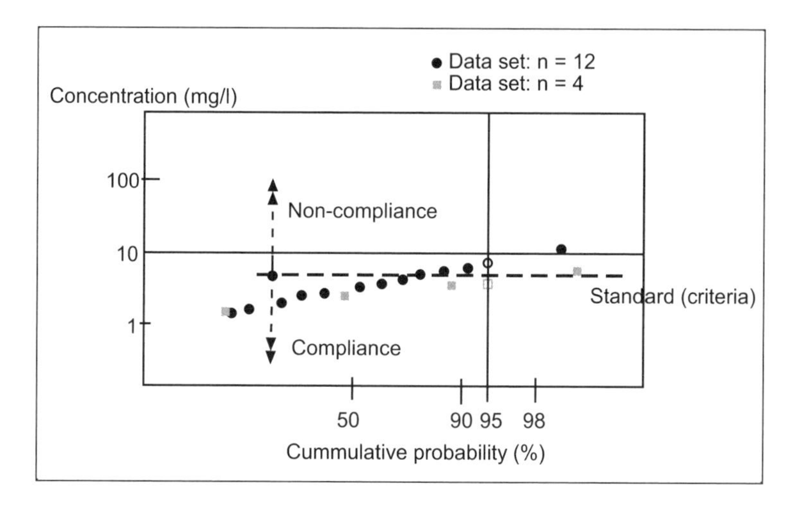

Figure 12.2. Graphical presentation of the cumulative probability of data sets with a normal distribution and a different number of observations.

$$T_r = (100\ \Delta t) / p \quad (12.2)$$

where: T_r = return period; Δt = sampling interval; p = probability of being exceeded.

The application of these basic statistical concepts to a specific data set reflecting the observed values during continuous and regular monitoring of water quality, would help in a correct interpretation of the results and correspondingly, a correct assessment of the water quality status at this specific location. For an example, if the results of an arbitrary monitoring program, measuring ammonia at a frequency of once per month for a duration of one year, the data set would be composed by 12 values. The value representing the ammonia concentration during the year would be the maximum concentration at 95% cumulative probability, which could be expected to be exceeded by 5% of the observed values only. Then the frequency of the expected occurrence of the extreme value at this specific point would be equal to 20 months (from equation 12.2), given that p = 5% and Δt = 1 month. This example shows the need to link the monitoring process and related assessments with existing regulatory instruments.

The case studies presented in previous chapters, usually use mean values for the evaluation and assessment of water quality. Due to the time and resources limitations, the number of observations in some cases were too low, and usually a high variability of the data sets were observed, deviating from the normal distribution. This would require the application of statistical transformations and the use of more complex statistical tools, which were not available. In addition, the existing regulatory instruments include specific limits and criteria as fixed numerical values, but do not prescribe requirements with respect to the frequency of observations and the statistical analysis of the data obtained, living open the question of how the monitoring data should be interpreted. However, in the future implementation of regular and continuous monitoring programs, the statistical methodology for data analysis and assessment should be prepared during the design stage of the program and should form an integrated part of the data analysis and assessment process.

2.3 *Designing the monitoring program*

The design of a sustainable and practically implementable monitoring program for surface water control, would require a good understanding of local conditions, possible sources of discrete and diffuse pollution and the expected pollution characteristics. It should consider the available data and historical information collected. The implementation of a monitoring program to match the water quality objective approach could become a serious challenge. In some cases, the program might be oversized, with the inclusion of numerous sampling points (end of pipe and natural water quality) and too many parameters tested, which could jeopardize its implementation. In other cases, the locations and parameters tested could not be enough to provide the necessary information. In order to design a cost effective monitoring program, there is need to achieve a correct balance between:

- The size of the monitoring network;
- The number of parameters monitored; and
- The frequency of monitoring.

It would require expert knowledge in the field and should be judged based on the specific conditions, available resources for the implementation, and the existing regulatory instruments. Unfortunately, in numerous cases, the monitoring programs are rigid routines, which have not been changed since the date of their establishment.

The choice of parameters to be monitored should consider the available capacity in terms of laboratories and human resources. The material presented in previous chapters show that usually the level of the technical facilities, which are available in the vast majority of the countries in the region, except for the Republic of South Africa, allow only for the testing of basic parameters. However, in many cases, the information needs require the testing of toxic substances or other specific pollutants, by means of more advanced methods and equipment. Examples of such cases include:

- Point and diffuse discharges from large urban population centers;
- Industrial and mining enterprises, discharging to natural waters, which would be used downstream for direct domestic and agricultural use by a rural population.

Such conditions require special monitoring attention and the expert knowledge of specialists during the design and execution stage. An important parameter, which should not be missed from the monitoring program, is the flow rate at the most important monitoring stations and corresponding link with rainfall data.

The frequency of sampling is another aspect, which would considerably influence the reliability of the data obtained and the operational cost of a monitoring program. It has already been emphasized, that the monitoring of diffuse pollution means the monitoring of natural water resources and requires an event-orientated approach. Specifically for the case of tropical countries with distinctive wet and dry seasons, the major pollution loads are expected during the beginning of the rain season; therefore, the frequency of sampling during this period should be higher. The intensity of rainfall events is also an important factor because of its erosive effect and the increased risk of the transport of pollutants and should be included in the data sets. A proper determination of the annual frequency of sampling should be done, based on a detailed examination of the rainfall patterns specific for the region and should provide enough data for statistically reliable results.

Storm water quality monitoring in urban areas, specifically in cases of large drainage channels, could be regarded as a point source of pollution and should be subject to a separate monitoring program. The regular monthly monitoring of urban drainage, as in the case of the city of Harare (Chapter 3), would not provide adequate information. The determination of EMCs and unit pollution loads from typical land use patterns, or typical unit pollution loads for selected areas, drained by one drainage channel, could drastically reduce the cost of regular monitoring programs and could be very useful for the analysis of impacts and future developments. This could be done by means of separate projects and the application of automatic monitoring stations, as the procedure requires a high frequency of observations. The data obtained during such projects would be used for pollution loads estimation and the evaluation of different scenarios of future urban development. If the data with respect to EMCs is available, then the regular monitoring programs could have a lower frequency and the data obtained from them could serve as back up information and data validation. Also, it should be noted, that the design of a monitoring program is not a "once for ever" exercise but should be viewed as a process. The regular periodical assessments of its effectiveness and the consequent improvements are mandatory.

The data quality assurance aspect should be mentioned as pivotal for the success in the implementation of any monitoring program. It is directly connected to the available financial, technical and human resources potential and if these resources were not provided, the basic objectives of the monitoring program would be jeopardized. The proper data validation and storage procedures should be executed on a regular basis to minimize errors occurrence. The quality assurance measures are even more important in the cases of diffuse pollution control, because they would allow eliminating gross measurement errors and would reduce the natural high variability of the data sets, due to the uncertainty of rainfall occasions and runoff quality.

Designing a monitoring program requires expert knowledge, a good understanding of local conditions and consideration of the operational costs involved. It could be recommended that it is better to reduce the number of monitoring sites, but to have a full set of all necessary data with respect to this site, than to expand to numerous locations, with scattered monitoring data. For locations in remote areas, or low cost urban developments, the incorporation of biomonitoring techniques and their integration in the monitoring process could be a viable option, which could help to considerably reduce the monitoring cost and to involve the community in the process.

2.4 *Biomonitoring*

Aquatic ecosystems are sensitive to changes in their environment, due to natural or anthropologic causes. The response to such changes might be drastic, resulting in the death or migration of different species. In most cases, it includes a decrease in the reproductive capacity and a decrease in the normal metabolic rate of aquatic organisms. Biomonitoring aims at identifying and quantifying such types of responses, with respect to particular organisms and the evaluation of the suitability of water to sustain aquatic life.

Different types of biological effects, caused by the whole complex of physical and chemical characteristics of the aquatic environment, might be used for assessment during the biomonitoring process, but some of the most commonly used ones include:

- Changes of the species' composition in the aquatic community;
- Changes of the dominant group of organisms in a particular habitat;
- A decline in the number of some species;
- A high mortality of sensitive life stages, e.g. eggs, larvae.

The biological assessment of an aquatic ecosystem include different types of methods, which estimate and quantify the above-listed biological effects:

- Ecological methods
  - Presence or absence of indicator species-invertebrates, plants, algae;
  - Analysis of biological communities (biocenosis)-invertebrates;
  - Analysis of biocenosis on artificial substrates
- Microbiological methods
  - Detecting the presence of selected bacteria;
- Bioassay and toxicity testing;
  - Tests expose organisms (invertebrate, fish) to different contaminants in varying concentrations in order to obtain their response.

With respect to biomonitoring techniques for the evaluation of the status of natural water bodies, the ecological methods are by far the most widely used. Bioassays and toxicity testing are methods applied widely in laboratory conditions. They are used to determine the guidelines and criteria related to toxic or other biological effects caused by specific contaminants, as well as the determination of maximum pollution limits (criterions) with respect to acute or chronic toxicity.

The ecological methods for biomonitoring are using biotic indices as specific assessment tools. The oldest one is the Saprobic index, developed at the beginning of the 20th century to assess the impacts of sewer discharges on river water quality in Central Europe. It identifies four zones of gradual self-purification capacity of the river body, where the identification of selected indicator species is made, in addition to the determination of several chemical characteristics (as DO) and the characterization of the general conditions of the water body. Contemporary biotic indices include a much wider list of indicator organisms, typical for the specific climatic and geographical conditions. In general, no single group of organisms can be used to assess all aspects of water quality and when using biological approaches for water quality assessment, a combination of techniques using different organisms usually gives a more accurate assessment. Numerous tools for biological assessments and the evaluation of the health status of aquatic ecosystems have been developed in different countries and described in literature (Novotny 2003, Uys et.al. 1996, Hellawell 1986). In the region the South African Scoring System (SASS) has found an application. It provides a rapid assessment of bentic macro invertebrates using a scoring system for African families to calculate an index of pollution, based on the number of family representatives (taxonomic groups), which are classified depending on their tolerance to pollution. It is a rapid assessment method. According to SASS4 the invertebrates are collected from all the principal river biotopes at a particular site. They are taken back to a laboratory/office, removed from the debris of the sample and identified down to family level. The samples are counted and recorded in their respective taxonomic groups. Each taxonomic group has a grade or ranking which is relative to its sensitivity to pollution. Families with a ranking of 1 are very pollution-tolerant while those with a rank of 15 are very sensitive. These grades/ranks are summed to give the Sample Score, which is divided by the number of families of invertebrates identified, to give the Average Score Per Taxon (ASPT). This gives the indices of water quality when taken in conjunction with other factors such as the Sample Score. The ASPT values usually range between 2 and 8. In general, the higher the ASPT value, the better the water quality. The amount by which this average sensitivity deviates from the number obtained from pristine reference sites forms the basis for a quantitative assessment of the ecological impairment of the area. The analysis to a family level does not require anything more than good taxonomic keys and some assistance from an experienced person in the field.

As with any system of assessment and evaluation, biomonitoring methods and techniques have advantages and disadvantages that need to be taken into consideration in order to decide on the suitability of the method chosen. A brief listing is presented below:

- Advantages
  - Selected methods, as biotic indices, are simple to perform.
  - No special equipment or facilities are needed.
  - Biological communities may be affected by human interference in ways that are not detected in chemical tests such as the removal of riparian vegetation, dredging and canalization (Chessman 1995).
  - Both point source and non-point source problem areas may be identified.
  - Collecting equipment is simple, inexpensive and lightweight making it easy to transport thus testing the maximum number of sites with the minimum amount of effort and expense.
  - Rapid-assessment predictive models are simple and based on sound ecological principles, which may be presented in a way, which can be interpreted by non-specialists, such as water board managers and the public. It gives a quantitative measure of ecological impairment.
  - It is easy to tailor this type of model to any country and it produces results, which are in themselves legally defensible and can lead to enforceable management options.
- Disadvantages
  - Their use is localized to the specific bio-geographical region and the comparison between different regions is risky;
  - The knowledge of taxonomy is required with respect to each region;
  - They reflect environmental changes after the damage has been implied;
  - They are not specific with respect to pollution sources and causes;
  - There is a danger that classification of sampling sites and habitat types may be subjective.
  - Down stream catchment areas are usually vast and situated in gently sloping fertile valleys, which are ideal for settlement and agriculture. The choice of control sites in such conditions is difficult, as usually the whole area has been affected by human activities. Upstream sites could not serve as controls because of the different site conditions and stream morphology. The definition of pristine implies the "natural state" of the aquatic ecosystem, which is a rare occurrence in downstream sites, thus, a compromise of "the best attainable ecosystem" would need to be done.(Hohls 1996).
  - Resh et al. (1995) points out that there is a difficulty in defining thresholds of variation, such as: how much is due to sampling inaccuracy or natural variation, what are the limits of change in species composition, if any?

3 REGULATORY ASPECTS

The evident difficulty to monitor and control diffuse pollution requires an objective orientated approach, where the main goal should be to control and maintain natural water quality in a sustainable manner. The development of regulatory instruments play a key role in this process and would lay the foundation for a sustainable practice. These documents should be able to allow a comparison with actual measurements, the detection and assessment (quantification) of pollution levels, as well as the possibility to penalize polluters.

3.1 *The regulatory approach*

The Water Quality Objective approach should be implemented in order to provide a wider range of protection of the water resources and to specify the natural (surface and groundwater) background quality for each specific region, as well as to include the detection, assessment and regulation of diffuse pollution sources. The formulation of sustainable and practically implementable objectives would require:

- A sound estimation of the present status of the natural water quality in the catchment basin, or sub basin, and its corresponding beneficial use;
- A good knowledge of treatment and disposal methods for wastewater and storm water at local and international level and a choice of the best alternative (BA), related to local conditions;
- A consideration of the conditions for implementation of BA in the light of the economic status of the region and availability of financial, technical and human resources.

The implementation of this approach does not necessarily mean that it should lead to a conflict for cases where the effluent discharge regulations are in place. It should be viewed as a broader approach to the regulation of water quality, which could include guidelines/criteria for natural water quality and incorporate the acting regulations for effluent discharges in order to achieve the prescribed objectives. In addition, it would allow in future the implementation of the waste assimilative capacity approach, which requires a more extensive database, more sophisticated equipment and a higher professional level of human resources, involved in the process.

The need for introducing the water quality objective approach has been recognized in the region and many countries have adopted it. The practical implementation of these instruments is still in its infancy and most probably would be a lengthy process, which would require a considerable administrative effort and a corresponding restructuring of the institutional arrangements. Also, it would require the revision of the existing monitoring programs and their upgrading to a new level, which would provide for the enforcement of the new legislative documents.

One important point, which should not be omitted with respect to the implementation of this strategy, is the fact that in numerous cases, in the countries of the region the rivers under consideration are ephemeral, as it is the case of Botswana. Correspondingly, they flow only during and immediately after the rain season. In such circumstances, we could not rely on dilution or any assimilative capacity. Correspondingly, the effluent orientated approach should be applied and enforced for such localities. This serves to emphasize the need for the integration and mutual interrelation between the two approaches, as well as the consideration of the specific site conditions, typical for the locality, country and region.

3.2 *The link between regulatory instruments and monitoring programs*

Regulatory instruments, which are not enough explicit in terms of quantitative criteria for their implementation, but rely on general statements or descriptive formulations could be regarded as "paper tigers", which would not help to control and improve the water quality. In cases where specified values with respect to listed constituents have been incorporated, providing a dimension of magnitude with respect to maximum permissible concentrations, it would be necessary also, to specify the method of monitoring, including sampling procedures and monitoring frequency. This is necessary, considering the fact that water quality is a random variable as it was explained in section 2. For a given monitoring station, and a period of observation of one year, two data sets are available, as illustrated on Figure 12.2. For the data set with 12 observations the 95% cumulative probability value would exceed the recommended standard and should be interpreted as non-compliant with the regulatory instruments. The data set with four observations only would indicate compliance, because the 95% cumulative probability value is lower than the recommended limit. This example does not mean that the lower frequency would necessarily mean compliance. It is possible that even when a few observations are available, the standards could be exceeded. In several case studies presented in previous chapters, all values of the data sets, with respect to certain parameter, considerably exceed the recommended values, and do not leave room for misinterpretation of the results. However, in most cases, the data interpretation and pollution status evaluation could be very subjective, if regulatory instruments do not specify the required frequency. Consequently, the regulations enforcement and the provision of penalties with respect to clear cases of identified polluters would be difficult to implement.

Considering the random character of natural water quality, it should be well understood that 100% compliance is not possible. The choice of confidence level would determine the level of the stringency of the prescribed criteria. More relaxed standards could have lower cumulative probability requirements, which means that the cases of non-compliance along the period of observation would happen more often. In cases when toxic or other human health related criteria are exceeded, an important factor, which should be considered and specified in regulatory instruments, would be the duration of the exposure of the aquatic environment to the respective contaminant. In other words, when the standard prescribes a specific allowance for exposure to higher concentrations than the prescribed limit, it should also specify the period of allowable exposures.

In general, the frequency of the prescribed monitoring observations would vary for the different locations and type of media observed. Ground water quality does not vary rapidly, thus a lower frequency would be recommended. With respect to surface water bodies, a clear differentiation in the monitoring methodology should be done between rivers and lakes (dams). In general, larger rivers have a much more stable water quality compared to small streams. In cases where a naturally high variability of the data sets is expected, regulatory documents could provide for an acceptable range of variation of specific parameters, rather than fixed values.

3.3 *Water quality regulation and models applications*

The application of the water quality objective approach, together with the integrated assessment of the catchment characteristics and the formulation of specific criteria with respect to beneficial water uses, leads to an extremely large volume of information to be considered during the preparation and implementation of regulatory instruments. The task to analyze and evaluate such information could be regarded as unattainable without the application of contemporary tools for information storage, retrieval and analysis, as well as modeling of different processes and simulation of different scenarios.

Considering the nature of diffuse pollution and urban drainage specifically, pollution loads' magnitude would be determined by the land use and rainfall patterns. Therefore, its correct estimation would depend heavily on the proper evaluation of the characteristics of the drained area, the use of reliable methods for runoff estimation, and reliable rainfall data sets. These factors, presented in the form of an information database, together with the information provided by the corresponding water quality monitoring programs, result in a massive information block, which is difficult to be handled and interpreted manually. For this reason, the estimation and control of diffuse pollution requires the development and implementation of models, based on GIS systems, which could provide the necessary means for data handling, storage and retrieval. Also, the application of such tools would improve the reliability of results obtained and their level of accuracy. Their application is even more relevant for the cases of urban drainage pollution, where the variability of land use practices per unit area is much higher, compared to cases of agricultural diffuse pollution. However, most countries in the region do not have the capacity to develop and maintain such tools at present. It means that the practical implementation of the water quality objective approach in the region should be viewed as a long-term task, which should be executed in a phased manner. The data collection processes, based on regular monitoring programs, and any additional information from specific research projects, in addition to the creation of GIS systems, would be the first phase of this process. At later stages, the accumulated information could be incorporated in higher hierarchy models, describing the pollution transport and conversion mechanisms in an integrated manner at larger catchment characteristics, which would allow the estimation of assimilative capacities of natural water resources, the allocation of possible pollution loads and the simulation of different scenarios. This could help in the revision and upgrading of existing regulatory documents in a systematic and cycled manner.

It has been emphasized that water quality standards and regulatory documents are site and country specific. This statement refers to the different conditions, which would determine different criteria in terms of parameters and permissible limit values. However, regulatory instruments should also incorporate indications with respect to the methodology of the monitoring process, laboratory testing procedures, data handling and interpretation analysis, etc. In this respect, an integrated regional approach is very important and could help to boost the process of the practical implementation of the water quality objective approach. The preparation of regional guidelines, to support the national and local ones, could help in the following directions:

- It would provide a basis for the comparison of the data obtained in different localities and would enhance the investigation of trends and variations in the region;
- It would enhance the application of compatible tools for data handling and analysis;
- It would provide a foundation for quality assurance and would increase the level of reliability and accuracy of the process;
- It would provide a common basis for data interpretation and pollution assessment/evaluation, which could help solve trans-boundary problems related to a sustainable water resources use.

4 INTEGRATING RESULTS FROM URBAN DIFFUSE POLLUTION ASSESSMENTS

Previous chapters present different case studies, related to the assessment of the impacts of urban diffuse pollution sources on runoff and natural surface/ground water quality. The reader should have noticed that selected cases have been studied but they are not sufficient for a fully integrated assessment at a catchment level. This fact reflects the current capacity in the study area, in terms of available laboratory facilities, transport, manpower, data availability, regulatory documents, as well as the time and financial constraints of each specific project. It was mentioned that a fully integrated assessment could be done, based on a comprehensive monitoring program and the introduction of contemporary tools for data handling, analysis, evaluation and assessment. It is expected that in future, the data presented in previous chapters could be incorporated and enhance such a type of analysis and evaluation. The following part of this section aims at a partial integration of results presented in the different chapters and points out directions for future development.

4.1 *Urban storm water characteristics*

When discussing problems of diffuse pollution in urban areas, it is necessary to make a differentiation between the water quality and characteristics of storm water runoff collected and transported by urban drainage channels and the one contributed directly to natural rivers and streams. Drainage channels are usually lined man-made structures, which are expected to be maintained and cleaned regularly. Thus the storm water quality in such channels directly reflects the corresponding quality of the drained area and its land use patterns. Natural streams represent the combined effect of the storm water, groundwater recharge, all upstream tributaries and point sources of pollution, as well as the accumulated pollution from sediments and eventually, solid waste depositions. This differentiation is important in terms of data interpretation and the unit pollution loads (per unit area or per capita) calculation, and a careful characterization of the study area should be done during the data analysis process. It has been discussed that discharges from storm water canals could not be classified as purely diffuse pollution sources, as they have a well-defined location, especially in cases of large conveying structures. In such cases, the diffuse character of the pollution is related to the random character of the qualitative and quantitative aspect of the pollution, rather than to its dispersed location. However, it must be remembered that storm water drainage structures have numerous discharges along the system at different locations, thus it should be regarded as a completely diffuse source of pollution at the level of the whole catchment basin.

The results with respect to storm water quality collected and transported by drainage channels in Harare, Zimbabwe represent an inconsistent data set of observations made during the period 1995-2003 (Chapters 3 and 4). Despite the fact that the measurements were done by different laboratories and information with respect to flow rates were not available for the vast majority of the cases, results show a trend of pollution with respect to phosphate and ammonia at the two major channels, exceeding significantly the prescribed limits at the Coventry Rd channel and in the CBD. Pollution with respect to metals concentration was found in respect to Pb mainly.

Storm water runoff from roads and highways is considered as a major source of diffuse pollution but has not been studied in details in the case studies presented. Indirectly, it has been included as a contributing element to the general characteristic of urban drainage. It should be noted that in general, the vast majority of the countries in the region do not have high traffic loads of the streets and roads, as in developed countries; correspondingly, no high pollution levels could be expected. Exceptions, which need special attention, are large highways and urban centers, mainly in the Republic of South Africa.

Previous chapters emphasize the need for event-orientated monitoring programs, which is most applicable for urban storm water. It could be recommended that research projects, orientated specifically towards storm water quality assessment, could help significantly to fill the gaps in existing monitoring programs and to obtain the necessary data base for a reliable determination of unit pollution loads, which could be incorporated in the integrated system analysis. Also, it is highly recommendable to include a proper technical backup as part of such projects, with the provision of automatic monitoring stations, collecting data with respect to flow rates and qualitative parameters, as well as a more extended list of measured parameters. Such data, linked to rainfall characteristics for the respective observations could form a sound base for the determination of typical unit pollution loads, which are essential in terms of an optimal choice of pollution abatement measures, and the general assessment, simulation and prediction of the catchment's water quality status.

4.2 *Urban streams*

A spatial integration of the results obtained in previous chapters could be done based on the Marimba River catchment characteristics, flowing through Harare, Zimbabwe. Figure 12.3 shows a scheme of the catchment, which is characterized mainly by medium and high-density residential developments. The River originates in Mount Pleasant, and in its upper reaches, collects the runoff from medium density residential areas, including commercial areas in the form of shopping complexes and schools. This type of development does not pose a serious danger with respect to diffuse pollution, as the runoff from roofs is discharged into surrounding gardens and infiltrates, recharging the ground water aquifer. Large green areas surround the schools and sporting facilities with a very low runoff potential. The areas, which contribute the major portion of the runoff from this land use pattern, could be associated with roads and parking slots.

The middle stretches of the River collect runoff from Warren Park and Kambazuma, as well as from Tynwald, which are high-density residential areas of larger plots than the original ones, described in Chapter 5. The Tynwald stream collects the runoff from the Golden Quarry landfill, which has been abandoned as a solid waste disposal site, but is of concern as a diffuse pollution source with respect to surface and ground water (Chapter 7). The drainage area is not densely populated, with large undeveloped open spaces, which in many cases are used for informal agriculture, a typical land use pattern for the low-density areas in the city. This segment of the River collects the runoff from the Coventry road drainage channel (results presented in Chapter 4). The channel collects runoff from a mixed land use pattern, including industrial areas, small to medium business areas, medium-density residential areas and open undeveloped areas. The downstream part of the River collects runoff from typical high-density areas, described in Chapter 5, namely Kuwadzana, Crowborough, Marimba Park and Mufacose. The largest tributary – Little Marimba River, flows mainly through undeveloped areas, but new high-density developments are expanding in this direction (Kuwadzana, Dzivrasekwa, Tynwald and Bluff hill). This stretch of the River receives the highest pollution loads due to diffuse sources (runoff from high-density areas and Crowborough farm), as well as from a large point source – Crowborough Sewage Treatment Works (CSTW). Chapter 8 has discussed at length the impact of the farm.

The results from the different studies could not be directly compared, as the observations in the different case studies were made during different periods of time. The maximum pollution concentrations were measured at Little Marimba River and ware associated with the runoff from Crowborough farm.

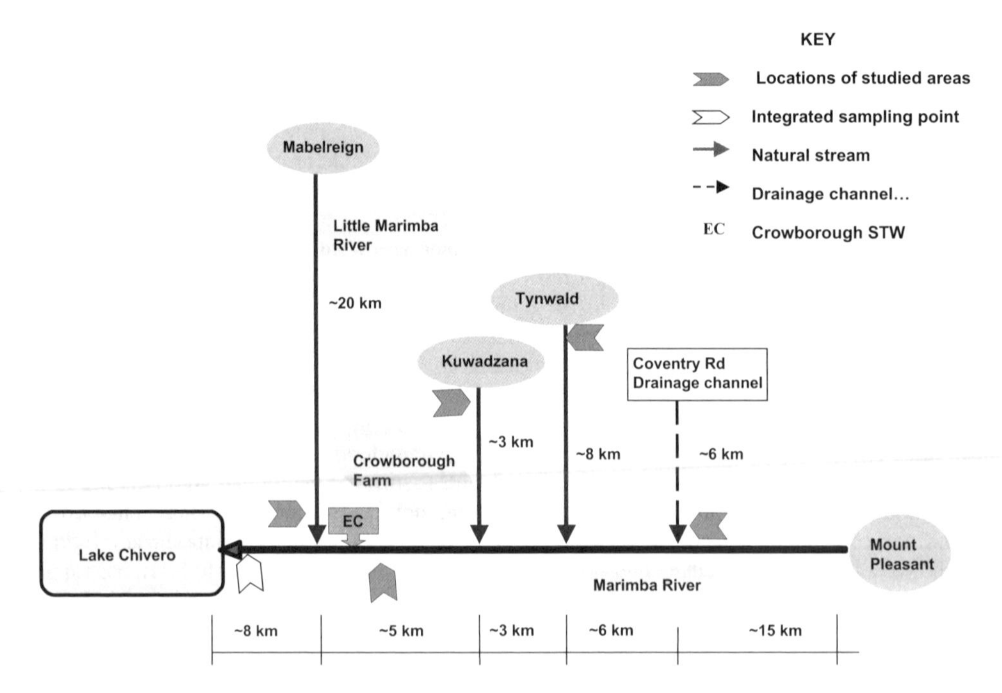

Figure 12.3. Scheme of the Marimba River catchment.

It could be assumed that the control point of the Kuwadzana stream represents the background quality for the Marimba River (Chapter 5), as it is similar to the background parameters in other points of the Manyame catchment (Hranova et al. 2002). The comparison with this point shows that the major polluting parameters are phosphorous, ammonia, dissolved solids, and selected metals as Pb, Ni and Cd. The comparison of the different sources with respect to metal concentrations show that their contribution is more significant with respect to storm water, but is not well pronounced with respect to the farm. The contribution of the farm with respect to surface runoff and impacts to Marimba River (Chapter 8) was significant during the beginning of the rain season with respect to Ni and Cd only.

In terms of an integrated assessment of the pollution load contributed from the catchment to Lake Chivero, the integrated sampling point shown on Figure 12.3 is very important, as it is provided with an existing discharge measurement device, and allows the simultaneous control of flow rates and qualitative parameters. Its inclusion in a future model for the evaluation and assessment of diffuse pollution in this relatively small catchment is essential.

The water quality variation in two other urban streams is presented in Chapter 11. The major pollution source of Lilongwe River was related to informal settlements and possible pollution from industrial enterprises. The water quality of Mbabane River was seriously affected by the malfunctioning of the municipal wastewater treatment plant. A more detailed analysis of the major pollution sources is made in the subsequent section.

4.3 *Groundwater*

Groundwater pollution is usually associated with diffuse sources and because of that, it has been omitted from monitoring and control under the effluent discharge approach to water quality management applied in the past. Chapters 6,7, 9 and 10 present different case studies related to the pollution of ground water resources in the Lake Chivero catchment. The diffuse pollution sources investigated are related to municipal wastewater and sludge reuse for the irrigation of pastures, and with specific urban sources – semi-formal settlements, cemeteries and a solid waste disposal site. At all studied locations, elevated levels of dissolved solids, nitrate, metals (Fe, Pb, Cd) and fecal coliforms were found. All examined boreholes at the solid waste site show that the pH levels were lower than normal within a range of 4 to 6. The highest concentrations with respect to coliforms were found at several locations in the cemetery and at all studied locations in Epworth.

The results obtained in Chapter 6 show that the case, needing the most urgent and immediate action, is related to the semi-formal settlement in Epworth, where the polluted shallow aquifer serves as a water supply source for the population, which generates the pollution. As a short-term measure, the people living in this area should be advised to boil the water used for direct consumption or direct body contact. The longer-term abatement measures have been discussed in more detail in the following section. The groundwater pollution with respect to metals is also of concern as it prohibits the direct consumption of groundwater for drinking or bathing purpose.

The spatial integration of the results with respect to groundwater quality and the assessment of the extent of transport of pollutants along the aquifer require the mapping of the available data with respect to the aquifer structure and its water quality. Localized studies, as the ones presented in this book, could serve as warning signals and should point out the need to take urgent or long-term measures. However, the successful solution of the problem and the process of recovery of the aquifer quality, as a result of implemented abatement measures, could only be achieved if such spatial integration is implemented in practice. The most appropriate tool for such an integration would be the implementation of a GIS, which could provide the spatial data integration with respect to rainfall, land use patterns, urban infrastructure, geological and climatic conditions in addition to surface and ground water resources in terms of quantity, levels and quality. The availability of information databases and their spatial reference could be used to analyze links and interrelations between the different components, surface-ground water interactions and the process of pollutants' transport. Such databases, if compatible, could be extracted for incorporation into higher hierarchy models, as explained earlier in this chapter.

5 TYPICAL POLLUTION SOURCES AND ABATEMENT MEASURES

Diffuse pollution is difficult to identify and assess. Its remediation could be extremely complex and costly, as it concerns the whole extent of natural water bodies. The recent practice of diffuse pollution abatement measures is related to the preliminary retention and partial treatment of storm water from densely populated urban developments and highway runoff, before it reaches natural water bodies (Debo & Reese 2003). In many cases, abatement measures need to be implemented with respect to the rehabilitation of the whole river system, which is a large scale and very expensive project. There are two possible scenarios in the cases, when water resources are already polluted. The first option is to implement structural abatement measures, which are complex and costly. The second option is to leave this burden to the future generations. Therefore, the best solution is to avoid at a maximum possible extent, the creation of diffuse pollution by implementation of source control measures. Such abatement measures should have a high priority. This is even more valid for developing countries, which do not have enough resources to implement large-scale pollution abatement projects. The source-control pollution approach requires the effort and involvement of the whole society and for this reason, public awareness and educational programs are one important activity to be envisaged. For this reason, the knowledge and expertise with respect to the nature and formation of diffuse pollution is an important consideration and has been stressed throughout this book.

Considering the implementation of source control measures, it should be emphasized that there is no need to monitor and control any possible diffuse source pollution, in order to prove that such pollution exists and is dangerous. Professionals in the field are well aware of the dangers associated with formation and spreading of pollution due to the specific sources, which are numerous and differ from one location to the other. For example, solid waste disposal sites are well-known sources of point and diffuse pollution and the implementation of source control measures includes the provision of well-protected sites. The seepage should be treated, before it reaches the natural water bodies and drainage structures should be provided to prevent runoff from reaching the site. The cases described in this book serve to obtain a partial assessment of the level of pollution and to form the base for a future integrated assessment at catchment level. This section discusses some of the most often met sources of pollution in the region and corresponding abatement measures.

5.1 *Semi-formal and informal settlements*

One of the most acute problems associated with diffuse pollution in the region is related to the recent trends of urban development, characterized by a very high rate of urban population growth and the formation of chaotic, informal and uncontrolled settlement patterns, resulting in the mushrooming of squatter camps and overpopulation of formal urban developments, mainly in the low-cost urban residential areas. There are no easy solutions of this problem and major efforts should be orientated towards regulating and limiting this practice in future, while providing basic water supply and sanitation facilities in the already existing ones. As mentioned in Chapters 6 and 11, most often, this practice is associated with bacteriological contamination of surface and ground water resources, which serve as a potable water source for the same population, which is generating it. The most imminent hazards are associated with public health and spreading of diseases. This is a burden for the corresponding local authorities, which are not able to prevent and regulate the settlement practice. They bear the consequences of pollution with respect to the locality, for which they are responsible, as well as, the one transported to downstream users. In many cases, political interference from different levels does not allow the implementation and enforcement of existing regulations, thus having an adverse effect on the successful implementation of source control and abatement measures.

Low-cost sanitation technologies, as pit latrines, which are widely used as an option in the region, both in rural and urban areas, often are the cause of diffuse pollution problems (Chapter 6). Local by-laws and rural sanitation guidelines should consider this problem and prescribe strict limits with respect to the application of appropriate on-site technologies in cases of high ground water tables, such as dry pit latrines and bucket latrines. The regulatory documents should provide for appropriate regular cleaning methods and their implementation should be backed up by public education with respect to operation and maintenance procedures. In cases, where water supply reticulation exists, the provision of a basic sewer system might be feasible. However, there are two possible problems, which might jeopardize the implementation of such measures:

- The first one is related to economic problems – the people, using the facilities might not be able to pay for the installation and service required. Such cases require well-regulated and administered system for subsidizing this practice by local authorities, governmental or donor funds.
- The second problem is associated with the number of people served by each facility. Even the best on site technology could not operate properly and will fail to serve a large number of people living in one place. In such circumstances, as well as in places, where no sanitation exists at all, the local authorities might look for the provision of ablution blocks, provided with clean water and sanitation facilities at convenient locations, which could serve a relatively small portion of the settlements. The system of ablution blocks could be supplied with clean water and a simple sewage system, and could provide for a medium-term solution in such areas. The construction costs could be funded by external sources, but the community should execute the operation and maintenance, with limited technical services provided by the local authorities.

5.2 *Storm water and solid waste management*

A proper maintenance of road structures and drainage channels, together with a sound solid waste management practice could lead to a considerable reduction of urban drainage pollution. Irregular solid waste collection, and spreading refuse over the streets and public places could be often seen in urban areas in the region, especially near market places and transport terminals. This problem has been acknowledged by official authorities and urban planners (Irurah et al. 2004). The corresponding pollution implications with respect to the drain systems and natural watercourses, which would collect the runoff and the washed litter, need no explanation or monitoring. The provision of a proper solid waste collection, transport and disposal system should be viewed as a priority measure of diffuse pollution source control, which should be provided by the local authorities. Diffuse pollution with respect to existing or abandoned solid waste disposal sites (Chapter 7), need special attention and the consideration of upgrading the facilities or the special rehabilitation projects with respect to abandoned disposal sites.

In addition to the above-mentioned source control measures, pollution abatement structures could be provided at the outlet of major storm water channels, such as wetlands, ponds, detention basins, etc. In places where land is available, the provision of natural treatment systems as wetlands before the discharge of drainage channels into surface watercourses could be an appropriate measure for the retention of suspended particles and reduction of nutrients (Kao & Wu 2001). The combination of different abatement measures with environmental corridors and buffer zones could be an attractive option. One important consideration during the design of such structures is the need for a correct evaluation of the quantity and quality of the storm water to be treated and the respective characteristic values, as design storms and corresponding pollution concentrations. This requirement emphasizes the need for a correct estimate of pollution loads, based on reliable information from research investigations or monitoring programs, as the basis for a reliable and cost-effective design.

5.3 *Industrial areas and illicit discharges*

Industrial pollution sources are regarded and controlled in most cases as point sources and cover a wide spectrum of problems. Excluding large industrial enterprises, which usually have provision for local effluent treatment facilities, the vast majority of the smaller scale industries in the region lack such facilities and often discharge their effluents without treatment into sewer systems (if they are available) or into the nearest watercourse. Also, many of the existing industrial effluent treatment facilities are in a poor state of maintenance and do not provide for the required treatment efficiency. The storm water from industrial sites reaches the drain system, collecting pollution from spreading raw and waste materials, oils, etc over the premises. As such, the industrial areas become a significant source of diffuse pollution through surface runoff, or through illicit discharges. Local authorities usually do not have the capacity to police and control the numerous smaller or medium size industries and workshops, generating pollution. From another point of view, governments try to encourage and support the development of industries as a means for economic growth and well being of the population. Such conditions create a difficult environment for the trade inspectorates and pollution management authorities, which enforce the existing regulations. In very rear occasions, penalties, even with respect to point sources of pollution, have been implemented and the polluting industrial enterprises penalized. Such practice actually means that the burden of pollution is transferred from the specific polluter, who is responsible for generating it, to the local authority, the tax-payers and the future generations, who have to deal and live with it.

In all cases of illicit discharges, from industries or from individuals, progress in the process of diffuse pollution abatement could be achieved only if the public and all interested parties realize the danger of such behavior and take their own responsibility for pollution prevention. In addition, a strong political will, and good management practices enforced by local authorities are imperative. The provision of a public involvement program, allowing for the control and reporting of cases of illicit discharges could help significantly.

5.4 *Diffuse pollution and water reuse*

Wastewater reuse is a promising strategy for a sustainable and effective use of the available water resources and is of significant importance for arid climates, typical for large areas in the Southern African region. The possible adverse effects of such a practice in the case of wastewater and municipal sludge reuse for agricultural purposes has been discussed in Chapters 8, 9 and 10. It should be noted that the problems of wastewater reuse and diffuse pollution are very often interrelated and there are several aspects, which should be considered.

In many cases, the cost for wastewater treatment could be significantly reduced if the subsequent reuse for non-potable purposes is envisaged compared to the cost of treatment, required for a discharge into an ephemeral river. Considering the economic benefits from the irrigated crops, wastewater reuse, if adequately applied, could be a cost effective measure for point source pollution prevention. But if not applied adequately, it could become a diffuse pollution source, as shown from the results in Chapters 8, 9 and 10.

The optimal and effective use of available water resources requires a detailed analysis of the existing sources of water related to different types of beneficial water uses. In arid countries, rainwater harvesting and utilization becomes a viable option for an additional water supply at household or small suburb level. At a larger scale, the urban storm water runoff could be regarded as a viable contributor to the local water supply source. Provided that source control measures are implemented and no major pollution from industrial or household origin is allowed to enter the drainage system, urban runoff could be successfully reused after partial treatment in detention basins or combined natural and man-made structures, in order to augment the available water resources. In such cases, however, careful consideration of the general urban planning and transportation structures should be done and specific goals explicitly formulated. Instead of applying diffuse pollution abatement measures orientated towards higher infiltration rates, attention should be given to source control measures and the increase of pervious pavements, which could generate higher runoff volumes.

One important aspect of storm water use is the scale at which this measure could be implemented and the corresponding storage structures required. This aspect, together with the wastewater reuse option, needs a more detailed examination from a systematic perspective. The introduction of decentralized sewer and drainage urban systems could become a more sustainable and cost effective option, compared to the currently applied centralized systems, especially for arid regions. The factors, which would determine the effectiveness of the decentralized approach are numerous and include not only purely technical and economic considerations but also social habits, the availability of water resources and a general understanding and willingness to implement a new and generally unfamiliar practice in the every day life.

6 MANAGING DIFFUSE POLLUTION

A sustainable approach towards the implementation of the diffuse pollution control practice requires that it should be viewed as a system of mutually interrelated activities – the monitoring, the data processing, the decision-making process and the pollution abatement measures, where the quality of the management and implementation of each one of them, separately, would reflect on the system outputs as a whole. The need for the integration of the specific activities into a cycling process has already been mentioned. In this section, more attention would be given to the decision-making process and the preparation and implementation of diffuse pollution abatement programs, with considerations of the social, economic and built environments.

6.1 *The diffuse pollution management program*

In order to achieve the cycled approach to the watershed management, the preparation and execution of diffuse pollution abatement programs is essential, as it provides the means for the formulation of objectives, and corresponding activities to achieve them, as well as, benchmarks for the control of their implementation and the final assessment of the achievements obtained. The preparation of a diffuse pollution abatement plan was discussed in Chapter 2, and it was emphasized that such programs should form an integral part of a holistic catchment management plan. Therefore, it follows logically, that the programs could be developed for different localities and at different scales, ranging from small urban or rural streams, to larger river basins, lakes (reservoirs) or large catchment basins. The subsequent steps of a program could be defined briefly, based on recommendations by Novotny (2003).

- Define objectives – requires the formulation of goals to be achieved based on detected imbalances between the desired status of the corresponding water body, with respect to a designated water use, and the actual perception of its quality. These are narrative statements, based on the desires of all stakeholders and the responsibilities of the institution, which will implement the program.

- Develop design criteria – this step provides for the conversion of the formulated objectives to numerical criteria, guidelines and standards, as prescribed by the existing regulatory instruments and engineering practices, such as prescribed water demand figures, population number and trends, design storms and corresponding return periods, etc. This would enable the quantification of the formulated goals.
- Use or develop numerical pollution criteria – the water quality objective approach would require different quality criteria for the different basins or sub-basins, which are site specific. They could be formulated in the context of an "ecoregion" – a geographical region with similar ecological, geological, geomorphological and land use characteristics. Most countries in the region have developed or are in the process of developing such criteria. One important aspect of this stage is the level of "attainability" of the criteria, which should be reflected in the subsequent stages of the program implementation and should provide for their practical implementation.
- Conduct water body assessment – this stage is most demanding, as it serves to quantify the imbalance between the desired and actual water quality. The need for continuous, regular and reliable water quality monitoring, of both point and diffuse sources, which has been emphasized throughout this book, including the determination of background pollution values, as well as the need to implement GIS for data handling, serves to backup this stage of the program and to provide reliable information for determination and implementation of the following stages.
- Formulation of an implementation plan, including activities to be undertaken, funding sources, timeframes, and other requirements is described in Chapter 2.

More complex programs, which are related to large catchment basins, lakes or dams of national importance or other high-priority water bodies, provided that the necessary resources are available, might incorporate assessments of their assimilative capacities and a corresponding allocation of maximum permissible loads. However, in numerous cases of smaller streams and rivers, such assessments are not necessary, as it is obvious that their assimilative capacity is limited. For example, in the case of the Marimba catchment, the river has very limited assimilative capacity, but a more detailed examination of the Lake Chivero assimilative capacity could be necessary for the formulation of a sound implementation plan. In all cases, the regular monitoring of the water quality status of the Lake would be necessary in order to control the effect of the implemented measures.

A very important aspect of the implementation of such programs is their cycling nature, where the timeframe of implementation should be specified within a prescribed period of time, most often several years. After the elapse of this period, the program should be revised and a new cycle started.

6.2 *Financing and economic aspects of diffuse pollution control*

Providing the necessary funding for any planned activity or program in the field of water resources is one of the basic requirements for its implementation. In many countries, such programs are financed by the government, as part of the whole system of natural water resources' protection and restoration. However, it is often argued that the burden for cleaning "your own home", or some local catchments should not be placed on the shoulders of the whole nation. As a result, many countries have adopted the "polluter pays" principle, which has gained momentum in its practical implementation and is justifiable for the point sources of pollution, because in this case it is easier to identify the polluter and the corresponding pollution load contributed to the water body. The same principle could be applied in cases of diffuse pollution from areas where there is a single owner, who could be identified to be responsible for the observed trends of increase in the pollution status of the water body. However, in the vast majority of the cases of diffuse pollution, the polluter is not defined at all, and could not be requested to sustain the cost for the cleaning.

Another important consideration in this direction is the fact that one of the best ways for diffuse pollution abatement lies in the implementation of prevention (source control) measures, before the pollution has been generated. Thus, another approach to sustain the cost of diffuse pollution control is to place the economic burden on the beneficiaries of such types of programs, which is known as the "benefits received approach". This approach has two problematic issues; the first one being "the willingness to pay" of the beneficiaries, and the second is their "ability to pay". Obviously in the reported cases of informal or semi-formal settlements "the ability to pay" of the beneficiaries is questionable.

Considering the nature of diffuse pollution generation in urban areas, and the much emphasized need for the implementation of prevention measures, as well as considering the fact that historically, local authorities have been responsible for the management and construction of pollution abatement structures, it could be recommended that the diffuse pollution control should be organized, managed and financed by local authorities. Campaigns for public awareness with respect to diffuse pollution impacts and the need for the public to pay for such services would play a very important role in this process. Also, the involvement and community participation is very important, as diffuse pollution prevention measures in many cases depend on the effort of the community as a whole, rather than on the efforts of the managing institution. Most probably, the combined approach for financing such activities will be most appropriate. It would require the direct managing institution (local authority) to organize and implement diffuse pollution abatement programs, and to provide for partial funding based on the "beneficiary pays" principle. On the other hand, the managing authorities at catchment and national level should acknowledge their responsibility in supporting such types of activities through different means, such as:

- Provision for a national framework of necessary regulatory instruments, to serve as a guide and direction for the local by-laws and regulations;
- Provision of financial support for approved local programs;
- Provision of technical and expert knowledge support during the preparation and implementation of the program.

One important point of the financing of any program for pollution control and abatement is its economic viability. In some cases, where pollution has already being identified and generated by upstream users or previous generations, it would require the implementation of costly abatement measures, and the program could have an adverse economic impact on the whole development of the area. In such cases, the designated uses of the water bodies should be revised and the corresponding regulatory documents put in place in order to provide for a staged, long-term pollution abatement measures. Very little research has been done with respect to the levels at which pollution could be allowed to happen for a certain period of time, and the widespread economic impacts of the different rate of implementation of pollution abatement measures. Present practice shows that these types of decisions depend on the discretion of the managing authorities at national and local level. Therefore, there is a need for high standards of expertise in the field of pollution control, which would allow for a proper assessment and evaluation of the specific conditions.

6.3 *Urban development and diffuse pollution*

Diffuse pollution is closely related to different types of land use practice and source control abatement measures could achieve a considerable reduction of this type of pollution in existing urban centers. The development of new urban areas should consider and envisage source control measures at the design stage. The implementation of this principle would require new philosophies and approaches in urban planning and development, as well as the introduction of regulatory documents and codes of practice, outlining specific measures, corresponding to the specific local conditions.

In the light of sustainable development, the introduction of measures as decentralized wastewater systems, and environmental buffer zones, would require specific planning and design approaches, where wastewater could be treated, reused or disposed locally, eliminating the need for wastewater transportation at long distances. The local treatment facilities could be linked with appropriate measures for diffuse pollution source control and incorporated into a dual-purpose environmental corridor or buffer zone, which could assimilate excess pollutant material and serve as a recreational facility. Such types of new urban developments, based on the decentralized approach to wastewater transport and treatment, including environmentally friendly solutions for the optimal reuse and disposal of waste material, could reduce drastically the load on the existing central transport and treatment facilities and could be the basis of a closed cycle of water utilization and disposal.

In arid or semi-arid areas, diffuse pollution source control measures could be incorporated in the integrated water resources management strategy of the wise use of available water resources together with runoff harvesting methods and wastewater reuse. In such cases, it could be admissible to increase the percentage of impervious surfaces in urban areas, in order to collect and use the runoff after partial treatment. Different options of combined wastewater reuse and storm water use at different scales could be envisaged and investigated. The scale of application of such measures, in terms of cost effectiveness and operational considerations, as well as, its regulation and control needs further investigation.

The studies presented in previous chapters show that in many cases in the region, the urban development takes the form of unregulated informal settlements and squatter camps. This pattern becomes typical for many African countries and efforts should be made to avoid and reduce it. Other negative examples of improper urban planning decisions, which could be classified as design errors, have been presented in relation to the selection of locations for solid waste disposal sites and cemeteries (Chapter 7). They show that decisions in respect to the choice of location of such facilities have been made without preliminary geological and hydrological surveys, and as a consequence, diffuse pollution of ground water could not be avoided. The same applies to numerous cases of use of pit latrines or soak-ways in areas of high ground water level. The importance of such planning errors could not be overemphasized, as once such developments are established, it is very difficult to correct the error, or to provide for pollution abatement measures.

6.4 *Social habits and diffuse pollution*

Social behavior and habits play a very important role in the process of diffuse pollution abatement and control. The link between littering and storm water pollution has been discussed, but also it is important to notice that no managing authority could cope with cleaning and maintaining a proper urban environment if the public has not developed the necessary habits and consciousness to use the facilities in a proper manner. Therefore, the process of diffuse pollution control and abatement should be viewed as a common responsibility of the public and the managing authorities. This could be achieved not only by educational programs, but also through the implementation of tough regulatory instruments and the enforcement of sensible penalties with respect to polluters, even in low cost urban population centers. A diffuse pollution program could meet its goals only if it is regarded as a task of the whole community, but not only of the managing authority.

Other aspects related to diffuse pollution abatement is the status of the operation and maintenance of sanitation facilities. Wastewater treatment plants, or on-site sanitation facilities could not operate on their own and require proper attendance and maintenance. Unfortunately, there are numerous cases of such facilities, which show full neglect and become a source of pollution instead of preventing it. Such practice indicates that even at management level, the need to operate and properly maintain an existing facility is not appreciated. Currently, water and sanitation programs have a priority for funding as a means for the provision of sustainable development. Such types of projects have been implemented in the past as well. They show that the project ends with the construction of the planned technical structure. After that, the local community is responsible for its operation and maintenance, and in many cases, the results are not satisfactory. Such type of projects should make provision for the proper training of the personnel involved and regular control of the level of operation and maintenance during a substantial period of time.

6.5 *Economic conditions and water quality management*

Water quality management in general and diffuse pollution control in particular, require extensive, regular and sustained efforts in terms of the provision of technical background, laboratories, transport arrangements, trained personal, institutional arrangements and financial support. During periods of economic and/ or social instability and decline, such types of efforts become unsustainable and the quality of the process is drastically reduced. Most often the water quality monitoring programs suffer most where the frequency of sampling falls down to very rare sampling occasions, or the number of sampling points is reduced considerably. In such cases, preference is usually given to the control of potable water supply only, in order to avoid imminent public health hazards. The diffuse pollution monitoring and control, in most cases, is left out of the scope of the monitoring process. Simultaneously, informal discharges of sewage from municipal or industrial origin would increase, due to economic hardships and lack of regular maintenance programs. Also, unfavorable economic and social conditions would enhance the formation of informal settlements or the overpopulation of the low-income urban residential areas, which would lead to an additional pollution load. Therefore, the successful implementation of water quality control and diffuse pollution abatement is very strongly related to the social and economic status of a country. This is most valid for urban areas. Thus, the successful implementation of programs related to water quality management and public health protection should be viewed as an integrated part of the sustainable and peaceful development of the whole society.

REFERENCES

Chessman, B.C.1995. Rapid Assessment of rivers using macroinvertebrates: A Procedure based on habitat-specific sampling, family level identification and a biotic index. *Australian Journal of Ecology*, 20, 122-129.

Debo, T. & Reese, J. 2003. *Municipal storm water management* 2nd ed. Boca Raton, Florida: CRC Press, Lewis Publishers.

Hellawell, J.M. 1977. Biological surveillance and water quality monitoring. In: J. L. Alabaster (ed) *Biological Monitoring of Inland Fisheries*, London: Applied Science Publishers Ltd, 69-88.

Hohls D.R.1996. National biomonitoring programme for riverine ecosystems: Framework document for the programme. *NBP Report Series No 1*. IWQS, DWAF, Pretoria, South Africa.

Hranova, R., Gumbo, B., Klein, J. & van der Zaag, P. 2002. Aspects of the water resources management practice with emphasis on nutrients control in the Chivero basin, Zimbabwe. Physics *and Chemistry of the Earth*, 27, 875-885.

Irurah, D., Malbert, B., Castel, P., Kain, J.H., Cavric, B. & Mosha, A. 2004. Challenges for sustainable urban development. In M. Keiner, C. Zegras, D. Salmeron (ed) *From understanding to action – sustainable urban development in medium-sized cities in Africa and Latin America.* Dordrecht: Springer SBM.

Kao, C.M. & Wu, M.J. 2000. Control of non-point source pollution by a natural wetland. *Wat. Sci. Tech.* 43 (5), 169-174.

Novotny, V. 2003 *Water Quality: diffuse pollution and watershed management*. New Jersey: John Willey & Sons.

Resh, V. H., Norris R. H. & Barbour M.T. 1995. Design and implementation of rapid assessment approaches for water resource monitoring using benethic macroinvertebrates. *Australian Journal of Ecology*, 20, 108-121.

Schultz, C. 2003. Decision support systems for equitable water sharing: Suggestions for consideration in the Water for peace Okavango pilot project. In: A. Turton., P. Ashton & E. Cloete (eds.) *Transboundary rivers, sovereignty and development: Hydropolitical drivers in the Okavango River basin*. Pretoria: AWIRU & GCI.

Turton, A., Ashton, P. & Cloete, E. (eds.) 2003. *Transboundary rivers, sovereignty and development: Hydropolitical drivers in the Okavango River basin*. Pretoria: AWIRU & GCI.

Uys, M.C., Goetsh, P-A., & O'Keeffe, J.H. 1996. National Biomonitoring Program for Riverine Ecosystems: Ecological indicators, a review and recommendations. *NBP Report Series No 4*. IWQS, DWAF, Pretoria, South Africa.

Index